Unrefined

synthesis

A series in the history of chemistry, broadly construed, edited by Carin Berkowitz, Angela N. H. Creager, John E. Lesch, Lawrence M. Principe, Alan Rocke, and E. C. Spary, in partnership with the Science History Institute

Unrefined

How Capitalism Reinvented Sugar

DAVID SINGERMAN

The University of Chicago Press
Chicago and London

The University of Chicago Press, Chicago 60637
The University of Chicago Press, Ltd., London

Published 2025
Printed in the United States of America

34 33 32 31 30 29 28 27 26   2 3 4 5

ISBN-13: 978-0-226-83737-6 (cloth)
ISBN-13: 978-0-226-83738-3 (ebook)
DOI: https://doi.org/10.7208/chicago/9780226837383.001.0001

Library of Congress Cataloging-in-Publication Data

Names: Singerman, David (David Roth) author
Title: Unrefined : how capitalism reinvented sugar / David Singerman.
Description: Chicago : The University of Chicago Press, [2025] |
Series: Synthesis | Includes bibliographical references and index.
Identifiers: LCCN 2024060626 | ISBN 9780226837376 cloth |
ISBN 9780226837383 e-book
Subjects: LCSH: Sugar trade—History | Sugar—Social aspects—History
Classification: LCC HD9100.5 .S54 2025 | DDC 381/.45641336—dc23/eng/20250319
LC record available at https://lccn.loc.gov/2024060626

♾ This paper meets the requirements of ANSI/NISO Z39.48-1992 (Permanence of Paper).

For Mary

Some friends of the Teacher were invited to tell the geography class the whole story of sugar. First came Mr. Chemist, who knew what sugar is made of, for he knows what nearly everything is made of.

Sugar, says Mr. Chemist, is made of three things, not one of which by itself is sweet, but when put together in just the right way they turn into a sweet substance which we call sugar. The names of these three things are carbon, oxygen, and hydrogen.

The Story Of Cane Sugar: As Told to an Imaginary Class in Agriculture (1928)

Water is H_2O, hydrogen two parts, oxygen one,
but there is also a third thing, that makes it water
and nobody knows what that is.

D. H. LAWRENCE, "The Third Thing"

So far, no chemist has ever discovered exchange-value either in a pearl or a diamond.

KARL MARX, *Capital*, volume 1, chapter 1

When you measure include the measurer.

M. C. HAMMER, on Twitter

Contents

Prologue: Outrageous Conduct in the Sugar House *xi*

1. **The Journey of Purification** 1

PART ONE 19

2. **Freedom from the New World** 21
3. **A Lever That Squeezes** 44
4. **There My Responsibility Begins** 67

PART TWO 95

5. ***Acarus sacchari*** 97
6. **The Unpracticed Eye** 121
7. **The Electric Apartment** 144

PART THREE 157

8. **Instructions Relative to the Use of the Polariscope** 159
9. **The Sum of the Errors** 172
10. **Ups and Downs** 195
11. **Final Receipt** 212

Acknowledgments *231*
Notes *237*
Selected Bibliography *281*
Index *307*

PROLOGUE

Outrageous Conduct in the Sugar House

Matanzas, Cuba, late April 1894, over the hump of the *zafra*, the cane harvest season, and nothing unusual to report. The sugar house at the Soledad plantation was bagging more sugar than usual, and of good quality too. The backlog of sticky second sugars from earlier in the month had been cleared out. A chemist from a factory in the province to their east paid a visit, eager to share his recent calculations of how his own production was going, and to peek at Soledad's own. The factory also welcomed the inventor of a novel evaporator, who was excited about retrofitting one of Soledad's older machines during the next grinding season. The island would produce a million tons that year, and though no one knew it yet, it was the end of the boom. Later that summer, the United States would raise the import duty on sugar, the price would crash, production would collapse, and a revolution would begin.[1] But for now the grinding ground on, and there really was nothing to tell the factory's owner back in Boston, Edwin Atkins, except that one of the sugar boilers had taken a swing at the head chemist.

The boiler had been drinking, though he was not drunk per se.[2] But it was a fight at work and over sugar. The boiler was responsible for crystallization, the most delicate stage of sugar manufacture. When the boiler felt the syrup was ready to crystallize in his pan, he would break the vacuum seal, the pressure would rise, and the sugar would begin to emerge out of solution. This was called his strike, and it took years of experience and trained senses to know when the syrup was ripe for striking.

By 1894 a factory's chemist was interested in recording the properties of every strike, looking for relationships that might improve the sugar that came out. This interest was part of the chemist's endless quest to monitor the factory's materials in motion and was a way to scrutinize the otherwise inscrutable

FIGURE P.1. Vacuum pans from the 1904 catalog of O. B. Stillman, who installed the equipment at Atkins's Soledad and many of the largest sugar factories in the Caribbean. O. B. Stillman, *Installations of Cane Mills, Multiple Effects, Vacuum Pans, Filters and Cane Sugar Making Machinery* (G. M. S. Armstrong, 1904), 73. Courtesy of HathiTrust.

work of the boiler. The chemist at Soledad, a Canadian named Wilfrid Skaife, thought that the boilers should have been keeping notes on the precise time at which they began their strikes. So Skaife asked one of these boilers, called Pat Leonard, to keep better records of when his strikes began.

Asked him "quietly," Skaife wrote afterward, but not quietly enough. The relentless grinding season dissolved their disagreements into the heat and noise and urgency of the factory. After all those eighteen-hour days, the sentiments of management and of labor were thoroughly supersaturated, and with his request Skaife broke the seal. The pressure rose, grudges crystallized, and Leonard, "partly intoxicated," right there in the boiling house, decided it was time to strike.

Skaife obviously dismissed Leonard from service in retaliation for his "outrageous conduct in the sugar house," as he put it to Atkins, but it was not clear whether he meant the punch or the insubordination about record-keeping or both. He promised his employer that Leonard's departure would not have an effect on operations, and that the only difference was that he had to work harder and longer. The *zafra* would be finished in a month, so for the moment Skaife and another boiler could divide up the extra work among themselves. But each week now brought news of some modification. The seconds had to be boiled on the light side to avoid the recurrence of last month's mess. Then the remaining boilers had trouble getting the liquid to form grains.[3]

Before the machinery was cool, Atkins had already contacted former boilers and begun looking for new ones to engage on favorable terms. The pair of Skaife and the other boiler couldn't keep up the pace for a whole season, and good boilers were not unemployed for long. Skaife advised Atkins to let him hire the new men himself, "especially as a more definite idea of their work ought to be instilled in them," as he delicately put it.[4] But none of this was good news for Skaife. He knew Atkins was looking for a new superintendent, and Skaife wanted the promotion even if it didn't come with a raise. He could deal with the machines, with the technical side of production, with the analysis. But now his claims that he could equally "manage men" rang hollow.[5] Leonard might have missed, but he had shown that a boiler's skills were indispensable. Making sugar was not as easy as the chemist thought.

1

The Journey of Purification

A Story of Cane Sugar

By the beginning of the twentieth century, Central Soledad was part of a sugar system spanning half the planet, bringing raw sugar to American refineries in order to bring refined sugar to American mouths. Consider a pamphlet from 1925 titled *The Story of Cane Sugar*, created and distributed by the Pennsylvania Sugar Company around 1925.[1] On the cover, rosy children with dark hair clamber over red boxes of Quaker-brand sugar. Let's unfold the pamphlet like an accordion, flip it open, and rotate the whole sheet 180°. Now we see a cartoonish map of half the globe in which the Americas are on the right, South Asia and East Asia on the left. Over the central Pacific is a block of text, framed within a shape that echoes the Pennsylvania's keystone-state motif. The map's title reads "Where Sugar Cane Comes From."

According to the map's text, sugarcane comes from between 22° latitude north and 22° south, "for it is here that a fertile soil, plenty of hot sunshine, and an abundance of moisture are likely to be found." Inset boxes highlight some of the chief sugar-growing places in that band: India, Java, the Philippines, Hawaii, Cuba, Puerto Rico, Louisiana. Look closely, and on each country or island you can see a crude and racist caricature of plantation life. Men wear straw hats, some stooping to cut cane with machetes, others riding carts toward the smoking furnace of a sugar mill.

This side of the pamphlet insists that where sugar comes from matters. It mattered to the Pennsylvania Sugar Company. By 1925, the firm had struck an exclusive deal with Atkins for his plantations' raw sugar, letting both companies float over the swells of the open market. The arrangement also allowed them to compete more effectively with other partnerships, like that between Hawaiian producers and Californian refiners, or between the Revere refinery

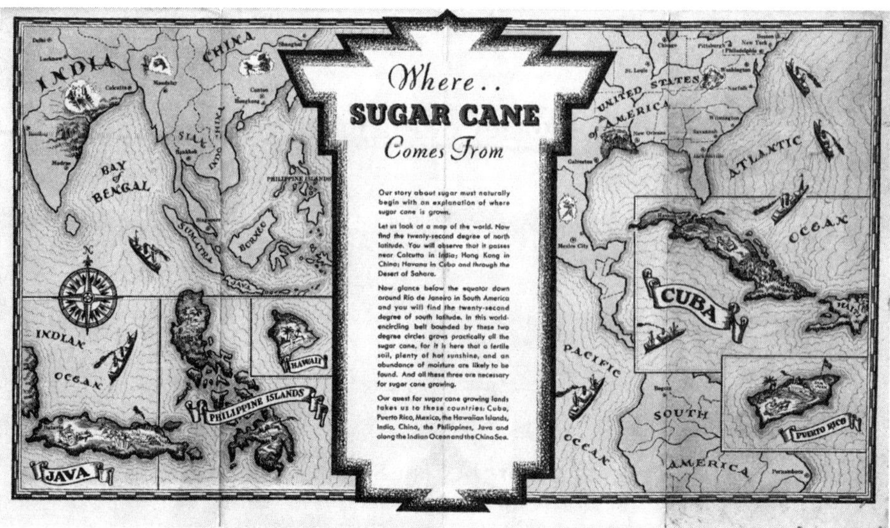

FIGURE 1.1. When customers of the Pennsylvania Sugar Company opened its promotional pamphlet titled "The Story of Cane Sugar" around 1925, this was what they saw: a map of half the world purporting to show where sugar came from. "The Story of Cane Sugar" (Pennsylvania Sugar Company, 1925). Personal collection of the author.

in Boston and the gigantic Cuban factories that were part of the United Fruit Company's hemispheric empire.[2]

The map tells a particular story of cane sugar's origins and its journey: Sugar grows in tropical islands and palm trees, in extractive agricultural economies, and flows to the modern industrial civilization of the United States. Notice that all the steamships here point north. Its center may lie over the Pacific, but the map as a whole displays a skewed metropolitan view of the world of sugar. The territories that it portrays as primitive held the largest and most sophisticated factories, and the most advanced knowledge about sugar, on the planet, as they had done for hundreds—and in some places thousands—of years.

Let's turn the pamphlet over now, so that we are looking at its entire reverse side. This view is dominated by a central map, which stretches vertically from the Caribbean to the United States' eastern seaboard. Below the map is a detailed aerial view of the company's Philadelphia refinery, long wharves jammed with ships. On the map itself, the city of Philadelphia is marked by a miniature of that same engraving. Down the left side, pastoral sketches depict the "brief story of the raw product before it reaches the United States." They show how cane is grown, harvested, and milled for its juice, the juice boiled down to crystals, until eventually the "raw sugar" is ready to be bagged and loaded onto vessels for Philadelphia. Ships steaming northeast across the map draw our eyes to Philadelphia and toward the upper right of the page. Down that side, more sketches show the mechanical guts of the refinery, where sugar is washed, filtered, and dried to make it finally pure and white. Now these crystals can be cubed and boxed for consumers. On the map, trains and trucks stream inland, carrying refined sugar to Americans everywhere east of the Mississippi River.

FIGURE 1.2. Fully opened, the pamphlet showed a map of the Americas, detailing the steps of sugar's production in such a way that made refinement and purity seem the inevitable outcome of progress. "The Story of Cane Sugar" (Pennsylvania Sugar Company, 1925).

The front extols the long history of cane as a crop and the specific history of each sugar cube. But the back seems to suggest that sugar's origins do not matter, that nobody should care where sugar comes from. At the bottom of the left-hand column, we read that raw sugar from the tropical factory is "brown in color and containing about 96% 'sucrose,' or pure sugar." This goes into the refinery on the top right, and what comes out of the refinery is "as pure as it can possibly be." The imagery depicts the North Atlantic and South Pacific as cogs in a great machine for turning cane plants into pure crystals.

On closer inspection, however, this side equivocates about the meaning and value of purity. Is the stuff in the Quaker boxes cane sugar, or is it sucrose? Sucrose, the molecule, is found in cane but also in thousands of other plants across the globe. If Quaker's product is one hundred percent this molecule, why mention the cane—why does the plant that the molecule came from mean anything? The business of a refinery was squeezing the past out of raw sugar, but to make money selling its product, the Pennsylvania Sugar Company found that it needed to mix a little of that past back in.

Sugar is one of the most important goods in human history. The search for transportable sweetness has shaped colonialism and imperialism around the planet for more than five hundred years. It has provided immense wealth for a few—and greater suffering for millions more. It has transformed the modern human diet. But what, exactly, is valuable about it, and who should get to say? Those turn out to have been surprisingly difficult questions, and the answers have their own history.

The Brief Story of the Raw Product

Where does sugarcane come from? The species that we group under the genus *Saccharum* all originated on the island of New Guinea, although at the time of "The Story of Cane Sugar," expert consensus held they had come from India.[3] Humans have been cultivating and breeding canes for so long that botanists keep discovering that supposedly wild lines are actually descendants of domesticated varieties.[4] Cane stalks have been chewed and eaten for as long as humans have been around it.[5] Getting crystals out of the sugarcane is harder work: extracting its juice to evaporate the water content, and then gradually and carefully heating and cooling and heating it again and again to encourage the growth of crystals.

The first people to have engaged in this process, what we would call sugar manufacture, do appear to have lived in what is now India, sometime in the first millennium BCE.[6] A few hundred years later, people in southern China were producing an "amorphous" sugar product too.[7] By the seventh century CE, a widely retold legend stated that a Buddhist monk had taught his neighbors to make crystallized sugar as compensation after the monk's donkey trod through a field of delicate cane shoots.[8]

Around the turn of the second millennium, merchants, entrepreneurs, and ambitious rulers brought large-scale sugar cultivation to the Mediterranean world. Until then, sugar had mostly been produced on a smaller and less centralized scale. In India or Egypt, the equipment for grinding and boiling cane was transported from field to field, so it was designed for easy disassembly. In the West, the mills were heavy units, built on site, owned by planters with more capital, and cane was brought to the machines rather than the other way around. This model of cultivation brutalized people and nature. Cane monoculture left soil depleted, grinding canes took enormous animal energy (though sometimes wind or water was available), and boiling the juice required whole forests to be cut and burned.[9] This was the beginning of a new form of "sugar capitalism," which, according to Ulbe Bosma in his global history, "started in the Mediterranean Basin, continued in the eastern Atlantic islands, and adopted its most radical shape in the small Caribbean islands."[10] Spain and Portugal brought sugar to Madeira and the Canaries in the eastern Atlantic, developed the sugar plantation model, and took it to the Americas in the middle of the sixteenth century.

In Europe sugar commanded high prices as a spice, an ingredient, and a medicine. As European colonies grew more and more sugar, the price fell, so European consumers demanded more of it, and the cycle renewed. As the anthropologist Sidney Mintz showed in his pathbreaking 1985 book *Sweetness and Power*, cheap sugar transmogrified elite rituals, in which displays of rare sugar flaunted status, into bourgeois and then proletarian ones, where sugar might sweeten a factory worker's tea.[11] The English ate two pounds of sugar per capita in 1670, twelve pounds in 1730, and twenty-four pounds in 1800.[12]

What made sugar cheap was slavery. Between 1500 and 1866, according to the latest and best estimates, over 9.5 million people from Africa were kidnapped by Europeans to work in the sugar colonies of the Americas, and another 1.5 million were murdered on the journey.[13] Some of those enslaved people also mined American silver so that Europe could pay for Asian goods to trade with Africa for the bodies to harvest cane in the Caribbean.[14]

The "sugar revolution" of the seventeenth and eighteenth centuries consolidated human slavery and other techniques into an efficient machine for making sugar: agricultural practices, mills for crushing the cane, boiling houses, and spaces for drying or "curing" or "purging" the remaining wet mass of crystals and molasses.[15] These plantations depended on the skills of enslaved people for successful harvests, for efficient grinding, for careful boiling without burning the liquid, and for the judgment to know when the sugar was ready.[16] The technologies at their core, especially mills and apparatus for conserving heat in the boiling process, took capital and skill to build and maintain. Plantations were some of the most coordinated and sophisticated spaces of production in the early modern world, more like modern factories than anything in Europe. It was for good reason that the sugar mill, in Spanish, was known as the *ingenio*, a word that connected engineering and ingenuity, and the name spread to the entire plantation itself. They were made visible to Europeans through countless illustrations, which generally elided enslaved people.[17] The plantation complex generated unconscionable profits by exhausting every ecology it invaded: Brazil in the sixteenth century, Barbados in the mid-seventeenth, Jamaica in the early eighteenth, then Saint-Domingue later that century, and Cuba and Puerto Rico early in the next.[18]

In the nineteenth century, a second sugar revolution, an industrial one, remodeled that plantation complex.[19] Railroads replaced carts for bringing cane to the mill. Wooden rollers driven by animals or wind were replaced by steam-powered mills that crushed more of the juice. Other new machinery helped extract more and cleaner crystals and dry sugar much faster. Planters imported scientific knowledge and experimental practices, employing chemists to control their production lines. Expensive machines and methods gradually demanded more capital and economies of scale, so enormous "central factories" sprouted throughout the sugar world. Large-scale sugar cultivation for export spread across the globe, partly to areas like Australia and South Africa, where sugarcane was a new species, and partly to places, like Taiwan and Hawaii, where what was new was only European-driven monoculture. It was powered, in different places, by free labor, by indentured workers, and, despite global movements for abolition and emancipation, by millions of the still-enslaved.[20]

Where could sugar come from? At the beginning of the nineteenth century, an unexpected entrant changed the nature of the question. Before 1799, sugar came from cane. Only a few grains of sugar had ever been extracted from other vegetables. Over the course of the following century, however, sugar from beets, grown in cold climates, rose to nearly equal cane's geopolitical importance. "The sugar question" between the two industries was central to European and then North American politics. By 1880, beet supplied half the sugar in world trade, and by 1890, nearly two-thirds.[21] Competition with government-subsidized beet producers drove cane sugar manufacturers to innovate, expand, and combine. But for sugar economies everywhere, the wealth from exports became a gilded prison. Beet sugar states bankrupted

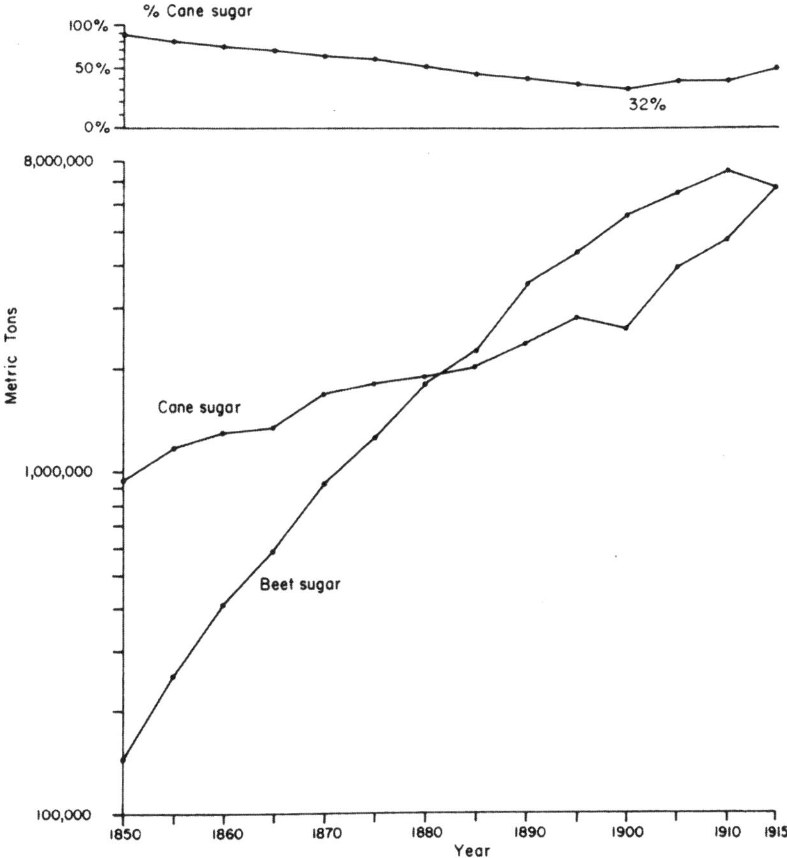

FIGURE 1.3. In the second half of the nineteenth century, the world's production of cane sugar exploded from a million tons a year to nearly eight million. Meanwhile, beet sugar rose from nothing to rival those levels and even, by the end of the century, outpace them. (Note that the vertical scale of this graph, reproduced from J. H. Galloway's 1989 work, is logarithmic.) J. H. Galloway, *The Sugar Cane Industry: An Historical Geography from Its Origins to 1914* (Cambridge University Press, 1989).

themselves with subsidies, while crashing cane prices sparked revolutions in colonial possessions.

At the beginning of the twentieth century, the global sugar machine yielded unprecedented quantities of sugar, sugar that was also qualitatively different from what had come before. It was purer, more consistent, and more stable; once refined and packaged, it could be sold to satisfy the world's enormous demand. The new sugar was, according to the great Cuban historian Manuel Moreno Fraginals, "as different from the previous product as the central was different from the old slave-run ingenio."[22] Sugar's properties, like those of other commodities of the age, indexed technological and economic progress.

Humbug

The long nineteenth century was an era of doubts and dislocations. Many of these worries were the product of new technological objects and new ways of thinking about the world. Locomotives, messages over electrical wires, photographs and then moving pictures, and even seed catalogs and glass boxes for transporting living plants across great distances: All of these contributed to a sense, as the phrase had it, that time and space were being annihilated.[23] It was also an age of confusion about where to place one's trust. People were dizzy from market revolutions and exhausted by knowing neither whom nor what to believe, now that everything and everyone seemed to come from over the horizon.[24] As the American diplomat and historian Henry Adams wrote, it was as though people used to the eighteenth century had been forced to keep up with the twentieth.[25]

The centripetal effects of global capitalism and global empires kept generating new things that needed to be measured, compared, and made the same.[26] Militaries, railroads, commerce, and communications networks all needed synchronizing and standardizing on a global scale.[27] Standards demanded and generated new institutions to create and implement them. So along with everything else, the pace of scientific meetings, conventions, and agreements accelerated during the second half of the nineteenth century. Toward 1900, all the industrial powers established new official organs for making standards in theory, practice, and law, and scientists themselves gathered more and more often for heated battles fought over—and under—national standards.[28] In one sense, the project to impose a universal order on the natural and human worlds was as old as modern science. "The causes of natural effects of the same kind must be the same," Isaac Newton had insisted. "Examples are the cause of respiration in Man and Beast, or of the descent of a stone in Europe and in America."[29] From the beginning, claims that people of European descent had special access to such universal knowledge supported claims that people of European (and later North American) descent had special privileges to rule the world.[30] In another sense, however, standardization was a frantic effort to stay afloat in the whirling vortex.

The nineteenth century's new commodities embodied, represented, and in large part caused the sense of dislocation and confusion. All along the world's "commodity frontiers," industrial capitalism was extruding nature's messiness into consistent forms more suitable for global exchange.[31] In 1800, no dyer had ever colored with artificial indigo. No butcher had ever sold frozen mutton. Certainly, no one had ever spread their toast with a substance called "extract of meat." Yet by 1900, natural indigo was almost gone from world trade, Britain imported more than half a billion pounds of chilled mutton each year, and beef carcasses yielded a savory extract.[32] Other goods were composed of familiar constituents but so reconfigured that they seemed unfamiliar. Gas burned instead of whale oil.[33] Coffee came in cans.[34] Cotton, tobacco, wheat, tea, and other familiar items moved not only in unprecedented quantities but also in unprecedented forms. Old objects had been reimagined and reinterpreted through modern scientific practices and ideas.[35] What truly differentiated the commodities of the nineteenth century from those of earlier periods was that many of these goods were produced, classified, traded, and taxed under increasingly standardized categories.

In economic life, standards for commodities were a revolutionary technology. They let buyers know what they were buying before those buyers set eyes on the goods, or even before the goods were drawn from the earth. The development of these standards allowed the commodities of the eighteenth century to become the mass industrial commodities of the twentieth and allowed the world of organisms to become the world of goods. In the environmental historian William Cronon's terms, "first nature" became "second nature."

In his book *Nature's Metropolis*, Cronon laid out an influential account of this process, by showing how the radical changes to the shape of markets in mid-nineteenth-century Chicago transformed the worlds of wheat, lumber, and meat. Take grain: It had once moved down the Mississippi River in individual sacks, and the value of a load of grain came from its individual character—its physical properties, but also the reputation of its grower. Around 1850, the railroad, the elevator, and the logic of capital made it far cheaper to ship grain in bulk than in bags. But now it couldn't be traced to its origin or priced on that basis in the way it always had been.[36] The solution was to invent what Cronon calls a "momentous" new legal technology: standard grades of grain.[37] The Chicago Board of Trade defined each grade's qualities, hired inspectors, and declared that one lot of, say, no. 2 white winter wheat was exchangeable for any other. Forward deliveries from Chicago to New York exploded: "No longer was it necessary to see a sample of any particular shipment, for all grain of a given grade was for practical purposes identical."[38] The grading system, Cronon argues, was a radical institution. Combined with new means of transport and storage, it gave rise to speculative dealing, futures exchanges, and ultimately the disconnection of markets from their physical products.[39]

In industry after industry, the logic of industrial capital pressed for consistent goods that could be stacked efficiently, processed automatically, packaged uniformly, and marketed widely. Each economic sector developed its own grades, and in each case, some institution had to assume responsibility for the defining. Standards did not emerge from uncountable independent acts of exchange, or by imposition from above, but from places where market actors came together to assign and delegate authority.[40] For Chicago grain, it was the Board of Trade; for Liverpool cotton, it was the Cotton Brokers' Association, which defined "fair" and "middling" and "ordinary" cotton, created samples that reflected them, and circulated these samples as far away as New Orleans.[41] These grades enabled cotton brokers to buy and sell promises of delivery and speculate on a crop before its seeds were even in the ground.

Of course, these categories were usually subjective. No one pretended that "no. 2 white winter wheat" described a species in the wild. The Coffee Exchange evaluated beans into categories from no. 1 to no. 8. Coffee was graded by an expert who swished it, spat it, and described it as "smooth, rich, acidy, or mellow," or "winy, neutral, harsh, or Rioy" (as in, from the area around Rio de Janeiro), or "musty" as opposed to "grassy." As Augustine Sedgewick writes in *Coffeeland*, "There were no fixed numerical standards, just a flexible new vocabulary," a competition for inventing adjectives.[42] Subjectivity, scholars agree, facilitated disagreements. Elevator operators, farmers, and merchants in Chicago challenged one another "over how to impose artificial boundaries on the world of 'natural' grain."[43] In 1871, the board sacked its chief inspector after he declared some no. 3 oats "mixed with Rejected barley" to be no. 2 oats, under pressure from a board director. The resulting scandal ended, Cronon writes, with the State of Illinois assuming responsibility for grading. "By 1874," Cronon reports, except for the odd shipment, "faith in Chicago inspections had been so restored that the city's grades were accepted without dispute."[44] The products of nature are diverse, the story goes, but however imprecise the standards, they were good enough to make things the same—to reconfigure nature for production and commerce.

Cotton grades were literally fuzzy, but the early attempts, writes Sven Beckert, nonetheless "formed the foundation upon which later enforceable standards would be built."[45] The course of standardization was bumpy but ultimately inexorable. "Without such standards," argues Beckert in *Empire of Cotton: A Global History*, "such a high-volume long-distance trade of bulk commodities would have been all but impossible—the vast diversity of nature had to be distilled and classified to make it correspond to the imperatives of machine production." Standards propagated upstream and reshaped the real world, from the kind of wheat planted on farms to the kind of people enslavers purchased.[46] Grading systems, in these accounts, are essential technologies of commoditization. Fit standard grades together with railroads, grain elevators, freezer cars, and telegraphs, and nature is transformed as if by an automatic process.

This is the standard story of standards, but it misunderstands how they work in an important way. To see why, we can turn to the history and sociology of science, and the difference between how scientists think about standards and how historians and sociologists of science think about them.[47] To scientists, the discipline called "metrology" is concerned with investigating, creating, and disseminating authoritative standard values to measure nature: its length, time, mass, electrical resistance, energy. A scientist would say that standards become widely accepted because they are precise and accurate. Standard values regulate scientific activity around the world and make possible a system of knowledge that pretends to universality, as well as all of the technological infrastructure that defines modern life.

But a standard value is a scientific fact, and like every piece of scientific knowledge, it is the product of humans, not gods. One of the great open questions for historians and sociologists of science over the last fifty years has been how, despite the specialness of the places we call laboratories, we moderns are persuaded that the facts produced in laboratories are applicable everywhere.[48] Even an ordinary laboratory is, compared with the rest of the world, an extraordinary place, full of specially trained people doing special activities. Standards are unusually unusual, developed in extra-special laboratories, under extra-controlled conditions, by extra-trained people. Yet in order for a standard value to seem like a self-evident property of nature, all of the very special work to create it must be hidden or ignored. Otherwise, a would-be universal value remains just another local creation by a peculiar set of humans, without any special claim on the rest of the world.[49] The sociologist Harry Collins compared standards to ships in bottles: We know a human worked hard to put it in there, but we are impressed because it looks like it got in on its own.[50]

But standards can't propagate themselves any more than the tiny ship can sail into the glass bottle under its own power. To get anything done in the world away from the laboratory where they are created, standards must travel accompanied by certain technologies, rules, and techniques, what the sociologist Joseph O'Connell called "the creation of universality by the circulation of particulars."[51] The symbolic climax of the nineteenth century's efforts at global standardization took place one festive day in 1889, when the platinum embodiments of the meter and kilogram were entombed in a special vault in Paris. "M" and "K" were placed underground, wrote the historian Peter Galison, because only by being protected from earthly concerns could they transcend their physicality and "function externally as universal measures."[52] It was not their painstaking production that mattered for their value as standards, but the elaborate ceremonies that declared them to be the most precisely engineered objects in the world. But to matter in the world, M and K had to be copied in officially sanctioned ways and distributed to special institutions where other things could be calibrated against them. Without all those rituals for its movement, a global standard is just a piece of metal in a French cellar.

Or, as the Alabama cotton sharecropper Ned Cobb put it to his biographer, "Much of it is humbug just like everything else, this grading business." This humbug is at the core of the story of the nineteenth century's new commodities.[53] Whether a meter, a kilogram, an electrical unit, or the grade of a commodity, it took power and persuasion to make the world accept a standard value. That power and persuasion is what historians of science mean by "metrology"—the work put in by a value's human advocates to make a unit look like an emissary from Nature itself. The prize, at the end, is elevating social relations—the power to make the rules—into the realm of natural fact. It is no coincidence that changing social relations into apparently natural ones is also the work of modern capitalism. What makes sugar different from other commodities, historians have argued, is that there were no artificial boundaries to impose. Sugar's grading system was defined by a measurable molecule, which was also the source of its value. In sucrose, its commodity form was already there, waiting to be freed from nature. But it turns out that this grading system, too, was humbug just like everything else.

The Power in Sugar to Produce Ideas in Our Mind

From the fifteenth century to the seventeenth century, sugar was the original commodity frontier. Europeans developed new economic, social, and ecological models of plundering the earth to satisfy their demand for sweetness.[54] At the very same time, philosophers of the period considered sugar an exemplar of a universal material that universally affected human beings.[55] John Locke, political philosopher and slavery investor, worried that not all gold was the same, that it was just "convenience that made men express several parcels of yellow matter from Guinea and Peru under the same name." But sugar was consistent enough for philosophical demonstration. When we taste the sweetness in sugar, reasoned Locke, it means that the sugar contains some essence that produces that taste: "For, if sugar produce in us the ideas which we call whiteness and sweetness, we are sure there is a power in sugar to produce those ideas in our mind."[56] The first words of the entry for "sucre" in Diderot and d'Alembert's *Encyclopédie* are *personne n'ignore*: "As everyone knows, sugar is a solid, white, sweet, and tasty substance."[57]

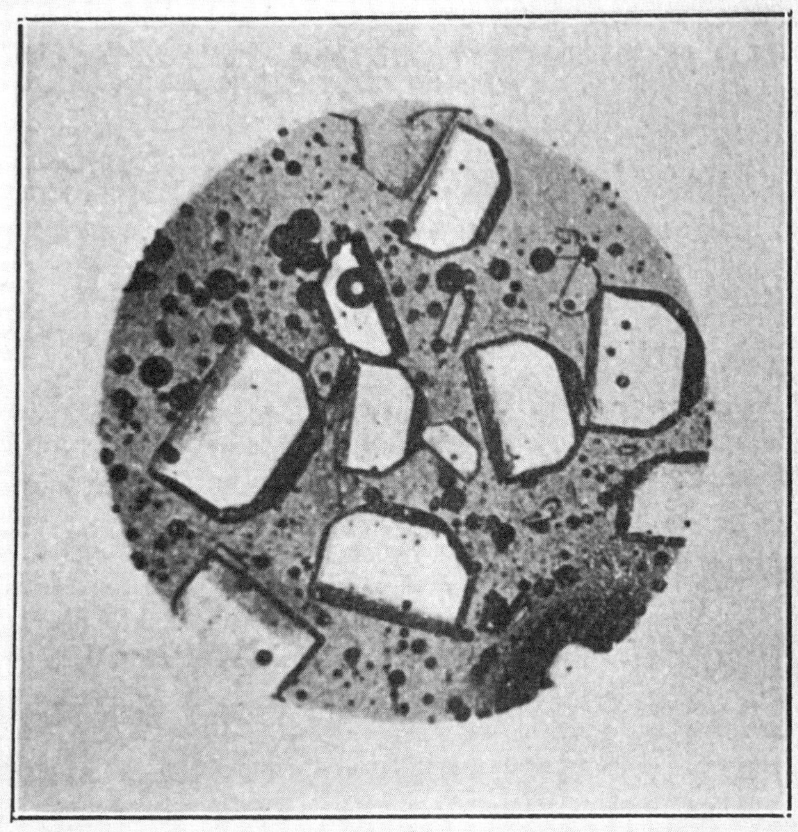

FIGURE 1.4. Sucrose crystals under a microscope, with black dots of molasses. University of Hawaii (Honolulu), *The Story of Cane Sugar, as Told to an Imaginary Class in Agriculture* (University of Hawaii, 1928).

To us would-be moderns, what matters in sugar is the molecule. "I am concerned with a single substance called sucrose," Mintz writes, "a kind of sugar extracted primarily from the sugar cane."[58] From New Guinea to the New World, historians write about the manufacture of sugar as if it were a constant quest for the purification and extraction of sucrose—a refinement of refinement. On the first page of his global history of sugar, Ulbe Bosma writes: "It requires ingenuity and patience to extract from plant material the complex sucrose molecule $C_{12}H_{22}O_{11}$. . . . The resulting white table sugar was a luxury 200 years ago and could only be produced in small quantities, through costly and time-consuming artisanal manufacturing."[59] It makes sense to view sugar this way. Sucrose is natural. People don't make sugar, the argument goes. Plants make sugar. "The plants know how to do it," one sugar-factory chemist marveled to his colleagues in 1888, "but we cannot."[60] In *Sweetness and Power*,

Mintz quotes the historian and policy adviser G. B. Hagelberg: "Though we speak of sugar factories," Hagelberg writes, "what actually takes place there is not a manufacturing process but a series of liquid-solid operations to isolate the sucrose made by nature in the plant."[61] The chemical identity of sugar ties its history together. As two scientists write in *Sugar: A User's Guide to Sucrose*, "The sophisticated chain of reactions that first occurred millions of years ago still goes on in the majestic laboratories of Mother Nature."[62] In this view, when humans claim to be making sugar, they are actually unmaking what the plants have wrapped around the sucrose.

Sucrose is, of course, very real. One sucrose molecule contains twelve atoms of carbon, twenty-two atoms of hydrogen, and eleven atoms of oxygen. Plants use sucrose to store photosynthetic energy, and an enzyme inside our small intestines breaks sucrose down into its two components, glucose and fructose, which our bodies metabolize into usable energy.

But sugar, the chemical substance, is not the same thing as sugar, the good that humans make, trade, and eat. And no matter how real sucrose is, it is not something that most people for most of the history of sugar could have known or thought about. The word "sucrose" was not used in English before 1857.[63] Like "carbohydrate" (1844) it belongs to the disciplinary idiom of modern organic chemistry, and, just as importantly, these words denote ideas from that field.[64] This way of thinking about substances only began to take shape in the early nineteenth century, and took decades to achieve its victory over other ways of thinking about matter, not least because many of those other ways seemed equally valid and more in line with human experience.[65] "In 1000 A.D.," Mintz writes, "few Europeans knew of the existence of sucrose, or cane sugar. But soon afterward they learned about it," and became "sucrose consumers." Yet they could hardly have thought that sugar was anything called "sucrose," nor conceive of it as a molecule, because they conceived of substances and their engagement with substances in other ways. Before the middle of the nineteenth century, when people spoke of "purity" or "purification" or "refinement," these were not simple translations nor antecedents of modern notions about the invisible structure of matter.[66]

The theories people hold about nature shape their interactions with nature.[67] To take just one example, we can look to 1675, when Barbados was the most advanced and profitable plantation colony. In that year, sugar featured in one of the earliest papers submitted by the Dutch draper Antonie van Leeuwenhoek to the Royal Society of London, Europe's most prestigious and well-connected learned organization. Van Leeuwenhoek surmised that the taste produced by any substance came from its physical effect on the surface of the human tongue, which contained thousands of infinitesimal "globuls," and described what he saw when he looked at grains of sugar and salt under his new invention, the microscope.[68]

Through his lenses he saw that each grain of sugar was an irregular crystal with many points and edges. It would therefore touch many globuls without exerting much force on any one of them, just as, said van Leeuwenhoek, you would be unharmed if you pressed a large pointed diamond into your palm. Make the diamond a thousand times smaller and apply the same pressure, and "it would not only cause smart, but doubtless, if no bones did hinder, run through the whole hand." If sugar were also a thousand times smaller it would cause smart to tongues, but instead it dissolved in the warm moisture of the mouth and touched many globuls with pleasant sensations. Meanwhile, crystals of salt stayed rigid enough "that they prick the globuls of our Tongue, though not so stiff, that they wound them."[69] That was van Leeuwenhoek's sensuous explanation for what his friend Locke called the "power in sugar."

The evocative and sensory terminology of the past, in other words, can't be easily mapped onto a periodic table.[70] People who dealt with sugar understood it through the way it tasted, felt, smelled, looked, and even sounded. On plantations, the highly skilled sugar master would slice one loaf of dried sugar into as many as sixteen different tranches of different qualities and prices. Consumers were no less conscious of its variation. One piece of French advice from the early eighteenth century was to "choose sugar that is white and dry, of an attractive grain, and a sweet taste almost like violets."[71] Pastry chefs in the courts of Spain began their cookbooks with the "calidades" and "especies" of sugar they would encounter. *Lisa, perla, soplo, pluma, caña, caramelo*: smooth, pearl, puff, feather, cane, caramel. Each had its own qualities.[72] English buyers asked their grocers for sugar from certain colonies and even from particular plantations.[73]

There were many ways of understanding sugar; none of them was ours. The actual number of people in 1000 BCE, in 1000 CE, or even in 1800 CE who knew of sucrose was zero.

The Equal Standing of All the Sugars of the World

Yet by the twentieth century, sucrose was all that mattered. To Edwin Atkins and his fellow industrialists, this conceptual revolution was the direct product of technological and economic changes. "Sixty years have passed since my first visit to Cuba," Atkins began his memoirs in 1926. "These years have seen such a revolution in the production and manufacture of sugar, and in all business methods, as probably, in a like term, will never be repeated."[74]

Over a decade later, in his book *Cuban Counterpoint*, the Cuban anthropologist and sociologist Fernando Ortiz contrasted the individualism of tobacco cultivation with the conformity of sugar's production complex. Every leaf was unique, every cigar a singular experience, but what mattered in sugar was quantity, not quality. Because "for both sugar-grower and refiner the aim is the most," he observed:

> The most cane, the most juice, the most bagasse, the most evaporating-pans, the most centrifugals, the highest crystallization, the most sacks, and the most indifference as to quality for the sake of coming as close as possible in the refineries to a symbolic hundred per cent chemical purity where all difference of class and origin is obliterated, and where the mother beet and the mother cane are forgotten in the equal whiteness of their offspring because of the equal chemical and economic standing of all the sugars of the world, which, if they are pure, sweeten, nourish, and are worth the same.[75]

This is the claim on the Pennsylvania Sugar Company pamphlet too. The refinery's job is to strip away everything but the sucrose. Above the map, a keystone emphasizes that Quaker Sugar is "100% Pure Cane Sugar, refined to the highest standards of quality and purity."

Purity should be a timeless property, yet it has a history. In the nineteenth century, Americans and Europeans gobbled up storybooks about commodities that took readers on a journey to the origins of the goods they bought and ate. Entrants in the genre promised to reveal surprising facts about a commodity, but they also enchanted their readers with a technological story that disguised the skill of non-European and especially enslaved people.[76] Scholars of these same commodities have fallen under a similar spell. By revealing the power of capitalism to put nature into boxes, historians of capitalism attribute that power to machines, techniques, and powerful institutions of authority. In doing so, they skip over how many different kinds of people it took to make diverse natural things seem the same: samplers, testers, weighers, chemists, customs officials, artisans. Nineteenth-century cheerleaders for sugar capitalism might have glorified sugar's "journey of purification," but historians shouldn't take them at their word. At the same time that Atkins wrote his memoirs and the Pennsylvania Company printed their pamphlet, one refinery engineer complained that chemical purity had become "a quantitative measure without just appreciation of the qualitative characteristics," whose simplicity concealed the need for a "just measure of comparative value" between any two sugars.[77] There was nothing inevitable, mechanical, or automatic about the process of commoditization. Every grain of sugar went on its own journey. Its qualities and properties had to be argued about, tested, and judged, alongside the even more

Raw Sugar Starting on Its Journey of Purification

FIGURE 1.5. Raw sugar from one of Atkins's plantations arrives at the Pennsylvania Sugar Company to begin its "journey of purification," according to a company history. Benjamin Allen, *A Story of the Growth of E. Atkins & Co. and the Sugar Industry in Cuba* (1925), 35.

fundamental questions of which qualities and properties should matter at all. The same was true with every parcel of grain or cotton.[78]

With its chemical value produced by plants, in what Ortiz called "those little natural sugar mills of the cane stalks," sugar seems the most objective of all commodities, the least open to dispute.[79] Yet Ortiz was playing on the seductive fantasies of commoditization. His real point was that, whatever Atkins and his allies might think, becoming a commodity did not erase sugar's history, any more than an enslaved person's history was erased when they were trafficked.[80] That insight is at the core of this book. If we can see how sugar, the most quantifiable of commodities, was as dependent on subjectivity and judgment as any other, we can start to understand the globe-spanning system that claimed to make things the same even as it depended on the ways in which they were different.[81]

This book is divided into three parts. Part one explores the changes in the global sugar economy of the long nineteenth century by recovering three of its untold stories: the novelty and weirdness of extracting sugar from beets, the difficulties of producing machines for making pure sugar, and the importance of skilled human labor in making chemistry look easy in sugar factories. Part two moves to the United States and shows how, from the 1870s to the 1890s, new kinds of sugar from industrial factories baffled people who relied on fixed relationships between substance and value but provided opportunities for others to exploit. Finally, part three shows how complex it was to actually create standards for sugar, how difficult it was to build a speculative commodity futures market on this unstable substance, and how both that complexity and that instability might be exploited.

By showing what was required to make sugar seem universally pure and indistinguishable, we can better understand what happens to all commodities as they come to be measured and valued with the tools, techniques, and language of modern science. There is never a point at which herds or harvests sublimate into something interchangeable. Instead, turning nature into commodities is a continuous process, and, as we will see, the people involved know how to reaffirm the similarity of different objects, or to instead object to their differences.[82] The idea that billions of natural objects can be said to be "the same" is a claim about nature just as a standard value is such a claim, and like all such claims, people have to make it travel. The machinery of commoditization appears to function on its own, but only because human beings are assaying the qualities of those commodities at every step. If we pay attention to the metrology, we can start to comprehend the machinery. The idea that all sugar is fundamentally the same is a product of the economic, political, and social order we call modern capitalism, just as the sugar itself on your table is a product of that order.

PART ONE

The three chapters in part 1 each address a global aspect of the sugar economy that shaped the process by which sugar was standardized as the chemical sucrose: the invention of beet sugar, the development of specialized machinery, and the rise and limits of "chemical control" in sugar factories.

In chapter 2, we explore the origins of the claim that all sugar was the same by looking at a moment when a specific form of that claim was intensely controversial. In 1747, a Prussian apothecary reported that he had extracted, from native European vegetables, a substance identical to cane sugar. This has become the origin story of the beet sugar industry, which by the end of the nineteenth century produced about half of the world's commercially traded sugar. In fact, earlier in that century, many forces resisted the notion that the two materials were equivalent, let alone identical, and they held different moral values too. Cane was reactionary, imperial, and violent; beet was progressive and nationalist, but it could represent a peaceful utopia or a corrupt industrial politics.

The new machines that transformed Caribbean sugar production are the subject of chapter 3. Some of them, like the vacuum pan, were the product of local knowledge, while others were adapted from European designs for beet-sugar production. But their design was only the beginning of the real story. Filling factories with these machines, keeping them working, repairing them, and customizing them required specialized companies, and nowhere on earth was more devoted to the manufacture of sugar machinery than the city of Glasgow in Scotland. To standardize a commodity on one side of the world, workshops and techniques across the ocean became increasingly idiosyncratic.

Chapter 4 shows the power and limits of "chemical control" in the Caribbean. The power of sugar machines has hidden the invisible skilled labor that made it look easy to transform sugar into a chemical commodity, but the industrial sugar factories of the late nineteenth century were a site of metrological conflict too. The skills of early modern sugar boilers are supposed to have been obviated and rendered unnecessary by the sophisticated processes of the new mills, but in fact they remained crucial to the success of the *ingenio central.* At the same time, supervising a sugar factory was hard chemical work, but the idea that sugar is sucrose has hidden this work too. Chemists did much more than ratify the natural perfection of sugar. On the one hand, chemists' rhetorical, intellectual, and physical labor rendered the role of the sugar artisan invisible. On the other hand, the more that chemists' expertise made chemical purity the standard measure of value, the more it obscured their own work.

2

Freedom from the New World

The Other Sugar Question

In 1844, the French proto-surrealist artist known as J. J. Grandville published a bizarre and beautiful book titled *Un autre monde*. In one chapter, Grandville imagines the citizens of the vegetable kingdom rising up against humanity, claiming their rights as living beings and throwing off the shackles of human empire. But it turns out that, despite their radical overthrow of the existing order, plants are no more immune to internal strife than people. A civil war among plants erupts, touched off by a brawl between a sugarcane and a beet.

Grandville's illustration is shocking, appalling, and amusing at once. His cane is dressed as a reactionary planter, leaf-hand poised to swing a switch of cane onto a defenseless beet. Before a background range of sugarloaf mountains, a frantic carrot tries to intervene, while a tobacco plant puffs a pipe and looks on. In this absorbing vision, Grandville imagined more than one plant killing its rival. He also imagined the forces of cane sugar obliterating a nascent world in which the powers of machinery and organicism combined to yield liberation rather than oppression.[1] "Combat de deux raffinés," the caption reads, the battle of two refined creatures, but these were very different meanings of refinement.[2]

Like seemingly every topic in nineteenth-century Europe, sugar took the form of a question.[3] The classic "sugar question" asked whether states should provide fiscal support to their domestic beet industries or continue to support their plantation colonies with tariffs on foreign sugar. The French sugar empire had once stretched across the Atlantic and Indian Oceans. Its greatest prize was Saint-Domingue, the richest and most productive sugar colony in the world until its enslaved people overthrew their French oppressors and defeated the British at the end of the eighteenth century. By the 1840s, France clung to its remaining sugar colonies, and viewers of Grandville's image

FIGURE 2.1. "Combat de deux raffinés." J. J. Grandville and Taxile Delord, *Un autre monde; Transformations, visions, incarnations, ascensions, locomotions, explorations, pérégrinations, excursions, stations, cosmogenies, fantasmagories, réveries, lubies, facéties, folatreries, métamorphoses, boomorphoses, lithomorphoses, métempsycoses, apothéoses, et autres choses* (H. Fournier, 1844), Getty Research Institute, Los Angeles (92-B7023).

would have recognized it as a parody of other political cartoons. An 1839 law that had lifted some of the taxes on cane sugar from French colonies had provided plenty of material to satirize the country's leaders and the plantocracy. The alternative was to abandon retrograde sugarcane colonies and support the new sugar of the future. That sugar, its advocates said, came from an unlikely and unlovely European plant: the beet.

The boosters of beet sugar industry portrayed it as a beacon of social, economic, and political progress. The future French emperor Louis Napoleon declared himself for this vision in 1842. Beet sugar alone, he wrote in his *Analysis of the Sugar Question*, fostered "the principles upon which repose the good organization of societies and the security of governments."[4] Beet sugar's boosters envisioned a Europe in which people bought and consumed the stuff in large quantities, an autarkic and patriotic replacement for sugar from the tropics.[5]

Beet sugar is a dark matter of nineteenth-century economic history, a substance that has little written history of its own and is only visible through its pull on the economic, social, and political trajectories of cane-growing regions around the world.[6] Yet its impact was enormous, as the chart in chapter 1 showed. In 1840, cane still accounted for 95 percent of world sugar production, but by 1890 beet took two-thirds. The quantitative aspect of beet sugar's history can overshadow its other elements, a distortion that is not helped by the scale eventually achieved by the global rivalry between these two crops. For the first decades of the nineteenth century, the continent's agronomists were preoccupied with the question of whether it was, or would ever be, profitable to produce meaningful amounts of sugar from beets. That was a matter of political economy, of the interaction of markets and the state. Historians studying the rise of the beet sugar industrial complex have mostly addressed this side of the matter by explaining the state ideologies and policies that sheltered and nurtured the industry until it was voracious enough that it almost devoured its parents.[7] But it was also a matter of knowledge and beliefs about nature itself. It required onerous taxation and immense subsidy, calculated the eminent German chemist Justus von Liebig, to make up for a simple agronomic fact he called "true natural economy." The cane plant was just better at making sugar than the beet plant was.[8]

Beet sugar did not possess a historical "momentum" that let it survive Napoleonic troops burning all its factories to the ground.[9] Its equivalence to cane sugar had to be affirmatively asserted rather than discovered. A few chemical investigators in early modern laboratories extracted one substance from root vegetables, hundreds of thousands of enslaved workers and artisans on tropical plantations extracted another substance from sugarcanes, and the identity of these two was not automatic or given.[10] And even if consumers could be persuaded that it was the same thing, it was another matter to overcome their doubts about its origins to the point that they would accept it as a substitute for their familiar food.

That substitution presented a contradiction. As Sidney Mintz pointed out, "Nature, precisely because it is natural, is not chemically clean. But scientific purity, exactly because it is chemically clean, is not natural."[11] The more chemists explained sugar from the beet, the less it might seem like sugar from the cane. Framing the "sugar question" in terms of pure political economy could only happen after an original sugar had been answered in the affirmative. Until at least the 1830s, it was not obvious that beet sugar was sugar at all, because sugar did not come from Europe. The process by which beet sugar became an established natural object was part of the larger change in chemists' understanding of the world, one that occurred from the middle of the eighteenth century to the middle of the nineteenth. Both that understanding and the one that replaced it were also inseparable from colonialism and its substances.

All the Rage

In eighteenth-century Europe, those who investigated the nature of matter divided their experimental materials into two kinds. There were "pure" substances, created in laboratories by chemists' own painstaking labors. Experimenters understood and described such substances through their composition, their affinities with and effects on other substances, and the means by which they could be analyzed or synthesized. But these were the small minority. For the most part, chemists, physicians, apothecaries, and others interested in natural history studied substances not made directly by humans, but by plants.[12] Thanks to centuries of colonial projects, imperial networks, and mercantile desires, a majority of the substances that an early modern chemist sought to understand were not animal nor mineral but vegetable.

One of these was sugar. By the middle of the eighteenth century, sugar was no longer the costly luxury used by aristocrats and merchants to flaunt their wealth, but neither was it yet the everyday ingredient of the later industrial diet. Beyond sugar, the range of vegetal products that lay at the fingertips of these mid-eighteenth-century Europeans would have been unimaginable a century before. Merchants' stalls and druggists' shops exploded with every part of the plant, "roots, leaves, flowers, fruits, seeds, woods, barks," and other raw plant matter.[13] From these arenas of exchange, one bought *simplicia*, substances extracted directly from plants. Many of these, including sugar, were classed as drugs, accepted to have medicinal and curative properties. Almost everything was a remedy for something. But in practical terms, for simple substances, one went to the market. More complicated products came from the sites of early modern industry, where primary plant matter was manipulated. Distilleries, coffee houses, weapons workshops, tanneries: From these places experimenters wrangled flours, oils, alcohols, salts, juices, tinctures, extracts, and other concoctions and decoctions.

This was the chemical world of the man who would invent beet sugar, Andreas Marggraf. He came from a pharmaceutical family as the son of the apothecary to the Prussian court, and he spent his early adulthood traveling around the German-speaking lands learning from and apprenticing himself to some of Europe's best-known chemists, pharmacists, physicians, and even metallurgists. By his late twenties, when Marggraf was appointed a member of the Académie royale des sciences et belles-lettres in Berlin, his contemporaries acknowledged his experiments for their care and skill.[14] For Marggraf himself, investigations into the makeup of matter went hand in hand with his work running his father's apothecary shop. Being a pharmacist, in the seventeenth and eighteenth centuries, trained you to be an empiricist.[15]

When Marggraf or one of his colleagues wanted to understand one of the substances they bought, sold, or analyzed, the provenance of a plant substance was at least as valuable as any other piece of information about it. Chemistry and natural history were close allies—indeed they often manifested themselves in the same person—and the taxonomic obsession that built cabinets of curiosity and *Wunderkammern* also shaped chemists' arrangement of the world of substances.[16] What plant it had been derived from and where that plant grew were important classificatory tools for a substance, alongside its more sensorily discernible properties, like whether it stank or at what temperature it would ignite. The world of the eighteenth-century chemist was sensory—or, to use Lissa Roberts's better and more evocative word, sensuous. Chemists had to learn how to deploy their taste, smell, sight, and touch, and how to train their bodies to be their most finely attuned instrument.[17] At the end of the eighteenth century, the rhetoric of "precision" championed by Antoine-Laurent Lavoisier and his allies would deprecate and then steamroll this form of knowledge into apparent obsolescence. But until then, all of these sensuous bits of information helped locate a familiar plant substance within existing taxonomies or told chemists where a novel substance should go.

Given the torrent of objects, specimens, and substances from across the planet, a major problem for eighteenth-century European chemists was to relate the classification of plants themselves to the classification of the substances derived from plants. Did similar plants always secrete similar substances, and, conversely, could substances help identify the plant? To Carl Linnaeus, and his fellow botanical system-builders, it was possible to discover general rules about both plants and their substances. Plant chemistry should map onto botany. "All plants of the same family always have the same virtues and properties," asserted an influential French contemporary of Marggraf.[18] But not everyone agreed with him that the plant kingdom should have natural laws, at least not rules that could be discerned by the human mind. This dispute was the important context for Marggraf when he began, in the early 1740s, to investigate why plants in his backyard tasted sweet.

As Marggraf wrote later, he was moved to the subject by the realization that the same "salts," as many crystalline substances were known, could be extracted from a variety of apparently distinct plants. He considered several candidates for his experimentations from the category of "species of plants manifestly endowed with a sweet flavor." He chose to work, in the end, on beets and chard. In particular, he selected the red and white varieties of the mangel-wurzel, a huge beet newly bred in the eighteenth century for use as fodder.[19]

The name literally means "root of scarcity." According to an English treatise on the vegetable from a few decades later, the mangel-wurzel, also known as the mangold, would feed animals in times of plenty and humans in time of famine. It was easy to grow when and where other crops, even its cousins among beets and turnips, proved reluctant. The author of the treatise declared that it should more justly have been called the root of abundance.[20] Not long after American independence, Benjamin Rush sent mangel-wurzel seeds to George Washington and called it an "extraordinary vegetable" and a patriotic resource, and Thomas Jefferson recalled decades later that it had been "all the rage while I was in France; but soon went out of vogue, I know not why."[21] A square yard of garden would yield fifty pounds. It had fat round roots and green leaves that tasted a little like asparagus. The mangel-wurzel was a miracle of eighteenth-century vernacular plant technology, but it was the opposite of exotic.

Marggraf rejected the facilities of the Académie and conducted his experiments in his family drug shop instead.[22] He chopped and dried the roots of his mangel-wurzels. As a good sensualist, he tasted them, examined them closely, dissolved them, and boiled them in liquor. After a few weeks, crystals began to appear, and like van Leeuwenhoek a century earlier, Marggraf placed these crystals under his microscope. (Despite van Leeuwenhoek's work, and depending on one's definitions of chemistry and laboratory, some historians have claimed this moment as the first use of the microscope in a chemical laboratory.)[23] The pale crystals, he wrote, looked just like those that one could derive from the cane plant. What he saw was "not merely a substance approaching sugar," he concluded, but was "in fact a true and perfect sugar that bears a complete resemblance to the familiar sugar extracted from the sugar cane."[24] It didn't just look like cane sugar, but really was cane sugar, only from beets. When he published his conclusions, in the *Mémoires* of his Académie in 1747, he placed this novel claim in the title: "Chemical experiments made with the objective of extracting a true sugar [*un véritable sucre*] from various plants that grow in our countries."

And in asserting his findings this way, Marggraf pushed beyond anything his predecessors had claimed. Olivier de Serres, for instance, a humanist and agronomist of the late sixteenth century, had noted a resemblance in passing between the products of beet and cane in his *Théatre d'agriculture*, a treatise on farming of 1599. Although the politics of beet sugar led some later French historians to backdate its invention to de Serres, he had claimed something less definitive and definitively less radical than Marggraf. All he had said was that when the juice of the beet was boiled, it acquired a beautiful red color and was "semblable à syrop au succre." Similar to sugar syrup, yet hardly true and perfect.[25] But then de Serres had predated the microscope by three-quarters of a century, whereas Marggraf had used it to show conclusively, or so he claimed, that the sugars were not just identical but truly the same substance.

A domestic vegetable mostly unloved except by cows was about as far from the tropical cane as you could travel among the edible provinces of the vegetable empire.[26] Could the same substance come from such different plants? In claiming that it could, in writing that the beet not only tasted strongly of sugar but actually contained it, "very copious and very pure," Marggraf attempted to advance a position about the interconnectedness and organization of the whole natural world.

Marggraf had been a prolific experimenter, and sugar was just one of his many inquiries. His publication itself provoked little contemporary reaction, although it was at least noted in France.[27] His work was part of a broader shift in the conceptual landscape of chemistry over the second half of the century.[28] First, chemists largely stopped differentiating between the *simplicia* plant products, like gum or resin or sugar, and the *composita* materials, which they extracted from plant matter through more elaborate processes. Marggraf himself, in his sugar experiments and in others, was one of many who noted the similarities between what they could produce in the laboratory from plants and what they could purchase. Then, toward the end of the eighteenth century, chemists again began to reconceptualize plants and their products around the theme of order. They distinguished "organized bodies" (plants were "organized" by their hydraulic systems) from the unorganized agglomerations they knew as minerals, and experimented with new relationships between the chemical and botanical worlds. By the last years of the century, even chemistry students as far away as Princeton were taught that sugar existed in almost all flowers and fruits, not just cane, beet, and maple, but birch, corn, carrot, and parsnips, among others.[29]

When Franz Karl Achard, Marggraf's protégé, assumed the directorship of the Académie's laboratory on his master's death in 1782, among his first projects was to pick up the work on sugar and redirect it.[30] The middle of the eighteenth century had been an era of variable but often high sugar prices.[31] Marggraf had hoped his discovery would prove useful beyond the laboratory.

He imagined that small farmers might grow and process vegetables for sugar in their own household. Since they could hardly be expected to boil their produce in alcohol as he could in his laboratory, he had proposed cheaper methods with cheaper ingredients, to free the peasant from the tyranny of purchasing imported sugar at ruinous prices.[32] And his revolutionary promise of independence was embedded into the title of his original 1747 publication: Sugar could be gotten from "various plants that grow in our countries."

When he picked up Marggraf's work, Achard focused not on the household but on production at scale. Drawing on his own training in botany, he used the grounds of his personal estate to experiment on the plants themselves and to investigate how to cultivate them. And where Marggraf had extracted sugar from roots in his father's shop, almost grain by grain, Achard took the laboratory he commanded at the Académie and turned it into a factory.[33] After thirteen years of work, in early January 1799, he communicated his findings to Frederick William III of Prussia. In exchange for exclusive rights to produce beet sugar and farmland sufficient to grow the beets he needed, he promised an independent Prussian sugar. As proof, he sent a loaf of sugar, shaped just the same as those from the tropics. This one even came wrapped in white silk. The flattery worked: Frederick declared sugar from beets a state project and ordered an accelerated program of official trials. These yielded Achard, by the end of November, a full-size factory built at the king's expense along with funds and land.[34]

Rumors of alternatives to cane sugar had already been in the air for a year or more when, in the fall of 1799, news of the beets reached France.[35] The letter relating Achard's success arrived just as Napoleon ascended to the position of first consul. Paris was not lacking in excitement, but sugar from beets broke through. Achard had, reports suggested, at last found a replacement for sugar in the humble animal fodder of the beet (although there was some dispute over whether this correspondent had gotten the species right).[36]

The Institut de France, the official scientific organ of the republican state, commissioned a panel to duplicate Achard's results. It hoped to demonstrate that sugar could be made from beets at an economical cost and profitable margin. But though the commissioners agreed that it was possible to tease a "beautiful" sugar from the beet, they reached only ambivalent conclusions about the economic value of the process. The American ambassador to Prussia met one of the directors of the new company, who "assured us it was impossible to make this sugar under double the price, which that from the West Indies amounts to." And even then, John Quincy Adams noted, its "appearance is equal to the very finest sugar from the cane; but it is neither of so close a texture, nor so sweet to the taste."[37]

News of beet sugar spread quickly to Cuba, where planters panicked that it would spell their ruin.[38] Thomas Jefferson pestered his French scientific contacts for the best beet intelligence on antislavery grounds. He already had the latest treatises but was not satisfied. "The bette-rave, which we can all raise, promises to supplant the cane particularly, and to silence the demand for the inhuman species of labour employed in it's [*sic*] culture & manipulation," he wrote an astronomer.[39] And to a botanist, whom he also begged for the recipe: "You will by example teach us how to do without some of the tropical productions. The bette-rave, I am told, is likely really to furnish sugar at such a price as to rivalize that of the Cane."[40] In this way, he was among the first in a two-century-long line of American strategists who dreamed of an alternative sweetener under the United States' control and who tied those dreams to complaints about forms of labor, especially foreign labor.[41]

Meanwhile, the project was shelved in France, even as the price of sugar rose. Wars, blockades, and self-declared embargos cut France's continental empire off from tropical sources of cane sugar, and Napoleon feared that sugar scarcity and cost might provoke popular uprisings. But the large-scale production of beet sugar seemed like a gamble.[42] What was possible in a laboratory might not be profitable at scale. Domestic fiscal and agronomic interests were far from reaching a consensus that it would be a good idea, even if it were technically and economically possible, and Napoleon himself doubted whether there was a future in homegrown French sugar, though he was alert for new sources of energy to feed his voracious armies.[43]

And the concept itself still seemed ridiculous. When, in 1803, Bonaparte publicly mused about the feasibility of extracting sugar from indigenous European plants, a royalist spy considered it evidence that the first consul had lost his mind.[44] To lend it credibility, in fact, French papers and politicians began to spread a rumor that the British had attempted to bribe Achard and suppress news of its existence. These reports emerged as early as 1811 and extended at least until the 1840s. The amount of the bribe offered to Achard to hide his findings was fifty times as much as he had been given by Frederick William III to develop the sugar. If he would publish a recantation, the British supposedly had offered him a hundred times as much. These purported sums spoke to the potential importance of an "indigenous" European source.[45]

Not until a second commission reported that beet sugar was economically viable, after a decade of quiet but busy improvement by French agriculturalists, did the emperor's policy change. By 1811, the price of cane sugar was twelve francs. Before the revolution it had cost just one.[46] So in March of that year, Bonaparte decreed that *l'état* would support launching a beet sugar industry: a million francs in loans, four years exempt from taxes, and forty square kilometers of northern France allocated to cultivation. It also created six "experimental schools for teaching the production of beet sugar."[47] Within a year, this crash program had yielded 3,400 tons of domestic sugar, and Napoleon hoped he would soon be able to banish all cane sugar completely.[48] By then, there were already more than a dozen factories operating in the German lands.[49] Beet sugar held out the promise that European rulers might unshackle themselves from those tyrannical colonial masters, cane planters. "Its manufacture is easier and more enlightened," wrote Jean-Antoine Chaptal, then director general of Commerce, Arts, and Manufactures, to a regional prefect in 1815, "and we are approaching the moment of freeing ourselves from the New World."[50] Stung by defeat at Haitian hands, here was France's chance for revenge.

The delivery of the first loaf in each country was a patriotic triumph, signifying that a "European sugar" had finally been extracted and purified from an "indigenous" crop. In one fanciful commemoration, Achard bows as he presents a white loaf to Frederick William III, though in reality this lowly chemist never obtained an audience.[51] In another, it is a minister presenting the loaf to Napoleon, who upon touring the first successful beet factory on French soil awarded a Legion of Honor to its proprietor and a week's bonus salary to every worker.[52] But after the collapse of Napoleon's continental empire, and the reopening of colonial trade, the price of cane sugar returned to earth and the prospects for beets collapsed. Most factories closed, the survivors shifted to adjacent products like syrup, and the industry receded for almost a decade and a half.

A Like Equality

A few years earlier, another sugar substitute had been all the rage in Europe, whose adherents also professed that consumers would be unable to tell any difference. This one also came from the New World and also promised to end slavery. The sap of the maple tree, like the pulp of the beet, held the possibility of a different economic and social form. Unlike cane sugar, it did not seem to depend on massive unfreedom and concentrated wealth and power.[53] To Thomas Jefferson and his fellow enthusiasts, one of the happiest characteristics of the sugar maple tree was that children could tap it. "What a blessing," he wrote in a letter meant for eyes in Britain's high offices, "to substitute a sugar which requires only the labour of children, for that which it is said renders the slavery of the blacks necessary."[54] Jefferson hoped that maple sugar would distract the wolf of slavery long enough to let him release its ear.[55]

One of the intended readers of the maple letter was the chief imperial naturalist, Sir Joseph Banks, and so Jefferson made an argument from economic botany. The trees required just women's and children's hands, yet they would last years and years, unlike cane, which demanded constant replanting. Jefferson's critique of cane was also partly ecological and partly civic. Colonists and now citizens had been distracted by cheap cane, careless about the bounty of their continent's own forests. But the chief argument was about quality: "This tree yeilds [*sic*] a sugar equal to the best from the cane" and, despite working the limited capabilities of the young and the weaker sex, "yeilds it in great quantity." He offered to send an example, but by a messenger rather than by the unreliable mail. The sample mattered, because as with every substitute, the worry was that maple sugar just wouldn't be the same.

European visitors to America had long disagreed about whether sugar made from maple sap was the same as sugar made from cane juice.[56] It also mattered whose hands, eyes, and minds were in charge of its making. Touring New France in 1721, the French envoy, historian, and priest Pierre de Charlevoix considered maple sugar curious enough that he brought some back to old France. Ultimately, he rejected the notion that crystalline crystals from cane juice would always be superior to sugar made from maple sap, and having rejected it, he pressed on to claim maple's superiority, under certain circumstances. "It is purer, and much better, than that of the Isles," he wrote, but only "when it leaves the hands of the Savages"; that is, only once the French, with their colonial sugar-making expertise, had shown the long-standing inhabitants of North America how to properly commercialize their familiar tree.[57] Armed with the samples he had carried back to Europe, he consulted a refiner in Orléans, who concurred that the sugar was as perfect as any he had seen.

A few decades later, the Swedish naturalist Peter Kalm leveled the odd charge that Charlevoix was "falsely informed" by his tastebuds. As we've seen, purity and goodness might be quite separate characteristics from taste, and in any case, whether the Jesuit found maple sugar tastier was surely up to him alone. And for maple boilers too, the art was in the eye: "Those accustomed to the process can easily judge when cooking and thickness are sufficient," Kalm reported to the Swedes, but no book could teach you how: "It is not as simple to learn from description and directions as from experience."[58] What was simpler in maple country was that when you needed to cool the syrup—for which Kalm borrowed the brewers' term "wort"—you could just carry it outside and leave it in the cold. This process made a "brown flour-like sugar" that Kalm compared twice to muscovado. But it behaved differently, forming lumps and taking the shape of its container. Kalm didn't like the taste, which was just "passable" in tea or coffee if one also added milk. Plus, maple sugar dissolved more slowly and had less sweetening power. But it was about the same for preserving, it did make better chocolate, and it could supposedly treat chest colds. If its deficits could be rectified through refining, Kalm too saw a bright future in global markets.

To Benjamin Rush, maple sugar could be "good grained" just like cane, and if American houses sported copper kettles the color would be "fairer" too. It could be clarified, could even be "clear," if other farm goods were added: eggs, lime, fresh milk. If a household was capable of making ordinary home products like "soap, cyder, beer, sour krout, &c," then its members already possessed enough knowledge to make maple sugar. Even better, the same northern ecology that produced hardy independent farm families ruled out the spiraling synthesis of industry and agriculture visible in Caribbean boiling houses. The trees were isolated, the sap had to travel a long way, and food had to be brought to the woods "in a season of the year in which nature affords no sustenance to man or beast." There was neither cause nor benefit to blighting the forests of New York with maple sugar factories. The ecology helped in other ways. The maple could boast that it was tapped during months "when not a single insect exists to feed upon it, or to mix its excretions with it, and before a particle of dust or the pollen of plants can float in the air." Meanwhile, bugs making up a whole "page in a nomenclature of natural history" feasted on West Indian sugar from before it was cut to after it was sealed in a hogshead.[59]

As always, there was more to purity than just purity. Opposing slavery did not, needless to say, stop Rush from paraleptically insinuating that the Africans who lived, worked, and died in the cane fields themselves contributed to the impurity of cane sugar, as contrasted with the virtues attached by autarkic Yankees. "I shall say nothing of the hands which are employed in making sugar in the West-Indies," he insisted, "but, that men who work for the exclusive benefit of others, are not under the same obligations to keep their persons clean."[60] If maple sugar seemed inferior, it was because it had been made "slovenly"; done properly, not even Alexander Hamilton could tell the difference.

The attention given to maple sugar in the early 1790s was largely the result of one hardworking booster, William Cooper. This energetic promoter of northwestern New York State hoped the sugar maple tree's output would benefit Euro-Americans of all classes and even ameliorate slavery.[61] It would lessen the trade deficit and bankrupt the planterocracy. But first, Cooper had to enroll the families and communities who did the sugaring—persuade them to work on a larger scale by buying new equipment from him, and to alter their comfortable domestic process to meet the fickle demands of faraway consumers. For to compete with cane sugar as "a Suitable Artical among Those of a Delicate Taste," he said, they needed to upgrade their technology and protect their products from other aspects of New York's nature. Both Cooper and his sugar were woodsy bumpkins who dreamed of entering urbane society.

Cooper teamed up with a Philadelphia manufacturer, but getting kettles to the middle of New York was hardly easier than getting mills to the Antilles from Scotland. Nor did the state's weather comply with Cooper's projections. Production fell far short of what he had promised to his subscribers in Philadelphia, who had agreed to pay 7 pence a pound. This price was a 30 percent discount from the prevailing price for cane sugar in the northeastern United States and reflected both the speculative quality of the project and the unknown quality of the sugar itself.[62] Some things were the same. It could rain in Cooper's Town just as it could rain in the Caribbean, and Cooper, inaptly, could not keep water out of the hogsheads into which he'd packed his sugar—a note of similarity to cane sugar, perhaps, for hogsheads were poorly suited to internal transport. But once the barrels made it to Philadelphia, the sugar, "superior in flavor to the best muscovado," was worth it. His ally Benjamin Rush pressed some into the hands of George Washington, and Cooper's samples spread across the ports and capitals of Europe.

Cooper learned some lessons. The next year, he handed out kettles on credit to more farmers, he built his own instructional works and even a refinery in the woods, and he hired men to tap trees for a mile in every direction. Already, in order to compel American ecology to satisfy a European market, Cooper had to sacrifice the yeoman vision Rush had promised. But Cooper was not the first and not the last to discover that sugar refining was harder than it looked: an "Extrem failure," he said. The weather still didn't cooperate, nor did the trees, nor his tappers. Just as Jefferson and Europe were getting hooked, Cooper cut his losses.[63]

By the time Cooper gave up, maple sugar was all the rage in Europe. Samples appeared in Amsterdam, that longtime capital of sugar commerce and refining knowledge.[64] A loaf of refined Philadelphia maple showed up in Paris, "occasioning mobs." The city's aristocrats-turned-revolutionaries were most pleased to show it off, especially if they counted themselves among an antislavery crowd.[65] When the anticlerical Republicans abolished slavery and replaced saints with plants on their new calendar, they gave the twenty-first day of Frimaire to the sugar maple (though its placement at the end of autumn suggests a lack of familiarity with the tapping season).[66]

Back in maple country, not even Thomas Jefferson could get his hands on any. He saw "maple sugar, grained" advertised in the papers and sent James

Madison questing around New York to purchase a hundred pounds for him.[67] But either the advertiser had jumped the gun or maybe he just knew how to juice a market, and Madison found the city's sugar buyers anxiously awaiting the shipment from Albany.[68] A Virginia lawyer was no match for Manhattan merchants, and by the time Madison had heard of the shipment's arrival, its cargo was gone, gobbled up at auction. Refiners, not just Jefferson, had caught wind of the demand among the "friends of humanity" in Paris.

After two weeks of hunting, Madison managed to secure a sample. The price of the "maple sugar, grained" that Jefferson wanted was still pegged to the going rate for muscovado; the maple tappers had not yet managed to pull that market inside out. Although the ravenous demand might have inflated Madison's expectations about the sugar itself, it was a bad sign that the only stuff he could find was a good deal cheaper than cane sugar. "The quality is so far below the standard formed by my imagination," Madison despaired, that he dispatched a sample so Jefferson could decide for himself. And indeed it was so bad as to be useless for Jefferson's goal, which was not to sweeten his coffee but "to send it to Monticello, in order, by a proof of it's [*sic*] quality, to recommend attention to the tree to my neighbors."[69] The enslavers of Charlottesville were not going to switch to maple on ethical grounds. "When double refined it is equal to the double refined of the Cane, and a like equality exists in every state of it," Jefferson insisted.[70] But not the stuff Madison had gotten his hands on.

Both Virginia's Piedmont and Parisian salons were minor territories in the imagined maple empire. From the rockily unpromising farmland of upstate New York, the maple sugar boosters of the early 1790s dreamed of toppling the whole brutal edifice of sugarcane. Benjamin Rush and his fellow visionaries—including his unsubtle Society for Promoting the Manufacture of Sugar from the Sugar Maple Tree—proposed not only freeing Americans from the moral peril of having to buy their sugar from enslavers but even bending and inverting whole oceanic triangles of trade.

If "sugar maple country" were sufficiently populated and developed, its promoters thought, its trees could not just supply the United States but could actually "undersell" cane in Europe.[71] Sooner or later slavery would end, and producing sugar on Caribbean plantations would no longer seem so economical, nor prices of cane sugar so affordable. During the maple season on farms across New England, they supposed, women and children were sitting around with nothing to do. Hence maple sugar would soon be abundant, cheap, and of high quality. But until then, when Jefferson wanted to send some to his admirers in France, he had to borrow some from President Washington himself.

Suck, My Dear, Suck

Despite Marggraf's initial declarations, the changes in beliefs among the chemical elite of Europe, and the enthusiasm among the physiocracy for the new beet sugar, it was not yet clear that it actually was sugar. Everyone already knew what sugar was, after all.[72]

So reactions had been mixed, in 1799, when news of Achard's project arrived in France. Some readers greeted it as news that the day was approaching when France and Europe could escape the British stranglehold on trade, but to others the whole thing was obviously an illusion. One refiner recalled laughter directed at the credulous, while skeptics "were not afraid to call [beet sugar] charlatanism."[73] Louis Napoleon's *Analysis of the Sugar Question* denigrated the reaction in order to exaggerate later success. "Parisian sarcasm greeted the precious discovery, and men who still doubt the unknown laughed at this new conception of genius."[74] He might have known, having rotted his own teeth on the sugarcane his grandmother grew in the greenhouses of Malmaison.[75]

The object of the sarcasm was a light brown powder, dry and resistant to absorbing humidity, "which tasted like sugar unaltered by any foreign tastes," according to the initial letter from Germany.[76] It is hard to know what the initial beet sugar tasted like. Some reports from the nineteenth century suggested that however much this first beet extract appeared the same as cane sugar, it still tasted earthy and acrid, like the beet itself. The beet's defenders retorted that this was an unsupportable slander.[77]

The natural world was full of substitutes, impostors, and outright forgeries. Given the ingenuity of plants, who knew what this powder really was? Perhaps it would disappoint, just as the "new sugar" from grapes had disappointed. A French chemist named Proust had isolated it from *les raisins* that same year, 1799, but as he announced this novelty he also admitted it was different from cane sugar and would not readily crystallize.[78] Grapes had been Robinson Crusoe's "agreeable" substitute for sugarcane when the fictional planter (and renegade enslaver) found them growing on his fictional pseudo-Caribbean island.[79] Sugarcane grew there too, uncultivated, but grapes provided a source of sweetness in his shipwrecked life. Unlike crystalline cane sugar, grapes wouldn't have helped Crusoe sweeten his tea, which of course he didn't have either.

The letter reporting Achard's news, therefore, reassured its readers that the powder was really sugar. Dissolve it in alcohol, carefully cool it and evaporate it as Marggraf had once done, and the "saccharine parts" would emerge as "a sugar very white and perfectly pure."[80]A decade later, Napoleon's 1811 decree, and his national crash program to build beet factories and gain independence

FIGURE 2.2. A cartoon supposedly from ca. 1811, mocking Napoleon's plans for beet sugar and tying them to the difficulty of persuading consumers to switch their allegiance. "Suck, my dear, suck! Your father says it's sugar!" Jakob Baxa and Guntwin Bruhns, *Zucker im Leben der Völker: Eine Kultur und Wirtschaftsgeschichte* (A. Barten, 1967), 135.

from foreign sugar, still prompted ridicule as well as urgent motion.[81] And not all pictorial representations of beet sugar were flattering. One cartoon in particular has survived, crude but effective. On the right, the emperor frowns as he wrings a few miserable drops from a beetroot into his coffee. On the left, his perplexed infant gums another beet. "Suck, my dear, suck," grumps the nurse to the dauphin in his cradle. "Your father says it's sugar!"[82]

At that very moment Achard himself was still protesting the quality of the sugar extracted from the beet, fending off claims that it could be distinguished from its competitor. He discussed the issue at length in his 1812 *Traité complet sur le sucre européen de betteraves*, describing how he could make a sugar "entirely similar to the sugar of the colonies." It looked worse, he confessed, because its crystals were covered by a thin film of beet molasses, but once that had been washed away, he promised that the sugar was "capable of receiving, by refining, all the shades of whiteness and the degrees of purity that one might desire." He also detailed another process, more elaborate but yielding better rewards in the form of a dry yellowish sugar with crystals like sand, “absolutely similar to raw Indian sugar, and can be compared to it for quality.”[83]

Within a few years France boasted sugar schools, beet factories, and countless acres under cultivation, yet Chaptal, the statesman and chemist, still felt the need to address repeatedly the question of whether beet sugar was the same as cane sugar. He reassured the regional prefect that rumors of any distinction were merely just a “prejudice” left over from the birth of the industry a decade and a half before. In his *Mémoire* on the sugar beet, the very first query he answered was “Le sucre de betterave: est-il de la même nature que celui de canne?” “Today, there is not the slightest doubt in the minds of enlightened men about the perfect identity” between the two, he replied. “And when they have been brought back, by refining, to the same degree of whiteness and purity, the most informed person can find no difference.”[84]

Of course, Chaptal acknowledged to the natural philosophers and engineers of the Institut royal,[85] the beet sugars of a decade earlier had tasted acrid or burned. But its consumers had understood, he said, even then, that the faults came down to the process and not to the sugar, and the enlightened had known the truth. In his own house he exclusively served beet sugar, and he asserted that guests complimented him on its beauty and suspected nothing about its origins—or rather, they suspected only on “a few days.”[86]

It did not help that the process for producing beet sugar was nothing like the process for producing cane sugar. The beets were sliced and soaked to extract their sweetness, rather than crushed, and intermediate products, unlike those of cane sugar or maple syrup, were unpalatable. By contrast, maple’s backers soothed potential customers in part by promising that the methods of production were similar to those for cane sugar; so similar, Rush said, he didn’t need to explain it. Furthermore, the method by which a plant substance was extracted from its original place in nature was part of its identity. The point was viscerally illustrated by another cartoon, far cruder than the swipe at Napoleon. This one survives from a German-language pamphlet dated around the time of Achard’s initial presentation to the Prussian king.[87] Four men surround a happy cow. On the left, one gingerly feeds the animal whole mangel-wurzels, and apparently smelly ones. On the right, a second man, even more gingerly than his colleague, receives the animal’s output. From its rear end are emerging whole sugar loaves, beautiful and perfectly conical, just like the ones presented to the crown. The third man milks molasses from her udder. And in the background, the man of science and calculation, observing and taking notes: “Is the sugar fine and the syrup pure? I’ll put it in the ledger.”

The image’s precise technical style apes and mocks the meticulous depictions of production found in contemporaneous works like the *Encyclopédie*.[88]

Wie die Krähwinkler Runkelrübenzucker fabriziren.

FIGURE 2.3. How country folk might think beet sugar was made. Jakob Baxa and Guntwin Bruhns, *Zucker im Leben der Völker: Eine Kultur und Wirtschaftsgeschichte* (A. Barten, 1967), 135.

A machine pointlessly belches steam. "Wie die Krähwinkler Runkelruben-zucker fabriziren" reads the caption; roughly, how the obtuse small-town bourgeois makes sugar.[89] The image ridicules not just Achard and his scheme—is this the scientific process that made the loaf for the king?—but also the idea that sugar could be extracted from the beet at all, and the provincials who might eat it. Perhaps it's even a banker's conspiracy: "The beets go up in steam and the interest charges remain," according to the supporting text. The French historian Daniel Roche, probing the invention of consumer culture, noted that sugar was urbane: "Sugar and femininity were linked with lightness and delicacy, in contrast to the rich and gross content of the menus of the people and the bourgeois."[90] Some novelists, philosophers, and critics were anti-sugar and pro-townspeople, and for them, beets were heroic; this cartoonist demonstrates that the opposite view had many adherents too. Only a rube would think of whatever Achard could make from beets as "sugar," and it would be just as fit for human consumption as manure. The mangel-wurzel was a vegetable that in Germany was given to cattle, never mind that in France it might have supported the people. Thus, the economic question about beet sugar—whether it could be produced at a cost to compete with cane sugar—was meaningless if nobody believed it was the same thing in the first place, and Achard and Chaptal needed to continually protest their equivalence.

Extract It from Everything

By the 1820s, cane growers could again ship their slave-made sugar to Europe; no amount of idealism could support the nascent beet sugar industry. "During the late troubles in the West-Indies," a children's book explained around the same time as Chaptal's insistent defenses, "when Europe was but imperfectly supplied with sugar [. . .] very good sugar was obtained from parsnips and from carrots; but the process was too expensive to carry on this enterprize to any extent."[91] The few factories that survived the end of the blockade did so either as hobbies for ambitious and independently wealthy men or as producers of syrup, not granulated sugar. They were helped, at least in France, by the protectionism of the colonial sugar lobby. Legally, what they produced was sugar. If they could export any of their wares, they still received a bounty, just like cane producers who reexported theirs. But unlike colonial importers of cane sugar, they owed no taxes on entry. And they continued to tinker with their methods. In technical terms the period from 1815 to 1830 was just as important to the development of beet sugar as the previous one had been.[92] A moment of discovery was less important than a process of extraction and acceptance.

By the 1820s, elite opinion in Europe, even among antislavery scientists, held beet sugar as still just another failed substitute. The Prussian polymath and explorer Alexander von Humboldt had traveled to Cuba in 1800, just after "some samples of beet sugar from Germany which people said 'threatened the existence of the sugar islands in America.' The planters had recognized with a sort of fear that this was a substance entirely similar to cane sugar, but flattered themselves that the scarcity of labor in Europe and the difficulty of separating crystallizable sugar from such a large mass of vegetable pulp would render the operation unprofitable at large scale." By 1825, when he wrote about this voyage, he noted that "chemistry has arrived to conquer these difficulties," but France had only ever produced a minimal percentage of its own sugar demand. The "well-informed" planters of the Caribbean "no longer fear beet, rag, grape, chesnut or mushroom sugars, nor coffee from Naples, nor Indigo from the south of France."[93] Humboldt, fiercely antislavery himself, was glad that the fate of New World slavery did not rest on the shoulders of peasants in the French countryside.

Within a few decades, the vitalist world of the eighteenth century, in which substances were intimately connected to the plants from which they derived, was in retreat, as seemingly every year brought new ways for chemists to breach the defenses of the organic world with substances brought from the inorganic. Cane sugar became just another assemblage of elements. Beet sugar became a domesticated type, and extracting it was the kind of science project that scientifically inclined children could attempt in their home kitchens.[94] By the middle of the century, sugar from beets defined reality to France's most celebrated author of realist novels. "It used to be believed that sugar cane was the only source of sugar. We can now make sugar from virtually anything," wrote Honoré de Balzac, "and the same is true of poetry. Let us extract it from everything, for it lies everywhere and in everything."[95]

One of the principal advocates of the new organic chemistry, Justus von Liebig, remained a skeptic of beet sugar's economic possibilities. Around 1830, he had been dispatched by the German state of Hesse to investigate French techniques for growing beets and refining sugar. He counseled the Hessen government to establish a beet factory, in which he invested. It went bankrupt, and he and his fellow investors lost all their money.[96] He grumbled that the beet sugar industry was staffed by dimwits: "Any ass can manage a sugar factory," he told a colleague.[97]

By the late 1840s, Liebig rejected the economic logic of domestic beet sugar, its "impolicy," yet he was still impressed with the technical achievement of beet sugar refining and extraction.[98] The humble beet, he marveled, out of which Napoleon had only squeezed a grim treacle, had been transformed by science and now yielded "the most beautiful white sugar." The people of Hesse were paying for the inefficiency of beet sugar. In economic terms, German taxpayers might as well construct a "prodigious hot-house" and just grow sugarcane themselves. A wiser course would be to grow other crops, export them, and buy sugar from someone else.[99] "The satisfaction of eating sugar grown on our own soil is therefore purchased by a not inconsiderable sacrifice." The beet sugar question had different fiscal and emotional answers. And Liebig refused to believe that sugarcane planters would remain benighted forever. They were already learning how to avoid losing 60 percent of the sugar in their cane juice. If they managed to get better still, and by just a few percentage points, the entire beet-sugar industry, as he saw it, would have its profits wiped out. The beet sugar industry was in fact already nurtured in a prodigious hothouse, one made of tariffs and subsidies rather than glass.[100] As soon as it was exposed to the cold winds of the marketplace, it would shrivel and wilt.

But already, as Liebig's mission to Hesse suggested, a new set of political and economic considerations had led to reinvestment in France, Germany, and other European states, and from there to experiments across the Atlantic as well.[101] In subsequent decades, these states taxed beet sugar in different ways, sometimes by output, but often by the pound of beet or by the size of the factory plant. Shifting systems of taxation directed and redirected the techniques that beet producers harnessed, refining their machinery as well as their sugar. When the weight of raw beets was the taxed quantity, for instance, manufacturers of sugar and machines poured resources and ingenuity into squeezing more from each kilogram.[102] At the same time, cane-sugar planters imported many of these beet manufacturers' machines as they sought new ways of controlling old forms of labor. As the historian Daniel Rood has shown, planters' initial enthusiasm for adapting some of them to Cuba was wrecked on the differences between beet pulp and cane juice, between temperate European air and humid Caribbean climates, and between a free labor regime and one of racialized slavery.[103]

The next chapter explores where those machines came from, how they got to where the sugarcane grew, and how they worked and stayed working while they were there. Before then, however, the beet-sugar question connects to the remainder of our story through an instrument called the polariscope, the closest thing to a definitive answer.

New instruments accompanied organic chemistry, answering old questions while raising new ones. One of the most enigmatic was the polariscope, sometimes called the polarimeter. It was developed in France by the physicist, colonialist, and counterrevolutionary Jean-Baptiste Biot, who refined his design over the first few decades of the century. Biot used it to analyze and document the weird ways in which light changed when it bounced off a mirror or refracted through prisms made of certain materials. Or, he later observed, when it passed through aqueous solutions of certain substances, including sugar.

His invention, we would now say, detected the direction in which light was polarized. Ordinarily, the waves of light oscillate in random directions, though always pointing perpendicular to the direction the ray of light was traveling. Today we understand that electromagnetic radiation, such as light, is the effect of moving electric charges upon a universal electric field, yet still the phenomenon can be difficult to picture. And as Theresa Levitt points out, in her book about Biot and his rival and friend François Arago, their contemporaries did not even have that electric model to guide them. A polariscope looked like more famous microscopes and telescopes. But those simply magnified their objects, making them normal size if they were arbitrarily too small or too far to see with human eyes. No matter how small or how close, nobody could see polarized light. Polarization was an unobservable thing, which the polariscope first claimed to make observable and then also claimed to quantify.[104]

Remarkably, and inexplicably at first, the polariscope also seemed to be able to distinguish between the products of the living world and the products of human artifice.[105] Substances from organisms tended to rotate light, while substances created in laboratories did not. Solving this mystery was one of the first successes of Louis Pasteur's career. The "optical activity" of these categories of substances seemed a way to distinguish the true from the false and the living from the dead. And in one of his early polarimetry papers, in 1816—a decade and a half after Achard's announcement, half a decade after Napoleon's indigenous sugar program—Biot used the polariscope to assuage doubts about beet sugars.[106] "Solutions of sugar from milk, sugar from cane, and sugar from beets have an effect upon polarized light that is detectable and is of the same nature," he told the Société philomatique in 1816. "This action is of equal intensity in sugar from cane and sugar from beets, which confirms the identity of these two substances."[107] Biot tested samples sent to him by colleagues, containing cane, beet, corn, and other sugars. As late as 1836, however, he was still leaving a crack for readers skeptical of their chemical identity. Two samples, including a beet sample, were subjected to polarimetric tests "afin de constater la nature du sucre qu'ils renfermaient": in order to ascertain the nature of the sugar that they contain.[108]

Biot was an enthusiastic supporter of France's cane sugar plantations. He dreamed that the polariscope might settle both sugar questions at once. It could answer for good the question of whether beet sugar was the same as cane sugar. It could also bury beet sugar as an economic rival by bringing cane sugar factories up to modern speeds. "In repeating the same test, after successive operations which have reduced the natural juice—evaporation, clarification, concentration, reduction into grains—one would know right away, and at every instant, the effect good or bad of each step," he advised colonial planters in 1840. And it would eliminate risk for metropolitan buyers and refiners, who still purchased raw sugar of uncertain qualities "except their being more or less white, or granular, and the knowledge of whence they come. All of this is very vague and uncertain, and exposes buyers as well as sellers to many disappointments."[109] Sugar was a paradigmatic commodity for nineteenth-century capitalism, and the difference between cane and beet sugar, as newly analyzed chemical substances, was a paradigmatic question for nineteenth-century science.

The wrestle for market supremacy between these two plants, and the humans profiting from them, warped the political economy of the entire planet for the rest of the century. Competition for market share in sugar led to European and American governments raising and lowering their tariffs and subsidizing domestic beet industries to the point of state bankruptcy. Politicians tried to civilize western North America by turning it into a giant beet farm, reshaping its economics and its ecology at the same time. Beets even led nations to begin to formalize international trade law.[110] None of that economic history would have happened without the quietly radical idea that the same sweetening substance could come from plants on opposite sides of the planet. At each stage in the creation of beet sugar, its promoters not only had to make the political, economic, moral, and fiscal case for their proposed industry. They also had to make the case, scientific and otherwise, that beet sugar really was sugar. Observing the revolution among the plants, one of Grandville's protagonists wonders if they should warn (human) civilization about the coming upheaval in its diet. No, replies another: "Better to remain silent, and start looking for an invention that will replace vegetables."[111] Beet sugar was already an invention that had replaced a vegetable.

In histories of sugar, beets often come at the end. This placement makes sense, since sugar has been extracted from beets for a sliver of the time it has been extracted from cane. But that placement also makes beet sugar seem like the climax of the story, implicitly the goal to which sugar knowledge was pointing all along. A precondition for the rise of the sugar beet as an economic and culinary force was that enough people had to be willing to eat it, and that meant European consumers had to accept that there was a comparison to be made in the first place. It was not up to chemists to "blaze the way to a new world industry," as sugar's most exhaustive historian summarized.[112] Those who spoke for the beet had to assert and reassert, assure and reassure, that what they produced was really the same substance and that it was an acceptable thing to eat in place of what the world already knew, however bitter and bloody was the aftertaste of the familiar.

3

A Lever That Squeezes

Prime Mover

The sugar plantation of the early modern period demanded extraordinary coordination between field work and factory work. In its precisely coordinated operation, it was, as scholars have remarked, more "industrial" than anything yet existing in Europe. Observers and promoters of the plantation compared it to "a well-constructed machine, compounded of various wheels, turning different ways," as the Antiguan planter Samuel Martin wrote, "and yet all contributing to the great end proposed: but if any one part runs too fast or too slow in proportion to the rest, the main purpose is defeated."[1] At the heart of the plantation was the milling equipment itself, called the *ingenio* in the Spanish-speaking world, but its name in every language served as a stand-in for the entire process.[2] The milling machinery crushed the canes and "expressed" the juice stored inside it, leaving a depleted remnant known as *bagasse* or *megass*. Richard Ligon, in his seventeenth-century history of Barbados, wrote that the components of the plantation "depend upon one another, as wheels in a Clock," and the mill was the plantation's primum mobile, its prime mover, like a clock or the whole universe.[3]

The most common and iconic form of mill across the Caribbean was made up of three wooden rollers, oriented vertically, with their axles arranged as the points of a triangle. Although a Sicilian artisan was believed by historians for decades to have developed this arrangement, more recently the vertical mill and its correct gearing have been credited not to Europe but to China.[4] From there, it was brought to the New World by Jesuit missionaries, who added a third roller that allowed greater and more controlled crushing. One enslaved person fed the cane between two rollers, and then another, on the other side, guided them back through the other pair. The mills might be powered by wind, animals, or even human labor. Planters preferred wind for

FIGURE 3.1. William Clark's series of illustrations from Antigua in 1823 was meant to reassure audiences back home that slave plantations in British sugar islands were stable and orderly, even bucolic. Behind the fantasy, his depiction of cane arriving at the mill shows the integration between the agricultural and manufacturing aspects of sugar plantations. This one was powered by wind, which was cheaper than animal power but not as consistent.
Courtesy of the John Carter Brown Library.

its power, but cattle could still push in calm weather.[5] Water wheels were appealingly steady, but channeling the right flow meant an extensive project of moving earth. Whatever their power source, these installations were heavy and expensive, and the brutal workload of the harvest took a toll on them.

The job of putting canes into the spinning rollers was legendarily dangerous and violent, as the turning mills crushed the thrashing canes under the enormous force of half a dozen draft animals. Putting the canes between the rollers was an art itself, known in francophone colonies as "donner à manger au moulin": feeding the mill. The man who fed the mill cane was sometimes called a feeder, and mills, notoriously, sometimes fed on him too. Sugar plantations were full of enslaved people who had lost limbs to the machine.

European writers described this horror with varying combinations of detachment, fatalism, and annoyance.[6] "If a Mill-feeder be catch't by the finger, his whole body is drawn in, and he is squeez'd to pieces," complained the Barbadian planter Edward Littleton. He wanted to rouse his readers' sympathies not with the dead person but with the image of an entrepreneurial planter punished unjustly. In fact, Littleton thought his fellow planters happier if the worker were simply killed rather than maimed, since the latter meant a burden to carry. "He must have more Negroes, or his Works must stand"—still, that is to say, as the planter watched his harvest rot in the fields.[7] Axes and machetes were as much part of the machinery as the mills themselves.[8] Even if a human body didn't break the rollers, running the machines cost almost as much as they earned. Littleton complained, "The Wear of our Mills (to say nothing now of the Tear, which is casual) is also a continual Charge to us." Plus, he complained, a new mill cost almost five hundred pounds.[9] As Caitlin Rosenthal has shown, the plantation was the origin of accounting, and few planters were as fastidious about their books as sugar planters.[10]

The centerpiece of the boiling house was a series of increasingly small kettles, called coppers or taches. The fire below each kettle heated the juice, and the particulates and scum that floated to the top were skimmed off to stop them from imparting undesirable flavors or impeding the desired kind of crystallization. By the beginning of the nineteenth century, sugarmasters were adding a quantity of lime to precipitate more of these substances out of solution. The art and skill of sugar-boiling lay in reducing the liquid in the open kettles without beginning to burn the sugar itself.[11] Some plantations used the so-called Jamaica train, in which all coppers were built into the same unit of brickwork and heated by the same furnace. As firewood disappeared, planters turned to the cane trash called *bagasse* for their fuel. The Jamaica train was more energetically efficient but trickier to control.[12] From copper to copper, slaves carried the thickening syrup. Finally, in the smallest kettle, the sugarmaster manipulated the heat and additives to bring the solution to the point where crystals began to form. Then he made his "strike"—removed the syrup from the heat and transferred it to troughs, where it was beaten with wooden paddles as it cooled to encourage it to crystallize further.

Once the sugar had been struck, it needed time to cool, solidify, and purge itself of molasses. The long room of the purgerie reminded one French traveler of a hospital ward.[13] A main aisle stretched down the center. To either side, smaller aisles branched off between the "beds." Into each of these beds workers laid dozens of conical forms, where the sugar rested until its molasses bled out. One space in the curing-house or purgerie was known as the "knocking room," where workers took the pots of cured sugar and knocked them hard against the ground, so that the sugar loaf fell out "as a bullet out of a mold." In Europe, sugar loaves on grocery shelves pointed upright like pyramids. On the plantation, those same loaves pointed down. The widest part of the loaf, "somewhat brownish, and of a frothy light substance," said Richard Ligon in Barbados, and its tip, "a much darker colour, but heavy, grosse, moist, and full of molosses," were severed for reboiling. It was the large middle portion, "of a bright colour, drie and sweet," that was sent to fetch a price in European markets as muscovado sugar. Clayed sugar, which was whiter and fetched twice the price, took four times as long to prepare.[14] A "moderate-sized" Cuban plantation of the early nineteenth century might have ten thousand conical molds.[15]

After some unsuccessful and uneconomical eighteenth-century experiments, sugar mills increasingly turned by steam in the nineteenth century.[16] Steam power obviated the need for the rollers to remain vertical, and newly installed mills were rapidly becoming three-roller horizontal models. These presented a much simpler feeding prospect. With one roller rolling atop the other two, the cane only had to be fed once to pass through both crushing gaps, and it was also easier to spread cane across the whole surface of the rollers, minimizing the wear to any one point. It took decades for the additional and more available power to improve the extraction of juices, and for engineers, technicians, workers, enslaved people, and sugarmill owners to work out technical details through long experience. For instance, the gap between the upper roller and the first lower roller needed to be set at around half an inch to crush the cane, while the gap between the upper and second of the two lower rollers needed to be just the width of a sheet of paper to squeeze out the most juice.[17] An additional advantage of horizontal mills was that they could be lined up in sequence, and the cane passed through multiple sets of rollers. By the end of the nineteenth century, some factories were equipped with milling units of over a dozen rollers that dwarfed anything a century earlier.[18] The roller in each position might have a special groove pattern calibrated to the state of the cane by the time it arrived at that spot.

What steam power did promise immediately was the freedom to dispense with herds of oxen for turning the mills and to lower the number of enslaved people required to tend to the animals. And as iron and then steel became the material for rollers, rather than wood, the additional energetic output of steam engines began to make a bigger difference.[19] Yet as Lizette Cabrera Salcedo shows, the steam engine raised as many technical problems as it answered. It needed ready access to fuel and to skilled repairmen, and to make the most of the investment, the rest of the plantation system had to be expanded to keep up with its speed and power. That meant more animals, more feed land, more enslaved workers, and ultimately other new machines as well.[20]

The vacuum pan began to replace open-air kettles beginning in the 1830s. If the juice and syrups could be contained in pressure vessels, and the air taken out, then the water would boil off at far lower temperatures, allowing greater control over the process and lowering the risk that sugar would burn.

As Daniel Rood writes in *The Reinvention of Atlantic Slavery*, this technique and its necessary devices were commonplace in beet factories, and the French engineers who brought vacuum technology to the Caribbean knew to adapt it for cane sugar plantations thanks to their own extensive experience in both sorts of sites. Yet making it work in its new context took creole expertise. In particular, it was a Paris-trained free Afro-Louisianan named Norbert Rilieux who figured out a design that was reliable, efficient, and economical for cane planters to adopt.[21] The vacuum pan itself was the name for the device in which the sugar was finally struck, but before then, the syrup passed through a "multiple effect evaporator," an efficient sequence of increasingly lower-pressure chambers, where the vapor from one pan boiled the liquid in the next.

A final crucial machine was the centrifuge, in the sugar trade more often called a "centrifugal," a word that also came to refer to the sort of sugar it helped produce. This was a giant suspended basket, like a salad spinner, which flung the water and molasses from the sugar crystals instead of letting gravity drain them. It could dry out tons of sugar on the order of minutes. This machine, which took shape in the 1850s, was an even more globally creole product. The contributors to its designs and modifications lived in Germany, France, Britain, Calcutta, South Africa and Hawaii.[22] It was the centrifugal that gave its name to the new kind of sugar the modern factories produced, and still does. By the time the Spanish poet Carlos Peñaranda visited San Vicente, the first modern factory in Puerto Rico, in 1878, the presence of the centrifuge was well established. He was dazzled by the centrifuge's product.[23] The sugar, he wrote in his *Cartas Puertorriqueñas*, "purified of all its molasses, which filters to another external vessel, washed by steam, and dried and compacted by the speed of the air that forcefully dismisses it to the sidewalls of the drum, forms tight masses, and is conducted on litters of white wood to the general warehouse, inspiring envy, with its purity and transparency, in crystal and snow."[24]

All these machines for sugar production came from industrial workshops in Britain, the United States, or France.[25] Above all, however, they came from Glasgow, in Scotland. The city rose to mercantile prominence in the eighteenth century with the West India trade, and became the heavy-engineering center of the nineteenth century after those merchants invested in mining, cotton, railroads, and shipbuilding.[26] Before the abolition of British slavery in 1834, as Stephen Mullen has shown, the city's sugar aristocracy grew rich from their involvement in the slave trade and plantations.[27] Through its people's engineering skills and its merchants' ties to sugar and slavery, from 1800 through 1914 Glasgow produced more of the world's cane-sugar manufacturing equipment than anywhere else, with some estimates as high as 80 percent.[28] Mirrlees Watson, perhaps the world's largest maker, dispatched over two thousand mills from its 800-person Glasgow workshop, judged "one of the most splendid and completely equipped engineering establishments" in Britain.[29] The city that built sugar machinery affords global historians an opportunity to link commodity frontiers with the places that made the tools for transforming nature.

For all the city's contemporary fame in the sugar economy—in 1878, the British consul in San Juan reported that "such well-known names as Tait and Mirelees and Buchanan are to be seen on scores of sugar plantations"—Glasgow has remained largely absent from historians' analyses of that economy.[30] When the city's manufacturers are mentioned, it is usually just to use the city's engineering reputation to add heft to claims about the technical sophistication of a new sugar factory.[31] The machines have fascinated scholars of sugar's second revolution, but they have arrived miraculously.

Glasgow's engineering workshops provide historians with an opportunity to study just what forms of knowledge it takes to produce "standard" commodities at a global scale. Historians of science have long accepted that factories and plantations, as well as a wide variety of spaces of exchange, are zones where knowledge can be produced, just like laboratories.[32] And the commodity status of nineteenth-century sugar depended more heavily on modern science than any comparable natural object. Yet the more that sugar became a chemical object, immutable and mobile, the more that machines for making it became customized and unruly. The circulation of standardized natural goods—superficially effortless—in fact depended entirely on deeper currents of people, objects, and knowledge that could not be standardized so easily.[33] The devices that enable and create precision only travel fitfully and after much social work has smoothed their way, but that travel is what makes it possible to claim that knowledge is universal.[34] The advanced machinery that filled sugar factories in the late nineteenth century enabled efforts there and elsewhere to standardize workers' behavior and divide it along racialized lines.[35]

These sugar machines, as we will see, were fundamental to a standardized commodity precisely because they resisted standardization themselves. Moreover, factory owners' efforts to de-skill labor depended on that individuality, on machines tailored by Glaswegian manufacturers to their sugar factory's exact requirements. Sugar machinery was designed and maintained through individualized systems of design, irregular channels of communication, and the ability to recontextualize people and things. In the production of sugar, wrote Fernando Ortiz, "it is always a machine, fundamentally a lever that squeezes. Sugar is made by man and power."[36] These objects, the instruments of de-skilling, in fact remind us that human craft, skill, and labor are impossible to eliminate in the infrastructures of commodity production, but can only be shuffled out of sight.[37] This chapter illuminates the history of Glasgow's people and their expertise, using the fragmentary records they left in account books, order books, and in traces of their own hands on diagrams across the whole period of sugar's second industrial revolution.

When You Expect to Come Home

Sometime around 1800, one of Glasgow's West India sugar merchants placed an order for a sugar mill with a flax millwright named James Cook. By 1805, Cook had so many orders that he was forced to move his workshop to a larger site on the south bank of the Clyde, and within two decades Cook's works were among the city's leading industrial operations.[38] So many engineers were trained there, wrote one industry observer a century later, that Cook's works became known as "The College." His firm anchored a cluster of sugar-engineering works on the city's south side.[39]

These firms, and the others that would dominate Glasgow's sugar-machinery business, all began as joint partnerships. Some partners contributed capital, others engineering expertise, and others familiarity with or connections to the West India merchant class.[40] In the mid-1860s, for instance, the young engineer Duncan Stewart was offered the position of chief engineer by a Glasgow merchant firm that owned seven estates in British Guiana. Although he had recently started his own London Road Ironworks, Stewart accepted the appointment in order to "master the process of making sugar from the cane," which he could not do from observing Scottish refineries alone. When he returned to Glasgow, a biographer wrote later, he brought "wider technical knowledge which proved to him of greater worth than the mere possession of capital."[41]

The "technical knowledge" that Duncan Stewart brought back from his years overseas incorporated the whole complex of sugar production, human as well as mechanical. His reputation was built on a series of patents for hydraulic rams that regulated the pressure a mill's rollers applied to the cane.[42] Early modern sugar mills were notorious for maiming or killing human beings, but in a steam mill it was far less likely that a slave's or factory worker's fingers or limbs would get caught. Instead Stewart set to work on his hydraulic cylinders to prevent damage from an overload of cane. The management of sugar factories invariably blamed a broken roller on the "injudicious and unskilful" workers. An "accident," in sugar parlance, switched from meaning an injury to a human to an injury to a machine. This explains why Stewart's patented hydraulics were called a “safety device”: not because they prevented a person from being milled, but because they ensured the safety of the capital investment in machinery.[43]

The engineers who traveled back and forth between the Caribbean and Britain played crucial roles as go-betweens, mediating the relationship between sugar estates and the firms that manufactured their machines.[44] Cheaper and faster oceanic travel meant that these engineers could easily spend the grinding season supervising a factory in the Caribbean, and the rest of the year elsewhere, sometimes in the cane fields of Louisiana, which harvested their cane on a different schedule, or in beet-sugar factories. Boosters of Glasgow industry argued that these engineers’ voyages provided unusually direct experience with the processes of sugar-making: “By their long connection with the trade, and especially from the circumstance that principals of several of the firms engaged in it have travelled in the sugar-growing countries of both hemispheres, they have come face to face with the planters upon their estates, and acquired an exact acquaintance with their wants.” Through such careful management of foreign customers, the sugar-machinery business “like the sugar plantations themselves . . . has been anxiously, intelligently, and enterprisingly cultivated.”[45] One engineer, for instance, “began his life of wandering” immediately upon joining Duncan Stewart as an apprentice in 1876, a life that took him to the company’s clients in Argentina, Barbados, and Cuba, as well as to its displays at international exhibitions.[46] Most of Glasgow’s machinery makers in other industries did not need to go to such distant parts of the globe to see their products in action.

Planters came to rely on traveling engineers as crucial intermediaries with the firms from which they wished to order new equipment or replacement parts. Purchasing agents who spent most of their time in Britain were better placed to communicate with engineering firms, but too far from the plantation to know what needed ordering. By contrast, itinerant engineers were uniquely positioned to understand how sugar factories actually worked.[47] While many such travelers were clearly employed by a particular British firm, plenty of others exploited oceanic distance. They might, for instance, maintain a consulting position for a Glasgow firm while still working for one or more sugar estates. Not infrequently an engineer would earn a commission from the plantation for ordering a machine from a firm and then get hired by the same firm to install that same machine.

W. H. Ross, an engineer who placed orders for many Cuban estates and was a confidant to their American owners, traveled back and forth among Havana and New York, Glasgow and Liverpool.[48] From Boston, the plantation owner Edwin Atkins, for instance, asked Ross’s “opinion of the approximate

FIGURE 3.2. Advertisements like these appeared at the beginning and end of handbooks, trade journals, and other publications read by the owners and managers of sugar factories. They emphasized a company's ability to supply all the equipment that such a factory might need, and, crucially, their ability to continue supporting that equipment after purchase. From H. C. Prinsen Geerligs, *Chemical Control in Cane Sugar Factories*, rev. ed. (Norman Rodger, 1917). Courtesy of HathiTrust.

value of a new mill del[ivere]d at Cienfuegos . . . made by Tate Mirless & Watson" and left much up to Ross himself to decide.[49] Shortly after he had purchased a plantation called Soledad, Atkins wrote Ross "giving order for a mill for regrinding on Soledad, you will note that we leave the matter in its detail pretty much to your judgement, as I am sure you will take [care to] give us only the best; and your long experience in these matters make your judgment of value." What made Ross valuable was not only his experience with sugar in general but with Atkins's new plantation in particular: "You know the kind of grinding we aim at doing, and trust you will govern your work in accordance."[50] To maintain himself in Mirrlees Watson's good graces, Ross asked Atkins for data about the performance of the sugar-boiling equipment, which Atkins allowed Ross to share with the manufacturer.[51] Those who set up shop in Britain, trying to manage Caribbean affairs remotely, had access neither to orders nor to the currency of Atlantic information, and they frequently found themselves without clients.[52]

In 1883, the Manchester merchants N. P. Nathan's & Sons purchased a full sugar-production arrangement from Duncan Stewart for a site in the Canary Islands. That price included everything needed for a factory: a horizontal engine with double gearing, a sugar mill with hydraulics, a fifty-foot cane carrier, clarifiers, filter presses, a triple-effect evaporator, vacuum pan, four centrifugals, and molasses tanks, plus piping, mounting, staging, spares, and tools for erecting, all of which cost £7,304.[53] After the parts had been produced in the works, the entire machine was assembled next door in the erecting shop, tested as much as possible without any sugar, and photographed. Then it was disassembled, packed, and shipped, along with erecting tracings and, later, photographs.[54] For the manufacturer, those documents served as insurance against future claims, and they helped reconstruct the machines after delivery.[55] These reconstructions of sugar machines were attempts to put together a fixed set of existing parts, not to build copies from scratch based on two-dimensional representations or textual descriptions. But clearly Duncan Stewart themselves believed paper insufficient, and the tacit skills of knowledgeable people necessary. One of the firm's engineers, Robert Gilbert, accompanied the Nathan's order to the Canaries to supervise.

Although he was Duncan Stewart's eyes and hands, Gilbert was legally an independent contractor. In November 1883, he signed a contract directly with Nathan's, witnessed by the manager of Duncan Stewart and the company clerk. The main task for which he had been hired, at £5 a week plus "suitable Board and Lodging," was to supervise "the erection of and fitting and putting into working order, of Cane Crushing and Evaporating Machinery." Once in Las Palinas, Gilbert was responsible for the machinery, including its unloading from the ship.

Individuals such as Gilbert, capable of supervising the construction of a whole sugar factory, were in short supply. The terms of his contract as a traveling erecting engineer emphasize the difficulty of the task. He was to obey orders from Nathan's, their manager in the Canaries, or their attorneys. At the same time, Gilbert was also obliged to "devote his whole time and attention to and employ his whole art and skill in the said erection and fitting up of the said machinery, and in getting the same in to proper and satisfactory working order."[56] Gilbert could not rely on quick or frequent communication with those who had designed the machines, and at the same time, his erstwhile employers were stifled in their efforts to understand what was going on.

Contracted erecting engineers sent as agents of a firm did not necessarily keep their principals apprised of their activities. In July 1884 Gilbert was still in the Canaries, and the personnel of Duncan Stewart were not even sure of how the work was progressing. He had not kept in close touch. The manager of the works in Glasgow asked Gilbert to "kindly write to us on receipt of this and let us know when you expect to come home." Not only did they want to know how his mission was going, but they needed him back in Glasgow so they could ship him off again—this time on a mission to Brazil, accompanying "a large Sugar Plant which will be ready for shipment early next month."[57] He missed that trip. He was still in the Canaries in mid-September, and the sugar factory was still not finished, awaiting replacement parts from Duncan Stewart. "We have sent as much of the details as you have ordered," the company wrote, "as we possibly could in the short time they gave us before the ship sailed." Distance tyrannized in both directions, frustrating both Gilbert's attempts to get his parts and Duncan Stewart's attempts to recall him.[58]

A year and a half after he had left Glasgow, in May 1885, Gilbert was still in the Canaries and still waiting for parts. They had been shipped, at least, just in time to be installed for the next grinding season. "We hope therefore all will reach you in good time so that you may be able to start the factory to [crush] the greater part of the crop satisfactorily," the firm wrote. Despite the delays and frustration, however, the presence of an engineer like Gilbert could serve a firm's interests. While twiddling his thumbs in the Canaries, Gilbert had been in contact with other planters and was on the verge of securing orders for two further factories.[59]

Such orders were capital expenses that plantations hoped to undertake only once or twice a century. Company order books show that plantation owners usually ordered just one mill-and-engine combination at a given time, and rarely more than two. Properly maintained, such a pair might still match the efficiency of new machines fifty years later.[60] One of Mirrlees Watson's first mills, shipped in 1850, was not broken up until 1927.[61] A three-roller milling unit installed in British Guiana before the American Civil War was still being used after the end of the Second World War.[62] The bulk of the work of the engineering firms was not on new machines but on extensions, additions, and repairs to old ones. The phenomenal cost and mass of these machines shaped both the plans of the sugar producer and the information demanded by the engineering firm.

Acquainted with Your Patterns

While mills lasted for decades in the Caribbean, their paper representations were carefully guarded back in Glasgow, where engineering companies were reluctant to part with original paper plans long after they had been turned into metal.[63] The transatlantic history of the equipment that filled the industrial sugar factory is thus not only a story of attenuated communication

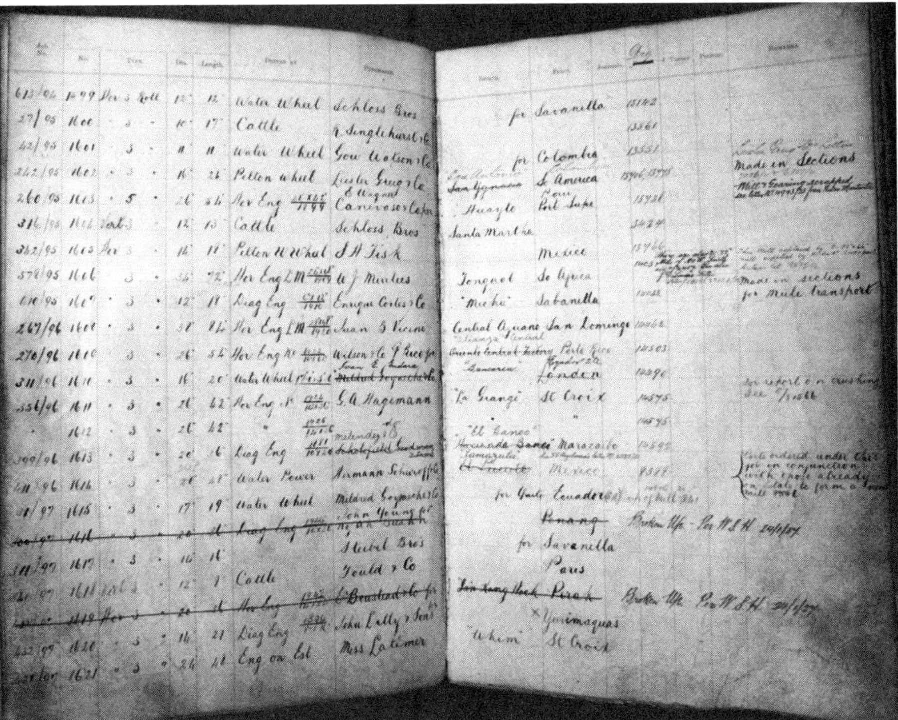

FIGURE 3.3. Order books, like this one from the Mirrlees Watson company, kept track of the purchases a company fulfilled. They were also where companies tracked their products after delivery. Clerks would annotate the books to add information as it arrived and record how they learned it, since news sometimes arrived from oblique and unexpected sources. From Mirrlees Watson order book, UGD118/2/4/37, Records of A & W Smith & Co., Ltd., Sugar Machinery Manufacturers, Glasgow, Scotland, University of Glasgow Archives and Special Collections.

but also a story of difficult maintenance at a distance.[64] Historians and other scholars are increasingly interested in studying maintenance and disrepair as ordinary states of affairs, rather than ones that are extraordinary or "pathological."[65] Sugar machinery spent little time at factory spec, as the hot, noisy, and humid factory placed immense demands on people and metal. During the six-month grinding season, a mill was run almost round the clock. Every day brought more things to lubricate and to tighten; every week or two everything had to be given a full inspection. At the end of each season the machines were partly dismantled and packed away to preserve them from the humidity, a curious combination of robustness and impermanence.[66] In such an environment, machines broke often, and when they broke, they were far from the tools and expertise that had made them. Historians of the sugar industry have largely paid attention to when devices were invented or, more usefully, when they were widely adopted. But for most of their lives sugar machines were "old, existing things" rather than new ones, and ought to be analyzed as such.[67]

In order to be able to fulfill requests for replacement parts, engineering firms made it their business to know the fate of their machines many decades after original manufacture. The annotations in the margins of the Mirrlees Watson mill order book testify not just to the kind of information the firm wanted, but also the process by which they acquired it. For example, Mill No. 360 was originally purchased for the Caledonia estate in Cuba in March 1859. A marginal note shows that in 1886 the mill was transferred to San Isidro, and whoever recorded the movement took care to note the letter that carried it: no. 9367 in 1904, fifty-five years after the machine was first bought. Or Mill No. 449, a horizontal mill sent to the Armonia estate in Cuba in June 1861. In 1903, the company received word that the mill unit had been transferred to La Reglita on the island. But the next year, they received conflicting information: Perhaps it was only the accompanying engine, No. 365, that had been moved. Other entries note the company's own modifications, like the one for Mill 723: built in 1869 for Demerara as a three- or four-roller mill, it was expanded in 1889 to five rollers, "reverted" to three some years later, and finally sold to the Bradford estate.[68]

These were fragile systems for acquiring knowledge. In December of 1857, the Cuban owner of the Soledad plantation purchased a mill with thirty-inch-wide rollers from Mirrlees Watson, No. 257. A few years later, the San Antonio estate purchased Mill No. 445, which sometime in the next two decades was sold to Soledad. In 1882, Edwin Atkins came into possession of Soledad by foreclosure and found its Mirrlees apparatus to be older than he wanted. "We replaced it with more modern machinery," he recalled, but "it was still in good condition and we sold it to another estate."[69] His new mill had significantly larger rollers—thirty-eight inches in diameter, the largest that Mirrlees then fabricated. The property of the estate when Atkins took it over also included Mill No. 1406, which Mirrlees had just shipped there in February 1883. It might take Mirrlees Watson years to find out about activity in this secondary market, despite yearly contact between Atkins and Mirrlees's agent. Or they might never find out. The last time Mirrlees's books recorded Mills 257 and 445, for example, they were at Soledad. If Atkins sold them, the engineering firm never knew.[70]

Rarely could sugar factories upgrade their entire production lines at once, so they cobbled together pieces of others. Sugar factories were agglomerations of machines from different makes as well as different vintages, so Glaswegian firms needed to know how their apparatus were being fitted together, not just

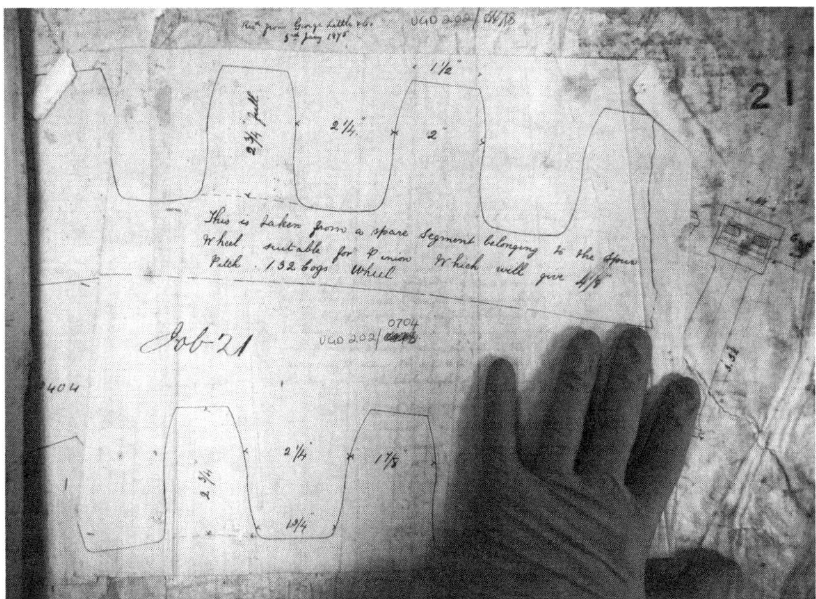

FIGURE 3.4. Full-size tracing of gear teeth from a sugar mill spur wheel, sent to Mirrlees Watson in 1875, with the author's hand for scale. UGD 202/0704, in Records of Mirrlees Watson Co., Ltd., Sugar Machinery Manufacturers, Glasgow, Scotland, University of Glasgow Archives and Special Collections.

who owned what. In 1907, the purchasing agent Victor Mendoza ordered Mill No. 1745 for Cuba's Central Mercedita. This was a standard three-roller unit, which the company recorded was "making with mills 1405, 1712 & 1713 a 12 roller train." The position of the unit in the "train" was important to know because it helped predict pressure and wear patterns and thus the likelihood of the unit's breakage. Similarly, Dos Hermanos in Cuba took delivery of Mill No. 1326 in 1882 and Mill No. 1555 ten years later. Each mill came with an engine and gearing, but in 1909 the company received word, via Mendoza, that the engines and mills had been switched, so that the later engine was driving the earlier mill. The earlier mill's rollers were both longer and wider than the newer, and thus would experience different pressures and volumes of cane, feature different patterns of wear, demand different amounts of power, and have different lifespans.[71]

When engineering firms heard from their clients, it often included a drawing or a tracing. Caribbean engineers used Glasgow firms' practices of record-keeping to test their reliability as manufacturers and credibility as suppliers of parts. In 1838, the locomotive engineer Peter McOnie, having fallen on hard times, wrote to a friend in the West Indies seeking work on an estate, but was advised to set up a repair shop in Scotland instead.[72] When this friend needed spares for his Cook-made mill, he needed to have "sent home such complete sketches with all sizes carefully marked" before McOnie could produce working patterns for the forge.[73] Later, when the plantation needed a new mill, the friend "made careful drawings of the set he required and sent them home to McOnie, who started at once to make working drawings from the sketches." The estate's owners were skeptical of McOnie's ability to complete the order, and only signed the deal "on Mr. McOnie explaining that he had working drawings already made to sketches from D. Cook's engine and sugar mill in Trinidad."[74]

This was true not only for new partnerships but long-standing ones as well. An engineer in Puerto Rico, at the end of the 1864 sugar season, sent Mirrlees Tait a request for two new mill cheeks on behalf of the estate of Señores Patxot and Polidura in Mayaguez, "as they have split the two this last crop across by the centre brass." The correspondent even told them where to look in their own books. "Knowing that *you have always your plans and models of the machines at hand in case any accident such as above should happen*, for that reason I only send you the number of mill 374, and year 1859, *thinking that is quite enough*." He attached only a "small sketch" to show where the original pieces had failed. But he also wanted the firm's guarantee that the new cheek would be functionally identical, "finished so that the same brasses will suit when necessary." A few weeks later the firm received another letter on the same matter from a different correspondent, identifying the estate as "Ysabel for which you have executed before this several orders."[75] According to the company's books Ysabel's mill was still being driven by a water wheel, and this client had required an unusually large number of drawings.[76] The engineer Thomas Dodd, writing in 1859 for Estate Florida near Ponce, on the southern coast of Puerto Rico, enclosed a drawing of the links of a driving chain of which he needed twenty-four feet. "I forgot to take the diameter of the pitch wheels to give the curve of the links," he wrote, "*but you will have it in your dimension Books*."[77] Dodd, apparently satisfied, was still ordering from Mirrlees in 1891.[78]

As a result of their importance, draftsmen could be powerful figures in a sugar-engineering firm, and drawing sugar machines required knowledge that counted as specialized even within the world of engineers. The ascent of Robert Harvey, who began as the best turner at James Cook's lathes, emphasizes these points. In the early 1830s, Harvey's skill as a portraitist led to a successful trial as a draftsman. After a year, Harvey too left in search of a raise, and was hired by Neilson, a general engineering works that had received an order for an engine and mill from Cuba. Neilson's manager, William Tait, told the partners that he could not build a sugar mill "as it was all strange to him," so he sought out Harvey for advice. The expertise acquired from working with Harvey later brought Tait into business with James Mirrlees, forming the largest and longest lasting of the Glasgow sugar partnerships, while Harvey returned to his old firm as managing partner in the 1850s.[79]

In the 1830s, when Harvey moved into the drafting office, self-trained draftsmen took techniques from many exemplars and sources, from drawing manuals to magazines. A decade or so later, however, Harvey began to teach mechanical drawing to supplement his income, first at home in the evenings, then in the Mechanics' Institute, and at the Government School of Design, which became the Glasgow School of Art.[80] The demand for these courses testified to the attraction of the potentially more "gentlemanly" nature of draftsman's occupation than that of even skilled factory work—as, indeed, did Harvey's willingness to take an initial pay cut.[81] At the same time, however, educational sites like the Mechanics' Institute, and Anderson's Institution from which it had split, helped render draftsmen the "invisible technicians" of engineering workshops, including those in sugar.[82]

By the latter decades of the century, sugar-engineering firms, like those in other lines of manufacture, boasted large drawing offices that included men of widely varying skills, training, background, and status.[83] In organization and discipline, the drafting office of a firm came to resemble the works itself.[84] Some draftsmen were themselves designers of machinery, while others worked to translate those designs into shapes that the works could manufacture. The largest group of draftsmen worked as tracers, skilled at making many reproductions of the same image. The more complex process of manufacturing itself now demanded multiple copies of drawings where just one had previously been necessary, and what counted as an adequate copy depended on for whom it was being produced, their place in the workshop, and what information they needed from a drawing.[85] Finally, some drawings needed to impress customers; as Frances Robertson points out, "Superhumanly neat inscriptions on paper functioned as a promise to deliver the goods in the material world."[86]

Outside the drafting office, the patternmakers were responsible for translating draftsmen's designs into wooden forms. A firm's patterns were, like drawings, its most crucial assets, and the ability of a company to classify, store, retrieve, and reuse drawings and patterns was key to its fortunes.[87] So patterns, like the machines they represented, were built to last. In constructing a pattern by gluing multiple cross-grained layers of wood, the patternmaker had to consider how useful the pattern might be in the future, and build it accordingly: cheaply to be discarded, or solidly to last, with layers of shellac to protect it from warpage and precisely designed joints that would not alter with age.[88] When the partnership that James Cook founded finally dissolved, the managing partner sold the works, but "bought all the patterns and drawings."[89]

The Puerto Rico engineers Robert Bennett and Thomas Dodd sent many orders to the Mirrlees firm in the late 1850s and early 1860s. "As I am not exactly acquainted with your patterns" for juice clarifiers, wrote Bennett in 1857, "I have sent you a sketch of the position they are to be placed with regard to the engine and you can make any other alterations that may be required to fit your patterns."[90] They wrote again on the eve of the 1861 grinding season to order a new engine, cane carrier, and shafts for Estate Vista-Alegre. "Cane carrier sides to be fitted to your mill No. 116," but "if the patterns have been altered since [the mill was made] let us know & we will send you a sketch, as we have not had time to do so."[91] The next year they ordered new cheeks for the mill, but trusted the company's records less, sending them the "shape of cheek . . . taken of casting." They added that "the drawing you have corresponds with these dimensions so you can make the gudgeons [to] your drawing."[92] Neither drawings nor patterns, however, dictated the construction of sugar machines. The cooperative processes of design and construction meant that shop-floor workers had to use their judgment to figure out how to translate paper to wood and then to iron and steel.[93]

Externally, despite the limits to what drawings on their own could do, Glasgow companies still made their drawing offices' competence into a selling point, by advertising how little information they needed to issue a spare. "To request spare parts it is sufficient to give us the number of the centrifuge, which is inscribed in the vertical axis of each one," boasted Watson, Laidlaw & Co. in a brochure sent to the Spanish Caribbean in the 1920s and retained by a Puerto Rican central. "We hold designs and plans of every centrifugal made by us, from the beginning of our firm in 1870 until the present, in order to be able to respond to any request for spares."[94] The drawing and pattern offices, no less than sugar factories or the shop floor, expose the constant process of maintenance and repair that underlay the surface of mechanically standardized sugar—a process that relied completely on the judgment, collaboration, and physical labor of human beings.

Glasgow's firms sought information about their products, and therefore about the state of sugar manufacture more generally, from across the Caribbean and the sugar-producing world. As such it may be tempting to think of the city's sugar workshops as centers of calculation, "mobilizing" and "accumulating," in the words of Bruno Latour, the world on paper through fixed inscriptions that travel easily.[95] In reality, however, these pieces of paper were severely recontextualized when they traveled.[96] A tracing of a gear wheel meant one thing in Puerto Rico, next to the broken machine from which it was drawn. It meant quite another in Glasgow, amid a company's ledgers and dimension books, patterns from the wheel's construction, and the metallurgical skills of its makers. Moreover, Glasgow's manufacturers were less central or coordinating than they were vulnerable, and their business depended completely on what they could learn from distant plantations.

Through the nineteenth century, sugar plantations were unquestionably where knowledge about sugar production would be generated. A century of production and maintenance by Glasgow's manufacturers enabled an industrial transformation in those plantations and in how sugar was made. At the turn of the twentieth century, that transformation began to enable claims that the Scottish city itself might generate knowledge about making sugar, just as the city's own importance began to waver.

Factories in Miniature

By the beginning of the twentieth century, the newest sugar factories in the world were many times larger than their predecessors of a few decades earlier. Glasgow firms were frequently called upon to design and construct these new, technologically sophisticated "central factories" from the ground up. The Harvey Engineering Company produced fourteen complete factories between 1905 and 1909 alone.[97]

The fall 1911 term at the Glasgow and West of Scotland Technical College (the descendant of the Mechanics' Institute) saw a new post of Lecturer in Sugar Manufacture, financed by "firms and individuals interested in this industry, including representatives of Glasgow's wealthy community of West India traders, estate owners and commission merchants."[98] More than 75 percent of the financing for the lecturer's salary, however, was given by the city's sugar-engineering firms, including the Harvey Engineering Company, Mirrlees Watson, A. & W. Smith, and Duncan Stewart.[99]

The centerpiece of the sugar school was a remarkable model of a sugar factory itself. Mirrlees Watson provided a vacuum pan, Duncan Stewart delivered crystallizers, and Watson Laidlaw furnished its specialty centrifuges.[100] The college's 1913 annual report claimed that "the equipment for the demonstration of all the important processes in the treatment of sugar juice is now complete."[101] The first lecturer, a former West Indian factory chemist named Thomas Heriot, wrote in the *International Sugar Journal* in 1916 that the college had constructed a "complete factory in miniature," one in which "every essential feature of the factory plant is reproduced on a smaller scale."[102]

Yet that emphatic sense of completion excluded crucial elements of the sugar factory. Most obviously, there was no mention of a miniature mill for grinding sugarcane, which could not easily be brought to Glasgow before its juice hopelessly degraded. Similarly, neither the workers who would populate a real factory nor the ability to manage them were part of Heriot's miniature model. "Driving" an Afro-Caribbean workforce,[103] Heriot wrote elsewhere, "requires no technical skill, and may be better performed by others" than by the trained chemist or chemical engineer.[104] He acknowledged the importance of "administration," but felt it could be delegated to "native foremen" without much trouble.[105] And, finally, the machines were not subject to cycles of assembly and disassembly, heat and moisture, that confronted actual factories in the tropics.

This was not the first such model to be proposed. In 1852, the steel innovator Henry Bessemer had proposed a new process for extracting and crystallizing sugar, including a drying apparatus that worked something like a gigantic crepe pan. He proposed that his new company would build "a Model Colonial Sugar-house, in which the most perfect arrangement of all the apparatus shall be carried out," and every month grind 150 tons of canes into sugar, which would be open to inspection and which it would then sell.[106] As a solution to problems of repair, he proposed to keep "a duplicate of *every single piece*" on hand in each sugar colony.

The Royal Technical College's model was derived from the products of machinery firms, and Heriot's miniatures represented the operations of a sugar factory as procedures subject to fine control.[107] With the help of the manufacturers and instrument makers, Heriot had installed "conveniences for exact experimental work which are lacking in the sugar factory," which allowed him to educate students "*in a more direct manner than by practical experience*."[108] He intended that an enrollee in the sugar manufacturing course would learn "all essential principles outside the factory, by means of lectures and laboratory experiments, so that, when he first enters the factory, he understands what he sees, and needs no other instructor than his own eyes and intelligence."[109] In other words, Heriot claimed that his school was in fact a superior environment for learning about sugar production than any actual factory.

And students came, first from the engineering companies themselves. The evening course was packed with almost sixty employees of the local sugar-machinery works. They heard lectures that followed sugar from its agricultural beginnings as "raw material" through milling, diffusion, clarification, concentration to syrup, crystallization, drying, and packing, all before "chemical control of manufacture."[110] Within a few years, the school added a new laboratory course that included "analysis of sugars, juices, &c."

FIGURE 3.5. A model evaporator at the School of Sugar Manufacture, Glasgow and West of Scotland Technical College, early twentieth century, with the school's instructor to show the smaller scale. OP4/145, Archives and Special Collections, University of Strathclyde Library.

The course began as a way to teach local engineering students how factories worked, but it became a model of a factory that taught colonials how their factories were supposed to work. Over the next decade, the college reported that its enrollment shifted to include many more West Indian students, who came in the hope "that their work [here] will do much to increase the knowledge of modern methods of sugar production and manufacture *which they have come to this country to acquire*."[111] For a hundred years, the infrastructure of sugar production had been characterized by just the opposite flows of people, paper, and knowledge. Glasgow's own success at industrializing the production of sugar had led to institutions like the "factory in miniature" at the city's engineering college.

It also set the stage for Glasgow's undoing, as rival centers of sugar-production expertise could emerge. A 1901 survey of "Local Industries of

Glasgow and the West of Scotland" painted a gloomy picture of the future of the sugar-machinery industry. The British West Indian sugar market had run dry, though this was partly compensated by sales to independent Latin American nations, and to imperial possessions in South Africa, India, and Queensland. The real concern for the Glasgow sugar-machinery makers was being shut out of two of the largest and richest sugar producers in the world. A quarter-century earlier, following Hawaii's signing of a reciprocity treaty with Washington in 1876, Glasgow had supplied the islands with their first "complete" sugar factory. But the archipelago, "once a good market, is now practically closed to British manufactures." Fortunately Cuba was "showing signs of revival, and may yet prove a fairly good market," according to the survey, "if it is not turned into an American preserve in a similar way to Honolulu."[112]

Far more than a "preserve," the Hawaiian capital would soon become home to the most advanced sugar-machinery firm in the world. The Honolulu Iron Works had been founded in 1853 by David Weston, who sold Mirrlees Watson the foreign rights to his patent for a more stable centrifuge. In another reverse creolization, the Glasgow company shipped one of Weston's "self-balancing" models back across the Atlantic and included it in their display in Philadelphia at the Centennial Exhibition of 1876.[113]

Like so many of Glasgow's firms, the Honolulu Iron Works began by repairing others' machinery. By the 1880s, however, its work was in high demand, "especially since the great distance from manufacturers in America and Europe made it extremely necessary to obtain what was needed right at home."[114] By 1909, the firm's management was congratulating itself that it had "been able to go into a foreign field, meet the best sugar-machinery manufacturers of the old world on their own ground, and give so good an account of themselves."[115] Honolulu Iron Works profited from the creation of Hawaii as a model sugar-production zone, free from the historical labor constraints that had plagued planters in the Caribbean. In this sense it was a scaled-up version of Heriot's "complete factory in miniature," complete not despite its lack of skilled labor but because of it. The personnel of the Hawaiian company followed similar paths as the Glasgow firms, from draftsman to superintendent to manager to technical director.[116] And they even profited directly from Glasgow's work. A plantation engineer named James Kennedy Wann Carmichael was born in Glasgow in 1888, educated at the West of Scotland Technical College, and then traveled to work at Hawaiian plantations and at Honolulu's sugar machine maker.[117]

The Hawaiian archipelago is so lacking in iron ores that Captain James Cook needed to forbid his sailors from swapping the *Endeavour*'s valuable nails for sex.[118] Yet the Honolulu Iron Works plant was six and a half acres, ringed by a commercial atoll ninety-five hundred miles wide: another machinery plant in the Philippines, an engineering office in Havana, designers and salesmen covering two floors of the Woolworth Building in New York.[119] Crushers and centrifuges bore the firm's name in a hundred factories from Japan to Louisiana. In Cuba, the Honolulu Iron Works could claim to have built the largest central in the world. The company offered three million of its own dollars to rebuild sugar production in Morelos after the Mexican Revolution.[120] In 1926, when the executives of the dominant American Sugar Refining Company needed to gut-renovate their great refinery in Williamsburg, they came to the Woolworth Building.[121] And aside from its plant in the Philippines, all the iron wrought under the company's name needed to come through Honolulu, and the firm carried that fact as a colossal chip on its shoulder. Company executives even advertised "the enormous handicaps which the Hawaiian manufacturers were compelled to overcome, not the least being the fact that they must import every scrap of material used."[122] For Honolulu, just as for Glasgow, the tyranny of distance begat thrift and ingenuity.

The status of Glasgow as the world's dominant heavy-engineering city diminished in realms beyond sugar machines. Following the boom of the Great War, when many sugar-machinery makers retooled for armaments, the crop of 1920 brought the highest sugar prices in a century. Mountains of unfilled orders rose in the sales offices of sugar-engineering companies, and so did profits. But those lucrative harvests were temporary, merely filling the vacuum left by European beet sugar. When the beet crop returned in 1921, prices collapsed.

Moreover, in return for tariff reciprocity with tropical sugar producers, American protectionism had raised the relative cost of Scottish machinery. The construction of new factories in Hawaii and Cuba depended on their sugar entering the United States tariff-free, but neither did American machinery face taxes on entry to those islands. British manufacturers hoped this tariff surcharge would not entirely outweigh their firms' long-standing relationships with engineers and owners. But as Cuba's central factories failed following the 1921 collapse, they often fell into the hands of American creditors, who strongly preferred to buy from domestic engineering firms. That year, only a single order of centrifugals arrived in Scotland. Glasgow newspapers now reported on their city's sugar firms in different tones: They spoke of "moderate activity," business that was "fairly active," with "work not so plentiful as in the years past." Some firms received no new contracts until 1923.[123] Meanwhile, Honolulu Iron doubled its business between 1920 and 1924.[124]

By the 1920s, by contrast, the Honolulu Iron Works were producing factories for Cuba, the Dominican Republic, Formosa, and the Philippines. The firm boasted not only a branch in New York City but a full office in Havana, including engineering staff—a far more substantial presence than the consulting engineers on whom Glasgow firms had historically relied, and one that gave customers the possibility of far closer coordination. Technology, expertise, and individuals could move through multiple overlapping circuits of the sugar economy. If it remains strange to think of sugar expertise in sugar-free Glasgow, it should be no less discordant to our received geographies of technology, knowledge, and capitalism to find a prestigious iron works in ironless Hawaii.

In 1921, the manager of the Honolulu Iron Works published an article in a trade journal called *The Trans Pacific*, and with industry as well as agriculture in mind, titled it "the future of sugar is in the Orient."[125] That same year, the board of A. & W. Smith dispatched the aptly named Martin Ironside to Cuba "for the purpose of gaining knowledge of the most modern practice" in sugar-making and in machinery. They judged that "such a visit would be advantageous" to the firm and, perhaps acknowledging the usefulness of Honolulu Iron Works' Havana office, "a similar visit on the part of a member of the Works Staff would be even more so." Ironside returned six months later, having toured not just "the Sugar Factories of Cuba" but also Trinidad, Louisiana, and elsewhere in the United States.[126] He may well have visited Louisiana's new and "radical" Adeline Sugar Factory, which had been "designed in its entirety by the Honolulu Iron Works Company of New York"—a company name that concisely expresses the geographic inversions and reversions of expertise in the early twentieth-century sugar economy.[127]

4

There My Responsibility Begins

Dangerous Characters

Conflict between management and labor over the quality of sugar and who got to measure it was a perpetual feature of sugar plantations. This book began with a strike about a strike: During the 1894 grinding season in Cuba, one of the men in charge of boiling sugar on the Soledad plantation, owned by Edwin Atkins, sucker-punched the factory's chemist, who had asked him to keep better records of his work when he struck the sugar. "Pan-boilers do not like being interfered with," overseers in Demerara were advised.[1] The incident at Soledad was one battle in a much longer war.

In his 1926 memoirs, Atkins outlined the "revolution" that had replaced the traditional mill and plantation estate, built on enslaved labor, with great steam-driven central factories that matched their beet-sugar brethren in technical sophistication. He recalled that "almost every year, some new machinery was installed at Soledad . . . which enabled us to do better work, increasing our production and decreasing the cost of labor." As we saw in chapter 3, he had bought new mills to crush cane for its juice, a new vacuum pan to more finely control the reduction of the juice to a mass of molasses and crystals, and a centrifuge to separate the two.

In the mid-1880s, the final years of Cuban slavery, Atkins had hired indentured Chinese contract laborers, whom he praised as "faithful men and never missing from their places." Faithful or not, they still needed supervision. Thus he was glad that Soledad's increased revenue meant that he was able to hire "higher-salaried men in the responsible positions." Most of all, he had found a trustworthy deputy in "our" chemist, a Canadian named Wilfrid Skaife, who "took charge of the sugar house and was the greatest help to me." For improving the factory, Atkins recalled fondly in 1926, "he was full of new ideas."[2]

As we saw in chapter 1, the language of revolution in the industrial production of Caribbean sugar in the second half of the nineteenth century has been consistent for a century and a half, both among the industry's participants and its analysts. Contemporaries of that process in the nineteenth and early twentieth centuries, as well as later historians, emphasized the dazzling chemical consistency of sugar that emerged from the new factories of Cuba, Puerto Rico, and other sugar territories. The means by which such sugar was actually produced, however, have been written about in mechanical terms. As these machines begin to appear en masse after 1850, so too do narratives about how sugar was made begin to change. Up to that point, they are stories about human skill. Afterward, they become stories about machines. And as the machines that filled sugar factories become more impressive, and as the factories themselves come to dominate the Caribbean landscapes, they overwhelm the human beings who still worked skillfully within them.

This chapter shows how the industrial sugar factories of the late nineteenth and early twentieth centuries remained sites of metrological conflict between workers and management, long after the factories appeared to run by themselves. Skillful human work in sugar factories persisted long after most observers have supposed otherwise. More important, that skill was irrevocably embedded within fundamental features of industrial sugar-making.

These factories were called "central factories," *ingenios centrales*, or just centrals, because cane was brought by railroad from far greater distances and in quantities greater by orders of magnitude than had ever been possible in a small plantation. No longer integrated with the landscape, they were now meant to be isolated from it.[3] By the late nineteenth century the people who made Caribbean centrals work came from across the globe. By choice and by coercion, global capitalism had created a workforce that largely replicated imperial hierarchies. "Our manager is an American," as Edwin Atkins wrote to one candidate for the job of chemist, "the chief engineer Scotch, the pan men are Germans, laborers in sugar house mostly Chinamen," and of course the cane was harvested largely by Afro-Cubans.[4] Historians have debated whether the mechanization of sugar production brought planters' industrializing commitments into conflict with their desire to profit from the forced labor of people of African descent. As Daniel Rood shows, the labor hierarchy was connected in subtle ways to new technologies and how planters viewed the alleged capabilities of supposed racial types. They sought, in Rood's words, "Anglo machinists, white creole overseers and sugar masters, Chinese skilled laborers, and an internally divided population of enslaved workers, with Cuban-born operatives often placed in skilled positions, and African-born bozales left to cut the cane." Yet, as he also notes, these were caricatures that crashed on the realities of running a mill.[5] Engineers, managers, and chemists rarely stayed put but might spend half the year in beet fields and half in cane factories, or alternate between academic employment and factory superintendence.[6] They read handbooks and publications boasting global lists of contributors and subscribers.

As we saw in the previous chapter, even machinery has a human history. In the context of these factories themselves, however, the new machines have themselves hidden two kinds of invisible skilled labor, both of which were important for sugar to become a commodity defined by its organic chemistry. The first of these forms of labor is that of the skilled workers who made the sugar at its most crucial stages. The sweltering, noisy, open-kettle sugar-boiling houses of the seventeenth, eighteenth, and early nineteenth centuries depended on the know-how of enslaved artisans. This dependence was so fundamental that even their own enslavers themselves could not ignore it. Theirs are the skills that are supposed to have been obviated and rendered unnecessary by the sophisticated, chemically controlled industrial processes of the sugar mill, but in fact remained crucial to the success of the central until the second half of the twentieth century.[7]

The second form is the labor of those responsible for the chemical control of the sugar factory. Here, the notion that sugar is sucrose has led historians astray. In scholarship on late nineteenth-century sugar factories, the presence of a chemist has largely been used to index the factory's sophistication, showing that owners and managers understood the need for scientific knowledge. But the case of mistaken identity of pure sugar and chemical sucrose has made these factories appear both revolutionary in technical terms and evolutionary in other ways, merely the culmination of centuries of development toward perfect technical efficiency. The more we think of their product as pure sucrose, rather than the complex substance called sugar, the more it seems that the job of chemists is merely to ratify this perfection. Paradoxically, the more successful chemists are, the more they disappear.

In fact, accepting the notion of sugar as a globally uniform chemical commodity means ignoring the activity of the people responsible for determining that uniformity. On the one hand, chemists' rhetorical, intellectual, and physical labor rendered invisible the role of the sugar artisan. On the other, the more that chemists' expertise made chemical purity the standard measure of value, the more peripheral and obscure became the work of chemists themselves. The evidence for the continued importance of skilled labor comes in large part from examining chemists' own handbooks, articles, and other documentation of how chemical control was to be implemented. In another irony, it is these same works that reveal the places that artisanal skills survived in the sugar factory itself, and attest to the kinds of knowledge that "chemical control" could not control.

An ideal sugar factory would run by itself, making sugar as the plants did it. "The leaves correspond to the factory buildings," one pedagogical text from 1928 explained, "and the tiny granules of green chlorophyll are the machines." The cells of the cane "need no attendants to watch or guide them, for each seems to know its part in the factory's work and does it without being told." This was the fantasy of managers: that machines "work busily and with no noise or fuss whatever."[8] But in fact, within the newly industrial sugar factory, chemists oversaw laborious regimes of sampling, testing, and analysis, all of which fell under the term "chemical control."[9] As with much metrological work elsewhere across the history of the capitalism, the work chemists performed in enforcing the uniformity of sugar has been hidden. And by divorcing industrial sugar manufacture from skilled labor, we have misunderstood the term: Chemical control could not mean the control of materials without also meaning the control of human labor.

More Perfect in Their Business Than Any White Man

Caribbean sugar plantations of the sixteenth through the early nineteenth century were famous sites of artisan skill, technical know-how, and technological progress. To Enlightenment thinkers, sugar was a "wonder," to be ranked among other glorious inventions like "Guns, Printing, Navigation, Paper." Credit for that wondrousness was a valuable thing, and as Eric Otremba shows, the followers of Francis Bacon wanted to claim it for their new natural and political philosophy.[10] The one thing Europeans could not understand was the crystallization stage at the center of the process.

No figure captured this paradox of sugar-making better than the enslaved artisan boiler. Thanks to years of "unintermitted practice" and close observation, according to one writer, "Negro boilers must be more perfect in their business, than any white man can pretend to be."[11] Different fields and terrains yielded different cane qualities, so an overseer in Jamaica might notify his boilers "when he begins a new piece of Ground, that they may be ready to remedy any inconvenience from the variety of soils," especially by mixing canes from different elevations.[12] Planters reminisced about happier days under their favorite boilers. Overseers hated them, because boilers sometimes felt empowered to complain about poor treatment, but overseers also knew that without the boilers' cooperation and good health they themselves would be blamed when the sugar suffered.[13]

In plantations that used the so-called Jamaica train, in which all coppers were built into the same unit of brickwork and heated by the same furnace, the temperature was "controlled by yells" from the slave at the kettle to the furnace-minder.[14] From copper to copper, slaves carried the thickening syrup. Part of the art and skill lay in reducing the liquid in the open kettles without beginning to burn the sugar itself. Finally, in the smallest kettle, the sugarmaster manipulated heat and additives to bring the solution to point where crystals began to form. Then he made his "strike"—removed the syrup from the heat and transferred it to troughs, where it was beaten with wooden paddles as it cooled as encouragement to crystallize further. Knowing when the crystallization point had been reached was perhaps the sugarmaster's most coveted ability.

In 1765, the Antiguan planter Samuel Martin complained that the "art of boiling sugar [is] generally least understood either by overseers or their masters; but . . . trusted wholly to the skill of negro-boilers, who indeed arrive by long habit to some degree of judgment by the eye only."[15] After the strike, the cooling semicrystalline mass was poured into great inverted conical molds. Over several weeks, the molasses, thick sweet liquid from which crystals could no longer be coaxed, drained out through the bottom and into troughs that collected it for future use, while the loaf itself dried in the sun. Molasses was fermented into rum, sold of its own accord, or returned to the syrup as it boiled, to extract a few more granules. The loaf could be sold whole—or, in "the most personal operation in the mill," it might be divided by the sugarmaster's own eye, wherever he judged a layer of one marketable quality to yield to another.[16]

As early as the 1790s, Cuban planters attempted to import men trained in European schools of chemistry. According to María Portuondo, "the new scientifically trained sugar expert was expected to inhabit the world dominated by the sugar master and bring control and predictability to the process."[17] They and their overseers had struggled and failed for hundreds of years to learn how to boil sugar, to explain it to others, and to convey it in writing. "To impart by words the knowledge of sugar-boiling by the *touch* is as impossible as to teach any other mechanic art, which can be attained by practice only," wrote Martin.[18] Planters had little ability to discipline their boilers when sugar was not to their liking. They could fire them, if free artisans, or demote them, if enslaved. John Pinney, who neurotically managed an eighteenth-century plantation on Nevis, would "borrow" another planters' boilers if he was dissatisfied. But he was also so beholden to boilers' skills that he would, according to the historian Richard Pares, shower them with gifts if the sugars made

FIGURE 4.1. The boiling process was the most difficult part of sugar making to get right, and it was one about which Europeans had the least knowledge. But in illustrations like this one, from William Clark's 1823 series of views of Antigua, they appeared as under the orderly command of the planters and overseers. Courtesy of the John Carter Brown Library.

him happy.[19] Caitlin Rosenthal has shown how slave plantations, especially sugar plantations, pioneered techniques of accounting, facilitated by their near-total control over the bodies of their workers. And among the vast array of books kept by plantations were books for monitoring the boiling house, some of which, Rosenthal notes, were portable enough to have been carried near the boiling itself.[20] But these books were limited to quantitative observations such as the number of hogsheads, and could say little about the quality of the sugar they represented.[21]

By the middle of the nineteenth century, planters had brought European chemists into their sugar houses, as they sought further means to quantify and measure production that remained in workers' hands.[22] They brought one instrument with them that promised to change everything. That instrument was the polariscope, Jean-Baptiste Biot's invention that could detect the rotation of polarized light. The chemist Noël Deerr concluded his two-volume *History of Sugar* of 1950 with a discussion of this instrument. Its placement was a sign of importance, for "on it depends the whole process of the determination of cane sugar and of the control of the factory."[23]

As we saw in chapter 2, Biot hoped it would allow sugar planters to discipline their workers and their machines. In his 1843 paper titled "On the

Application of Optical Properties to the Quantitative Analysis of Liquid or Solid Mixtures in Which Crystallizable Cane Sugar Is Joined with Uncrystallizable Sugar," he again advertised his hope that planters could recover more sugar now that they had access to his creation. Even molasses from the best-run refineries, he had found, still contained large quantities of crystallizable matter. "In repeating the same test, after successive operations that have reduced the natural juice—evaporation, clarification, concentration, reduction into grains—one would know right away, and at every instant, the effect, good or bad, of each step," he had written.[24] Calculating the sugar concentration from the measured rotation required a fair bit of multiplication. It was easy with logarithms, which could just be added. Otherwise the fractions involved would make the experience "very painful." An analgesic was available: Standardize the masses of the samples. Biot refused to swallow it. "I have not sought to exempt [my method] from this necessity [of calculations], by restricting it to operations by fixed dosages, as has been done for some other scientific methods, by which it has been hoped to render them more vulgarly usual." He predicted endless difficulties standardizing the weights: "This alleged fixity of dosages," he snarked.[25] (He would be proved right, as we will see in part 3.)

Biot's true reason for not fixing dosages he saved for the end of his article: It would toughen up the planter class. Planters' enemies back home, and many of their friends, worried that owning people and making money in hot climates left them spendthrift and indolent.[26] The mathematics demanded by the polariscope would turn sugar planters and even refiners into better men. "It seems to me more advantageous, I would almost say more honourable, to raise [*élever*] manufacturers up to using such rapid and simple logarithmic calculations than to hold them down to a vulgar routine that keeps their results far from rigor and generality."[27] As owners and planters modernized their operations by using the polariscope, the polariscope would modernize the people who learned to used it. But for all that it could measure intermediate products, there were operations in the new factory into which it could not peer.

Division of Labor

By the middle of the nineteenth century, as first the grinding, then the boiling, then the drying became largely mechanized in Cuban and some Puerto Rican sugarmills, their conditions of life and work reached what Moreno Fraginals called a "super-barbaric stage." The parts of the process that remained manual became bottlenecks to the smooth flow of the whole mill, and pressure was applied across that barrier to force people to work ever harder. The beginnings of industrialization "increased the traditional barbarism of the mill by demanding synchronization of manual work with mechanical processes." Thus machines intensified the violence of slavery rather than ending it, and sugar houses became "huge grinders which chewed up blacks like cane."[28] At the same time, the *hacendados* who owned "semi-mechanized mills" also sought to quantify and measure those elements of production that remained directly in workers' hands.

Mid-twentieth-century historians who extolled the progressive industrialization of sugar production in the West Indies suggested that "wastefulness and slovenliness" were the predominant characteristics of sugar production before the central factory and its machines.[29] In fact, planters tried extensively to monitor that which was not "understood either by overseers or their masters," as Samuel Martin had worried. Precisely because labor on a slave plantation was in theory subject to round-the-clock oversight, planters found it useful to implement systems of financial and productive accounting. Their power meant they could both enforce the collection of information and felt that they could calculate its utility. A blank form from Cuba's Ingenio La Ninfa, printed during the first decade of the nineteenth century, shows the information its experimentally minded owner demanded from thirteen divisions: the mill, boiling house, purging house, drying room, and sugar and molasses stores to the distillery, sawmill, carpenter, forge, grain mill, and butcher. And slaves—bought, sold, born, died, or escaped—were enumerated in identically formatted boxes.[30]

La Ninfa's categories themselves indicate the integrated agricultural-industrial workings of the plantation. The later central-factory system was built on a different principle: on the separation of the field from the factory.[31] Central factories posed problems for control because of their sheer size and complexity. A strict distinction between their interior and exterior, both in reality and in rhetoric, proved to be important as chemists sought to conceive of the space within the factory as one that was like a laboratory. This was true even as controlling a factory meant a vast web of disciplinary and measurement practices that extended beyond the bench, out of the laboratory, past the gates of the central, even to distant outposts in the canefields.

Watching Soledad from Boston, Edwin Atkins demanded both weekly laboratory reports and monthly statements from his chemists.[32] By the end of the century, 96 percent sucrose measured by the polariscope had become the trade's target for raw sugars. After Soledad had begun grinding its 1890 crop, Atkins was pleased to hear of how much it was extracting but ordered his chemist to go no further: "95½ avg. test is as high as I wish it is useless to

FIGURE 4.2. Images of "modern sugar factories," like this one from Stillman's catalogue, highlighted their separation from agriculture. Cane arrived by rail cars, where it could be automatically sampled and weighed. O. B. Stillman, *Installations of Cane Mills, Multiple Effects, Vacuum Pans, Filters and Cane Sugar Making Machinery* (G. M. S. Armstrong, 1904), 73. Courtesy of HathiTrust.

make it higher."[33] In observing, measuring, and calculating the chemistry of sugar processing, chemists' employers hoped to exert coercive control over their once autonomous labor force. Conversely, dictating the behaviors of their workforce was to be the means by which factory owners could control parts of the sugar-making system that defied calculation and analysis.

The work of the sugar industry's most widely respected sugar chemist, Noël Deerr, on "the control of the factory" can teach us much about the nature of chemical control as it was envisioned by its most widely read proponent. His books were written over the course of decades as a factory supervisor, largely in Cuba, Puerto Rico, and the British Caribbean colonies.[34] Even as he depicted, in many ways, an idealized version of laboratory and factory, he also explicitly acknowledged and implicitly revealed the way chemical practitioners had to negotiate with the other human beings who populated the cane-sugar world, and with the material limitations and environmental realities of sugar production. "A modern factory must be conducted from the mill to the distillery as a huge chemical experiment," he began, "and efforts be made to account for every pound of sugar."[35] But in many ways the rest of his textbooks were devoted to showing where sugar chemists should compromise and where they should not. "No hard and fast rules can be laid down for the determination of yield and losses," Deerr admitted: "A great deal depends on the skill and ingenuity of the chemist." The very vagary of chemical administration of a factory was why the chemist "should alone be responsible for this department."[36]

The most important number when accounting for sugar within the factory was the amount that entered it in trapped inside the cane. Where factories received cane from a network of weighing or assembly stations, however, "the question of loss of weight of canes between cutting and milling is very serious." How—which is to say where—the weight was defined, either at the "outlying balance" or at the gates of the factory itself, could yield radically different intermediate and final measures of purity, yield, and efficiency. Juice in the cane would evaporate, sugar would ferment, and the calculations on which the factory's experimental efficiency depended would become increasingly imprecise.

In large factories, canes were weighed when they arrived on balances that were designed to make large but accurate measurements. Only two "precautions" were in order: confirmation of the "exactitude" of the scales themselves and calibration of "the (in general) native operator." Ideally, "automatic" balances that were "self-registering" put any potential manipulation "beyond the control of the attendant, whose functions are [now] merely mechanical."[37] Atkins, for instance, was disturbed by weight that sugar cargoes added between Cuba and Boston, "against the usual gain of ¾ of one per cent" that might be attributed to the absorption of moisture. He pressed Skaife to "inquire carefully into the weighing and see if our weigher is giving us full weight."[38] Skaife duly reported that he found nothing amiss—that he was satisfied with the regime of numerical control he had established.[39] The goal was to turn the sugar worker, whether in the field, in the factory, or in the places they intersected, into a piece of machinery, but one whose potential for malfunction or malfeasance was limited by the system of which he was a part. "Automatic" balances that were "self-registering" put any potential manipulation "beyond the control of the attendant, whose functions are [now] merely mechanical."[40] The arrival of the cane by rail facilitated control, since railroad scales could weigh entire cars against their tare set at the beginning of the season.[41] At the large Puerto Rican central of Guanica, the general manager was an ex–railroad employee and ordered checking of the cane, juice, and sugar scales with standard weights at least twice a week.[42]

Deerr emphasized and reemphasized that the fundamental source of most losses was human error on the part of the sugar factory's own workers. Losses that could not be accounted for he attributed to three sources. One was errors in the chemical analysis of the "molasses and low products" that contained less sugar and greater amounts of other substances, which made such analyses more complex. Another was loss to inversion of sucrose to uncrystallizable glucose and fructose, especially where acids were used to brighten or whiten the sugar to make it attractive for direct consumption. The third, "entrainment losses," mostly meant sugar that had been lost through workers' activity that was perceived to be avoidable, even if "very hard to determine directly." These included sugar that fell out during direct handling; "sugar lost in boiling over in the pans and triple [effect]," which theoretically, under proper manipulation, should never happen; and "sugar eaten and stolen by the workmen."[43] Likewise, various automatic juice-weighing devices were again intended as checks against the feared inaccuracy or worse of the "operators" of the discrete and continuous processes.

At the same time as chemists expressed mistrust of the workforce, however, Deerr's treatises on chemical control also revealed the degree to which the actual practical work of a chemist in a large factory was dependent on the workers themselves. Consider the way Deerr suggested determining the sugar content of canes. In one of the factories in which he had worked, which, he recalled, took in and ground sixteen carloads of cane per hour, he retrieved one cane shoot from each car. His method of sampling the canes themselves was elaborately fastidious. Of sixteen canes each hour, he cut the top sixteenth of the first cane, the second sixteenth of the second, and so on, down to the sixteenth part of the sixteenth cane. These he then had cut with a large knife into halves, then into quarters, and eventually into eighths, producing a final sample of between two and three hundred grams. "Proceeding systematically on these lines," he wrote, "as many as 100 canes can easily be analysed in a day of eight hours." But cutting sugarcane was not easy, and the chemist in Deerr's example had done no actual analysis yet. For the chemist, the analysis of 100 canes a day was very hard work—and this was only the beginning of his supposed "control."[44]

Likewise Hubert Edson, who began a long career as an important factory chemist working for Harvey Wiley as a Department of Agriculture liaison in Louisiana in the late 1880s, remarked in his memoirs on the enormous task he had faced. "Starting with a calculation of the sucrose in the cane," he wrote, "I planned to follow through with analyses of sucrose in each succeeding step through the factory." He wanted to test the mill juice, clarified juice, evaporator syrup, massecuites after boiling, and the sugar and molasses themselves. "Each of these several products had to be accurately measured or weighed and, with sixty or seventy analyses a day, I had to spend an average of eighteen hours a day in the laboratory. . . . I came out of it tired but happy." Yet he considered the list of tasks he was asked to accomplish as a young chemist almost comically excessive. Even Edson, "with my limited experience," he recalled, "fully realized that it would be impossible for one chemist to do even a minor part of the program presented."[45]

So it was no surprise that Deerr admitted that in many cases "the necessary control work of a sugar factory is performed by unskilled assistance," even as, with his choice of homophone, he effaced the actual assistants.[46] He acknowledged that "by simple analyses, such as can be performed by any member of the estates' staff, combined with attention to correct measurements and sampling, valuable and reliable data can be obtained."[47] Other chemists, in their own handbooks, textbooks, treatises, and manuals, likewise spoke of how "samples of materials at essential process points were collected continuously," without mentioning by whom.[48] Guilford Spencer, who worked for the Department of Agriculture before supervising chemical control at the Cuban-American Sugar Company, asked in his *Handbook for Sugar Manufacturers* whether one chemist could perform all that was necessary for factory control. "If the chemist is located in a house provided with crude measuring apparatus, where he must superintend his own sampling and keep a watchful eye on his employés, he will require several assistants," Spencer wrote. "If careless work is the rule instead of the exception, his labor will be largely increased." He recommended finding "a reliable man" who could be entrusted with supervision of the sampling, leaving the chemist himself and his assistant "to study the work of the sugar-house, and the resulting improvements will be of great advantage."[49] The sugar chemist's control over the factory workforce depended on collaborative activity with and trust in that workforce itself.

And more often than not it was chemists who were in charge of that workforce. "The manager, engineer, or chemist must not only know what has to be done," wrote Thomas Heriot, the West Indian factory chemist who organized a school for cane sugar manufacture in Glasgow, "but must be able to direct those who do it, and to see that their instructions are being carried out."[50] At Soledad, for instance, where in the 1880s the chemist earned $1,500 a year plus expenses, the man who occupied that position was also the superintendent of the human operations of the sugar house, while the mills themselves and the machinery were overseen by the chief engineer.[51]

When Atkins's general manager departed in 1894, Skaife, his chemist, pushed for complete control. He suggested that "all technical questions concerning the mill house and boiling house ought to be referred in the first instance to me." He argued that the chemist's domain was coterminous with the factory itself, both for reasons of efficiency and by right. "Let it be understood

SUGAR HOUSE NOTES AND TABLES. 47

JUICE.

Normal Juice: tons.	Juice % on Canes.	Tons of Sugar.	Tons of Glucose.	Sugar % on Sugar in Canes.	Density.

Total Solids %.	Sugar %.	Glucose %.	Ash %.	Sugar: Glucose.	Purity.

MEGASS.

Tons.	Tons of Sugar.	Sugar % on Sugar in Canes.	Sugar %.

Glucose %.	Ash %.	Fibre %.	Water %.

SCUM DIRT.

Tons.	Tons of Sugar.	Sugar % on Sugar in Canes.	Sugar % on Sugar in Juice.	Sugar %.	Glucose %.	Water %.

SYRUP.

Tons.	Tons of Sugar.	Sugar % on Sugar in Canes.	Sugar % on Sugar in Juice.

Density.	Sugar %.	Glucose %.	Sugar: Glucose.	Purity.

MASSECUITE I.

Tons.	Tons of Sugar as Crystals.	Tons of Sugar in Solution.	Total tons of Sugar.	Sugar % on Sugar in Canes	Sugar % on Sugar in Juice.

Crystallised Sugar %.	Dissolved Sugar %.	Glucose %.	Ash %.	Sugar: Glucose.	Purity.

48 *SUGAR HOUSE NOTES AND TABLES.*

SUGAR I.

Tons: gross.	Tons: net.	Per cent. on Canes: gross.	Per cent. on Massecuite: gross.	Sugar % on Sugar in Canes: gross.	Sugar % on Sugar in Canes: net.

Sugar % on Sugar in Juice: gross.	Sugar % on Sugar in Juice: net.	Sugar %.	Glucose %.	Ash %.	Water %.

MOLASSES I.

Tons.	Tons of Sugar.	Sugar % on Sugar in Canes.	Sugar % on Sugar in Juice.	Density.	Sugar %.

Glucose %.	Water %.	Ash %.	Purity.	Alkalinity of Ash, as K_2O %.	Glucose: Alkalinity of Ash.

MASSECUITE II. CURED.

Tons.	Crystallised Sugar %.	Dissolved Sugar %.

MASSECUITE II. ESTIMATED FOR CURRENT WEEK.

Tons.	Sugar %.	Glucose %.	Purity.	Tons of Sugar.

SUGAR II. CURED.

Tons: gross.	Tons: net.	Percentage on Massecuite II.	Sugar %.	Glucose %.	Ash %.	Water %.

SUGAR II. ESTIMATED FOR CURRENT WEEK.

Tons: gross.	Tons: net.	Per cent. on Canes: gross.	Sugar % on Sugar in Canes gross.	Sugar % on Sugar in Canes: net.	Sugar % on Sugar in Juice: gross.	Sugar % on Sugar in Juice: gross

FIGURE 4.3. Two pages from a widely circulated model for a factory chemist's weekly report in 1900. Each box represented a great deal of labor in taking, preparing, and testing samples. Noël Deerr, *Sugar House Notes and Tables: A Reference Book for Planters, Factory Managers, Chemists, Engineers, and Others Employed in the Manufacture of Cane Sugar* (E & F. N. Spon, Ltd., 1900). Courtesy of HathiTrust.

that the cane is delivered to me [outside the factory] and there my responsibility begins."[52] A few weeks later he tried again: "There would be little or no difference as regards the routine work but I could put at your disposal a good deal of knowledge and experience which is now rather wanted, for I have been a long time in the business and can manage men," he claimed, even as he wrote with vitriol in favor of European racial dominance.[53]

The number of surviving daily chemical reports of the form shown in figure 4.3 does not reflect the number which must have been created. Those completed reports that are available, however, contain a wealth of information of use to historians beyond the production statistics they enumerate. The extant chemist's reports for the huge Central Aguirre in Puerto Rico, owned by the New York–based South Porto Rico Sugar Company and founded immediately after the island's acquisition by the United States, demonstrate the way that the supervision of labor and the supervision of sucrose production were intertwined, and provides a sense of the intensive enumeration at the center of chemical control of a large and recently built factory.

George Rolfe, who on May 19, 1908, signed the report for that year's *zafra*, or harvest (from December 12, 1907, to May 6, 1908), had come to Aguirre as chief chemist from his position as instructor in sugar analysis at the Massachusetts Institute of Technology.[54] At the top of his sheet of yields and losses, Rolfe recorded that Aguirre had taken in 25,875,994 pounds of sucrose that year "entered by Cane," and an additional 136,700 pounds of remelted third sugars from 1907 and prior crops, for a total of 26,012,694 pounds of sucrose. Of this, 21,405,165 pounds had been recovered, and 4,607,529 pounds "lost" in manufacture. Of that loss, a tiny fraction had vanished with the mud or filter-press cake (211,212 pounds), in the molasses (3,360,369 pounds), and, worryingly, 1,035,948 pounds to what were called "undetermined" losses. This last category, Rolfe conceded elsewhere, was always to be expected within the range of 0.2 percent to 0.4 percent of the weight of cane, and it marked the conceptual limit of the chemists' power over what went on within the central's walls. And, ultimately, there was the sucrose that had never made it through the mills into the factory, 3,347,060 pounds of which had been burned in the furnaces with the bagasse. But all of these numbers were simply presented as sucrose—and what was bagged and shipped from Aguirre was not sucrose but sugar. So the final lines of the report, above Rolfe's signature, read "Commercial Sugar: 22,483,990 lbs., Polarization 95.20%," or 9.733 percent by weight of the cane. And then, "Note: Nothing left in tanks."[55]

On subsequent pages Rolfe detailed the links between sucrose and work. The central had ground 115,497 tons and 410 pounds of cane in the course of 109 "Days Work in House," or, more precisely, 2,776 hours and fifteen minutes: an average of 1,059.6 tons a day, or 41.6 tons an hour. But these had by no means been evenly split among Aguirre's three mills.[56]

Instead Mill No. 1 had barely operated all season, for 7 days of just over 12 hours each, while No. 2 had worked for 72 days, though only averaging 9 hours and 17 minutes per day. Only Mill No. 3 had run for 108 of the 109 days, for 18 hours and 45 minutes on average. Rather than periodically but completely breaking down, however, Nos. 1 and 2 seem to have been less than fully functional even when they were running. The former, for instance, during its brief week, had managed to grind only 24.05 tons of cane per hour, or barely half of what No. 3 was able to crush. And its bagasse was 5.72 percent sucrose rather than No. 3's 4.82 percent—in other words, Aguirre was losing 17 percent more of the substance it had paid for. Indeed, although daily and weekly reports and other documents from this period in the central no longer exist to confirm it, it seems likely that the alarmingly high sucrose content of

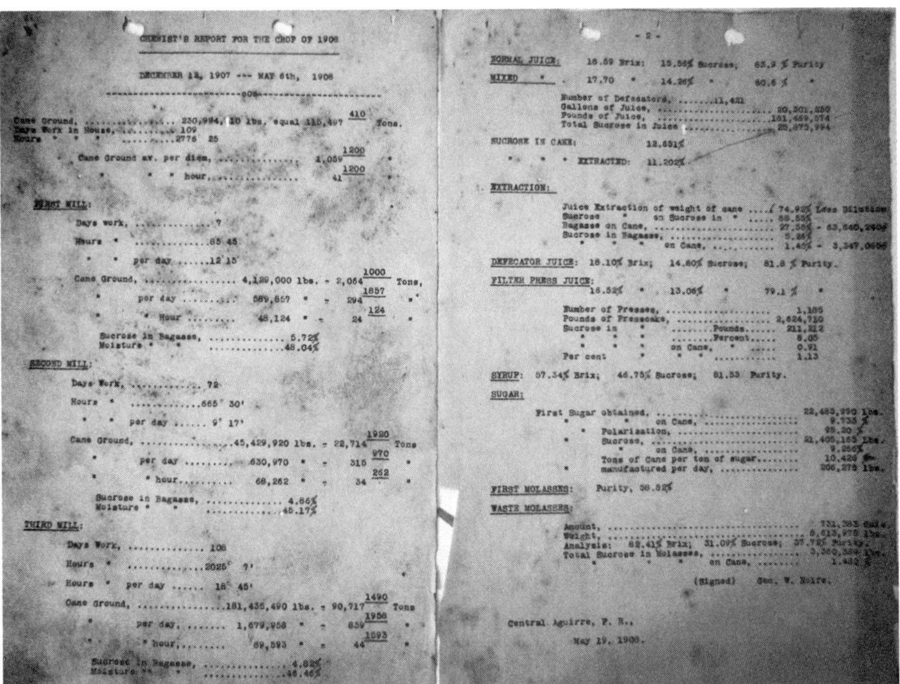

CHEMIST'S REPORT FOR THE CROP OF 1908

DECEMBER 12, 1907 --- MAY 6th, 1908

---------------------------o0o---------------------------

Cane Ground, ... 230,994,820 lbs. equal 115,497 410 Tons.
Days Work in House, ... 109
Hours " " " 2776 25
Cane Ground av. per diem, ... 1,059 1200 "
" " " hour, ... 41 1200 "

FIRST MILL:

Days work, ... 7
Hours " 85 45
" " per day ......12 15'
Cane Ground, ... 4,129,000 lbs. = 2,064 1000 Tons,
" per day .... 589,857 " = 294 1857 "
" " Hour .... 48,124 " = 24 124 "
Sucrose in Bagasse, ... 5.72%
Moisture " " 48.04%

SECOND MILL:

Days Work, ... 72
Hours " ....665' 30'
" " per day ...... 9' 17'
Cane Ground, ...45,429,920 lbs. = 22,714 1920 Tons
" per day .... 630,970 " = 315 970 "
" " hour.......... 68,262 " = 34 262 "
Sucrose in Bagasse, ... 4.86%
Moisture " " 45.17%

THIRD MILL:

Days Work, ... 108
Hours " ....2025' 7'
Hours " per day ...... 18' 45'
Cane Ground, ...181,435,490 lbs. = 90,717 1490 Tons
" per day, ... 1,679,958 " = 839 1958 "
" " hour, ... 89,593 " = 44 1093 "
Sucrose in Bagasse, ... 4.82%
Moisture " " 48.46%

- 2 -

NORMAL JUICE: 18.59 Brix; 15.58% Sucrose; 83.9 % Purity
MIXED " 17.70 " 14.26% " 80.6 % "

Number of Defecators, ...11,481
Gallons of Juice, ... 20,301,230
Pounds of Juice, ...181,469,074
Total Sucrose in Juice 25,873,994

SUCROSE IN CANE: 12.651%
" " " EXTRACTED: 11.202%

EXTRACTION:

Juice Extraction of weight of cane ....(74.92% Less Dilution
Sucrose " on Sucrose in " 88.55%
Bagasse on Cane, ... 27.58% - 63,640,240#
Sucrose in Bagasse, ... 5.26%
" " " on Cane, ... 1.45% - 3,347,060#

DEFECATOR JUICE: 18.10% Brix; 14.80% Sucrose; 81.8 % Purity.

FILTER PRESS JUICE:
16.52% " 13.06% " 79.1 % "

Number of Presses, ... 1,186
Pounds of Presscake, ... 2,624,710
Sucrose in " ....Pounds...... 211,212
" " " ....Percent..... 8.05
" " " on Cane, " 0.91
Per cent " " " 1.13

SYRUP: 57.34% Brix; 46.75% Sucrose; 81.53 Purity.

SUGAR:

First Sugar obtained, ... 22,483,990 lbs.
" " on Cane, ... 9.733 %
" Polarization, ... 95.20 %
" Sucrose, ... 21,405,163 lbs.
" on Cane, ... 9.266%
Tons of Cane per ton of sugar.......... 10.426 "
" manufactured per day, ... 206,275 lbs.

FIRST MOLASSES: Purity, 58.82%

WASTE MOLASSES:

Amount, ... 731,383 Gals.
Weight, ... 8,613,978 lbs.
Analysis: 88.41% Brix; 31.09% Sucrose; 37.72% Purity.
Total Sucrose in Molasses, ... 3,360,384 lbs.
" " " on Cane, ... 1.45 %

(Signed) [illegible]

Central Aguirre, P. R.,

May 19, 1908.

FIGURE 4.4. The annual chemical report for 1908 from the large and modern Central Aguirre in Puerto Rico. Central Aguirre Chemists' Reports, Colección Central Mercedita, Archivo General de Puerto Rico.

the bagasse was at least part of the reason that the mill unit was shut down in the first place.[57]

The situation of the mill deteriorated. The next year, in the crop of 1909, which ran from November 21, 1908, to May 27, 1909, the first mill worked no days at all. Apparently, it never worked any again: the Aguirre chemist wrote "FIRST MILL: NOT WORKING" through the 1919 crop, after which the report only bothered to list the second and third mill at all.[58] The 1910 crop reported a slight increase in the polarization of the first sugar—95.36 percent over 95.2 percent. The following year, however, the Aguirre chemist rather remarkably reported that the average polarization of the 20,799.91 tons of "commercial sugar" which the central had bagged was exactly 96.00 percent. In 1912, the sugar was 96.09 percent sucrose, and in 1913, 96.24 percent. The much-remarked-upon uniformity of sugar was always, on close inspection, illusory. In later chapters we will see why this illusion mattered so much.

The 1914 report was different, however, listing not only the bagged sugar, but its polarization "at Aguirre" of 96.43 percent, and its equivalent at 96 percent—in other words the tons of sugar that would have been bagged had the factory managed to manipulate its processes to produce sugar of the idealized market standard of polarization. And indeed over the following years the factory made no attempt to keep the sucrose content of its sugar rising. Instead its operators seem to have strived to maintain the polarization of their output within one- or at most two-tenths of the 96 percent target. The year-end reports do not yield much information about the practices of the factory's laboratory. The chemist and superintendent who followed Rolfe toward 1910 had begun to note, in their annual submission, that the sucrose in the molasses was being measured by the time-consuming but more precise technique of the Clerget method of double polarization, rather than a single "direct" polarization. The Clerget method accounted for the fact that a small number of the sucrose molecules might have split into the simpler sugars glucose and fructose, which had different polarizations in different directions that would throw off the whole reading. The fact that Aguirre's chemists began to note "Sucrose (Clerget)" in their analysis of the central's waste molasses may suggest that the South Porto Rico Sugar Company had assigned enough labor to the laboratory to permit the procedure to become routine.

Marvelous Secrets

Chemical control, as this chapter shows, was always about command of labor and not just of machines. Despite the efforts of central-factory owners and managers to eliminate artisan skill from sugar production, however, there was one crucial point of the process at which they were, to their frustration, unable to do so. This was the "strike" from the vacuum pan.

The notion that the artisan sugarmaster, enslaved or free, was rendered obsolete by new technology encapsulates the same paradoxical juxtaposition that renders the chemical sugar factory both revolutionary and evolutionary at once. If the centrifugal was taken to represent the new qualities of sugar, the vacuum pan has been most closely tied to the notion that craft knowledge was no longer required to make it. The old sugarmaster's expertise at discriminating among the qualities of the loaf was allegedly rendered irrelevant by the uniformity of centrifugal sugar. Moreno Fraginals concluded the first volume of *The Sugarmill* with an elegy on the chemical dissolution of the sugarmaster's authority: "His loss of supremacy came when vacuum pans were introduced and his judgment was replaced by physical measuring apparatuses."[59]

Yet the polariscope, for all Biot's promises, could not replace this judgment. The advocates of chemically controlled factories made two claims that served their rhetorical purposes while not being quite true. The first was that the old *ingenio* had been unsystematic and uncontrolled, when it was in fact controlled by the enslaved and free artisans who made the sugar, along with the early accounting methods of the planters. The second claim was that the new factory, stuffed with chemists and chemistry, actually was under complete control.

The new mechanical apparatus and systems of chemical control could not quantify the "secrets" of the sugarmaster. Sugar chemists themselves, in their own words, were more circumspect and consistently reluctant to make promises of control over a process that was mostly opaque and from which visual access did not yield much insight. Particularly after the end of slavery, owners and managers of sugar factories no longer felt sure they could trust the labor force they had once been comfortable controlling.

The legal status of workers in the Spanish Antilles was in question just as their technical milieu was changing. In 1861/62, when the government of Cuba conducted its first census in a decade and a half, it found an island that was becoming increasingly populated by people of light skin born in the New World. Of the 1,359,000 inhabitants, nearly 730,000 were white, and most of those were Creole, though *peninsulares* held a disproportionate number of powerful positions. About 225,000 were free people of color, manumitted or descended from freed ancestors. There were only slightly more slaves than there had been in 1846: 369,000 as against 323,000. Half of them worked on sugar plantations. There were, however, over 34,000 indentured workers from China, who had begun to arrive the year after the previous census. These were the first of 141,000 who would be induced to travel to Cuba, and of 125,000 who would survive the voyage, between 1847 and 1874, when the Chinese government banned the deceptive tactics of the procurement agents in its port cities and interior towns.[60] The labor force of Chinese people subjected to indentured servitude, like South Asians brought to the British Caribbean colonies, was intended to replace the manpower of slaves after the effective end of the slave trade in the mid-1840s. Edwin Atkins credited his Chinese laborers as being much of the reason that his sugar factory at Soledad, full of new vacuum pans, crushers, mills, and centrifuges, "was thoroughly organized and everything went like clockwork."[61]

The rebellion, based in the eastern provinces, that waged the Ten Years' War against Madrid's government in Cuba between 1868 and 1878 offered legal freedom to slaves who joined it as well as abolition if the rebels were victorious. The crown's response was the Moret Law, which in 1870 freed children and the elderly, at least nominally, but planters and their allies in the island's government continued to use their hold on power to limit slaves' practical freedom where it would interfere with their own economic interests, at least in the short and medium term. Between the 1861 census and the end of the war, Scott writes, "where sugar prospered, slavery persisted." Planters found various ways to incorporate Chinese and free workers while maintaining a labor force composed mostly of slaves. Slavery was extensive even in the largest and most technically sophisticated centrals. But the planters and factory owners did so at the cost of opening the eyes of the enslaved to alternatives that were within their grasp. Within a decade of the Moret Law, the *patronato* law established ten-year indentures for those currently enslaved, at the end of which would come the full abolition that the law in fact was intended to delay. But within barely half that span, the state abandoned its planned gradualism and abolished slavery entirely.[62] And during the 1890s, Cuba was racked by a war for independence and brutal counterrevolutionary violence in response.

In addition to the chemist's reports and yield summaries that he signed, George Rolfe also produced a treatise titled *The Polariscope in the Chemical Laboratory* in 1905. "In most first-class cane sugarhouses," he wrote, "the chemist is the superintendent of sugar manufacturing, *as he should be*." Chemical control had as one of its purposes "the guidance of the daily routine of the house."[63] It was the part that was not routine that posed the most problems for chemical control. Chief among these was the vacuum pan, whose operator, Rolfe emphasized, required "special skills."

The skills of the sugar boilers or "pan men" remained the most crucial ingredient to a successful crystallization. An index of their importance is that Atkins made time among his many other responsibilities to hire these men himself. He spent a great deal of time every off-season evaluating the performance of his current pan men and negotiating their salaries or granting them bonuses, even as he evaluated job applicants for the position, scheduled face-to-face meetings with potential new hires in person, and arranged training to complement the skills they already possessed.[64]

As industrializers discovered across the world, replacing human-performed tasks with machines did not end the need for human skill but merely displaced it onto a new type of worker and a different set of abilities.[65] Running particular machines well took its own special skill, and centrals were vulnerable to strikes by the men who managed the unloading of cane, who supervised the centrifugals, or who maintained the power station of a central.[66] The pan men were the most critical of all because, despite decades of efforts to automate skill in sugar production and transform it from a discrete to a continuous process more suitable for control, there was one crucial point of the process where they were frustratingly unable to calculate skill away.[67]

Like their predecessors, the skill of vacuum pan men was described in magical and supernatural terms. A pan man held "marvellous secrets," he was "looked upon as a wizard," his subject "regarded almost as a black art." He was a "boss conjurer."[68] Even a team of MIT management consultants in the 1920s was reluctant to claim how much progress had been made, only noting carefully that "science has put this process on a somewhat firmer basis."[69]

The amount of syrup that the "pan man" allowed into the pan, called the charge, had to be of just the right volume so that when sugar crystallized out of the syrup it would "just fill the pan full," no more and no less. Here, the economy of the sugar factory, and the maximization of the capital invested in the expensively fragile vacuum pan, depended unavoidably, in the labor historian David Montgomery's phrase, on the brain under the workman's hat. Predicting the quantity of syrup that could be nurtured into the desired volume of crystals of the appropriate size was, Rolfe wrote, "obviously . . . a matter requiring considerable experience," being "conditioned on the quality of the juice, as well as the size of the crystals required, and the polarization." Once the pan operator felt he had charged the appropriate initial quantity of syrup, or meladura, he began the process of crystallization, a task Rolfe conceded was "also a matter requiring much skill and experience." This was because further "drinks" of meladura had to be allowed into the pan in order to regulate the growth of the crystals. "On this will depend to a large extent the quantity of finished sugar of the required polarization, as well as the size of the crystals themselves." Yet while the latter might be observed by drawing samples of liquid out through the stop-cock, the polarization of the sugar crystals themselves could not be determined during the crystallization process.

Once the pan operator made his strike, the polarization of the resulting massecuite could be tested. The polarization of the crystals themselves, separate from their viscous medium, could be measured only once the crystals had been spun dry in the centrifugal and freed from their molasses—not before. So the pan operator had to manipulate the input of meladura and its temperature in ways that related to the nature of the initial juice. But it was not just juice of "poor quality" that needed the unquantifiable expertise of the pan operator; the task was no less mechanical nor reducible in the ideal case of juice from healthy canes, produced on the central's own land, under the supervision of its management. The "subsequent additions of meladura must be made skillfully" in every case, "so that the crystal growth will proceed regularly and continuously."[70] Chemical expertise could produce no formula that took in all the factors that composed the juice's "quality" or "character"—such as its purity, acidity, and albumin—and told the boiler the ideal temperature and flux to use in growing perfect crystals while mechanically pointing to the desired temperature and flux of syrup on the other. Quality or character

FIGURE 4.5. Long into the twentieth century, boilers took samples from the vacuum pan using a "proof stick," which let them examine the sugar without breaking the vacuum, but they still had to examine it using senses honed by experience. Bookers Sugar Estates, Ltd., *Bookers Sugar: Supplement to the Accounts of Booker Brothers, McConnell & Co., Limited* (Georgetown, British Guiana, 1954).

might make sugar "very wet & a little sour," or render one of Atkins's lots "lose in color and turn red" while leaving the next "as bright as the day they were made, free from grain and have a hard gritty feeling like dry sand."[71] Whether the juice was of poor quality or from healthy canes produced under the central's own supervision, the task was never mechanical nor reducible to calculable quantities.

What the "skillful pan men" knew also how to avoid was the "'false' or 'second grain'" by which sugar crystallized and then separated into smaller particulates. Between tiny and tinier grains, it was not clear which was worse. If the false crystals were not small enough to escape the centrifuge, they made purging the sugar of its molasses more difficult. Moreover, because a mixture of differently sized particles packs together more closely, their presence in the final sugar paradoxically increased its capillary capacity to carry moisture and thereby decreased its reading in the polariscope. But if the second grain was fine enough to escape the basket and drain away with the molasses, then it raised the sucrose content of the molasses while decreasing the yield of the factory's first and most valuable sugars. Rolfe reiterated the priority of expertise in this particular labor, which lay at the center of the sugar factory's process. "Skill engendered of long experience and familiarity with the workings of his pan are essential qualifications of a first-class sugar boiler. Such men, in raw-sugar work, command high pay and have much responsibility." By contrast, "an unskillful man may cause a loss of hundreds of dollars a week."[72] Maintaining the syrup at the proper temperature, pressure, and water content to grow crystals to a large enough size took serious knowledge. "By skilful pan-boiling, the crystallised sugar is obtained in a form permitting of easy separation from the molasses," Deerr wrote, while "the presence of fine crystals may cause considerable losses." Hence "the determination of the crystallised sugar" in the massecuite "affords a valuable check on the pan-boiler," his skill, and his ability or willingness to deploy it.[73] Measurement of the sugar content of massecuites thereby afforded another opportunity in the chemist's eyes for a check on theft.

As a result, the position of the pan operator was fundamentally a precarious one. A 1911 book on sugar for young readers, part of the *Peeps at Industries* series, invited those readers to "sympathize with the anxiety of the pan-boiler as he scrutinizes samples of the boiling . . . for you realize that with him rests the responsibility of deciding when the masse-cuite is ready to leave the pans."[74] In a raw-sugar factory, though a "native operator, [the boiler] holds a most responsible position; a trifle too little or a trifle too much boiling, and the contents of his pan are spoiled—moreover, his reputation, very likely an excellent one of long standing, has gone for ever."[75] Unexpectedly, for such a paean to imperial industry, the book struck a note of international solidarity among a skilled class of workers. The precariousness of the operator of the vacuum pan was not dependent on his position as a "native" within a sugar colony. The men who operated the vacuum pans in metropolitan sugar refineries, whether in Liverpool or in Brooklyn, were under the same surveillance and "anxiety."

This is the context for the occasion when, in the middle of the 1894 harvest, Skaife had to report to Atkins that he had fired a sugar boiler: not just the punch, but "because I asked him very quietly to put down the time of starting molasses strikes which he was neglecting."[76] What was at stake in these fisticuffs was whether knowledge about sugar-boiling was the property of the artisan or the chemist. In subsequent weeks Skaife had to work unsustainably

FIGURE 4.6. A boiler in British Guiana in the 1950s, recording his work in front of a bank of dials monitoring the crystallizers, where the massecuite cooled before being dispatched to the centrifuges. Bookers Sugar Estates, Ltd., *Bookers Sugar: Supplement to the Accounts of Booker Brothers, McConnell & Co., Limited* (Georgetown, British Guiana, 1954).

long hours to compensate for the absent employee. But he did not directly operate the pans, leaving that task to the boiler who remained.[77]

Sweetening the Rates

The separation of the factory and the field, however, which both defined the central and was important to chemical control, did not mean that chemists would not attempt to exert control over the places where sugarcane was grown. In the early twentieth century the work and expertise of chemists became the focus of conflict between the *colonato* farmers and the owners of central factories.

The analysis of the relationship between central and colono has been an important question for historians' understanding of the patterns of economic development in the large sugar islands of the Spanish Caribbean after emancipation and particularly after 1898. Ramiro Guerra y Sánchez argued that the railroad, by extending the mill's range, allowed these outposts of American imperialism to exert control over a larger zone and eventually to absorb smaller landholdings into huge latifundia.[78] The extensive and lengthy data available for Cuba allowed the economist Alan Dye to show that that differences in relationships between colonos and centrals contributed to the pattern of investments made by American capitalists in the late nineteenth and early twentieth centuries. In the west, the oldest and most densely settled part of Cuba, colonos tended to be former planters themselves, while in eastern Cuba the landowners were less wealthy, less well established, and therefore less powerful. The central factories built in the west, therefore, tended to have more difficulty negotiating with *independiente* cane suppliers, and less frequently owned the land while subcontracting out the operation. But problematically for a crop that depended so significantly on coordination between agriculture and processing, the scattered ownership of the cane fields made it more difficult to coordinate planting and harvest schedules. Thus, Dye argues, centrals were forced to attempt to exert control through contracts that more strictly defined how and when the cane was to be grown, harvested, and delivered.

By contrast, the eastern provinces of the island, especially the far east, had remained relatively undeveloped in the preceding decades and centuries, and it became easier to establish a new central and to accumulate large tracts of land under its own auspices or legally possessed by of closely interrelated companies.[79] These could then be rented or leased to colonos who would plant and manage and harvest the cane itself. The central's ownership of its cane land also helped, from its management's perspective, to solve the problem of its strategic vulnerability to a colono who refused to deliver his cane.[80] The need for the cane of a particular field at a particular time meant that a colono who chose his moment could throw the grinding schedule into disarray and potentially extract significantly higher payments from a central's management, who feared letting their expensive investment begin to rust.[81]

An alternative was for the central not merely to own land but to manage it as well and closely supervise the growing of the cane with its own personnel. Such land was known as "administration land," and the cane produced under the central's direct ownership as "administration cane." Yet at least through the early 1930s, the share of cane supplied to central factories by lands under their administration and the share supplied by *colonos independientes* both declined, replaced by *colonos del central,* managing land owned by the mills themselves.[82] In Dye's analysis, it was the logistical and biological problem of coordinating the reliable delivery of the sugarcane that spurred the growth of the large latifundia, not merely a local form of miniaturized imperialism.

Yet though the central could, in its contracts, mandate the agricultural techniques used by *colonos del central* or *independientes*, the quality of the cane delivered and its suitability for grinding was harder to specify and hard to measure. The purchase of cane, even from those working on a central's land, meant more work for the factory's chemists and assistants to do in a shorter period of time.[83] And in the early years of the twentieth century, there took place a crucial shift in the terms under which those contracts were assessed.[84]

In 1928, the new Brookings Institution devoted one of its first reports to a general exploration of the economic and social difficulties of Puerto Rico. Its appendix on "The Sugar Industry" was republished, at the request of the island's Association of Sugar Producers, under the new title "The Sugar Problem of Puerto Rico." Contracts between growers and grinders were a subject on which the Brookings investigators focused. They noticed that these documents, "essentially crop loan agreements," did not in fact stipulate that the central was required to grind the cane it purchased. As they put it, "the mill's interest in performing this service is so obvious that the obligation is not usually recorded." But given the disruption to a mill's operation when a roller or other piece of machinery broke—from the consequent degradation of the juice in the cut stalks waiting to be crushed, and the subsequent rush to make up for lost time—this was, in all likelihood, not a neglectful omission. Rather it was, on the part of the factory itself, a wise way to avoid a legal obligation to grind unprofitable cane. If forced to mill cane that had fermented into unprofitable territory, a central would then find that the newer cane was piling at its gates, its juice inverting by the minute.[85] The centrals also might use their contracts to stipulate the cane variety, irrigation, and fertilizer, and to cover those expenses frequently advanced cash to colonos until the grinding season, expenses of a sum that was often not enumerated but rather expressed as "sufficient to raise the crop." This meant that banks lent to colonos at a higher interest rate than to coffee or tobacco farms. A central did have a strong interest in grinding the cane of its own debtors first. This was not only when the central was pressed, in case of a logjam, but so as to maximize the return to those farmers who would then be obligated to it.[86]

Contracts between the colono and the central had, historically, been denominated in pounds of sugar per hundredweight of cane. Ordinarily, between five and seven pounds of sugar per hundred of cane were returned to the grower, which he could sell himself or ask the central to sell on his behalf (for a fee).[87] Centrals routinely tried to squeeze their colonos and shirk the terms they had agreed.[88] A deeper source of instability was a conflict over who deserved credit for increasing sugar output: the colonos who were investing in agricultural methods, or the centrals who were improving their manufacturing. More fundamentally, centrals and their chemists complained that these sorts of contracts yielded cane that was liable to be poorer quality and filled with "cane trash," such as leaves, or even with sticks thrown in to boost the weight.[89] If the colono were paid based on the measured juice purity of the cane he transferred to the central, they argued, then he would have an "incentive to raise a high quality of cane and to deliver it in the best possible conditions."[90] José Solá has found that at least one Belgian-owned Puerto Rican central paid colonos based on the density of their cane juice, a simpler measurement to make than polarization, as early as 1906.[91] After 1921, as sugar prices remained low in the wake of the post–world war crash, and as chemical forms of knowledge became entrenched throughout the sugar factory, such contracts came to be denominated in terms of sucrose content. By 1928, for instance, the insular government of Puerto Rico registered 605 weight contracts and 178 sucrose contracts.[92] But this system placed obvious advantages in the hands of the central, as the colono delivering to a mill "could not *know* whether he was fairly treated or not."[93] The juice of the cane could only be analyzed after it was crushed, and therefore all the power to decide the value of cane rested in the hands of the mill and its laboratory. For the colono, as one complained to a pair of US researchers, the system was a "funnel in which his end is the little end."[94] The funnel's squeeze was strengthened by the frequent requirement "that the cane shall come up to a minimum standard of purity and sucrose content, subject to penalties" if violated.

The authority thereby vested in the chemists of the centrals led, in Puerto Rico, to calls for legislation that would establish a force of chemists under the auspices of the government itself, in order to oversee the analyses of juice and even chemical control in the mills themselves. The proposed laws would also have mandated that a colono was entitled to at least 65 percent of the sucrose content of the cane he delivered. After several attempts to mollify these concerns in the 1920s and early 1930s,[95] the still dissatisfied colonos finally succeeded, in 1934, in achieving legislation that specified and regulated the details of the practice and place of sugar chemistry. "An Act to create the

Division of Inspection of Chemical Laboratories of Puerto Rico in the Department of Agriculture and Commerce" defined a "chemical laboratory" to be any "place or premises where chemical analyses or syntheses are carried out."[96] The new island-wide division of inspection was to be staffed by five chemists, who would be authorized to obtain and analyze "any sample of the cane juice of any *colono*, pertaining to cane ground in any sugar factory in the island, in order to contrast the analyses made at the laboratory in such factory." More generally they were to serve as an arbitrator (*árbitro*) of each "controversy" between colono and factory, to register all chemists and chemical engineers, and to dictate "the rules and system for taking samples and analyzing the cane juices" in factories' laboratories.[97]

By this point, in the 1930s, each American company dictated its own particular terms of payment to its colonos. These varied considerably and were never straightforward mathematical or chemical calculations. Central Aguirre, for instance, paid 65 percent of the expected recovery of sugar, calculated at New York prices minus the cost of freight. Fajardo paid 62.5 percent minus freight. Eastern Sugar Associates paid 65 percent, but at Puerto Rican prices, and they took a minimum of 3.5 percent by cane weight for the mill—a system designed to further reduce the incentives to deliver what the central considered lower-value cane.[98] On top of that, the colono selling to Eastern received not sugar but warehouse receipts, and then had to either market his sugar himself or pay Eastern to do it for him.

Central Cambalache, in the north near Arecibo, took a different tactic.[99] It had tried analyzing each colono's cane, but by the mid-1930s returned to the simpler basis of paying 7 percent by weight. Cambalache drew upon a particularly large number of particularly small growers, and its managers decided against the expense of chemical labor required to draw samples from each shipment. Yet they felt this would not reduce the quality of sugarcane Cambalache received, since the incentive effects of chemical control paled in comparison to the "aid in cultivation rendered by the mill's field inspectors."

The power of the centrals to control the specifics of those formulas had long been a source of contention among sugarcane growers, and indeed among mills themselves. Some mills were upset that "some competitor 'sweetened' the rates or the chemical analyses, to lure growers to its mill," reported a Columbia University study in the 1930s, while colonos complained that mills seemed to find an excessive number of deductions and demerits to reduce their payment. Cane weighing and analysis were judged by these researchers the least important points of contention in mill–colono relations. Although they argued that it was impossible to know whether the power of the companies to measure weights and conduct tests was being abused, the potential for abuse was real, "and its very existence can be the source of suspicion of unfair dealing and discrimination." Individually, each central defended its own idiosyncratic formula by pointing to an increasing body of chemists' evidence. But together, the effect of multiple, contradictory, supposedly rational formulas paradoxically reinforced a widespread sentiment among colonos that those same formulas were never purely "non-economic and technical in character." Their multiplicity "furnished a basis for popular suspicion . . . that they did not always operate to do strict justice to growers," and regardless of whether this was the case, it remained a severe irritant to the kind of relations between mills and colonos that both wanted, albeit on their own terms. If a decision among such formulas might and ought to be made on the basis of "careful technical study," it therefore seemed that any differences among them, on the part of centrals who claimed access to reserves of chemical expertise, were in fact the intentional result. How else but malfeasance to explain the differences among them, on the part of centrals who claimed access to reserves of objective chemical knowledge?[100]

Thus, the 1934 legislation was followed three years later by another bill, one even more explicitly intended to regulate contracts for cane purchases and, specifically, to establish "the procedure to be followed by the centrals of Puerto Rico in determining the crystallizable sugar per cent in the cane delivered by the colonos for grinding."[101] It defined a colono as any individual or corporation "delivering sugar cane for grinding to a central," and a central as any entity operating a sugar mill. And this new act attempted to further regulate the relationship between the mill and the field. The colono and central were to agree on the variety of local cane to be cultivated from those "which, because of their adaptation and noble qualities, may have become generalized in the region." The colono was responsible for his cane until delivery to the central or its agents. In return, the central was now obligated to grind the cane within twenty-four hours. And in a concession to the centrals, all of the cane that it processed, whether from its own lands or purchased, was to be "free from trash and earth." When the grinding season came to an end, each central was to file "a true copy of the final report on manufacture," including the mean analysis of the cane juice and the yield of 96 percent commercial-grade sugar.[102]

But it was not just the inspecting engineer, as an agent of the state, who was granted authority as arbitrator to enter and inspect the analytical premises of the central. Colonos had, as we have seen, long complained that the mills had total control over the analyses, "both in adopting formulas and in administering them," and that a colono delivering cane to the mill "could not *know* whether he was fairly treated or not, so the suspicion of evil had a breeding-ground."[103] Now cane growers themselves were granted the right to inspect their purchasers as well. This power was not, within the bill's text, temporally tied to the status of a contract: It could be exercised before, during, or after the grinding of a colono's cane, and both inside and outside the factory, at its scales in the field or incorporated into the railways, and in the "laboratories of the central" itself. Finally, and most remarkably, the law authorized them not just to "observe" but also to "intervene" in all the operations of chemical control: "the taring of cars, weighing of sugar, examination of scales, weighing of cane and juice, taking of samples and analyses of cane." For its part, the central was "obliged to permit access of such personnel to said dependencies of the factory, and to offer them reasonable facilities for the performance of their duties"—which might presumably include the facilities of the central's laboratory itself.[104]

Legislation to standardize the formula for the remuneration of cane growers all over the island was intended to clarify that the clash of interests between mills and colonos was an economic one, fought over the fraction of the price of sugar that each should receive. It was one final way in which the supposedly mechanical power of the central to transform sugarcane into pure sucrose actually depended on various kinds of skilled work and the battle to control it. The conflict had been present in the formula all along.

PART TWO

The following three chapters narrow the focus to the United States, and in particular to the port of New York, between 1875 and 1900. New York was the center of America's powerful sugar refining industry in the nineteenth and early twentieth centuries, and its custom house was more important than all the others put together, in an age when the tariff was a central political issue and the collection of import duties was the chief source of federal revenue. Both knowledge and deception about the nature of sugar took place here, in the conflicts among participants in the sugar trade and between the government and private interests.

The confusing properties of this new sugar destabilized how Americans understood what they were trading and eating. At the same time, as discussed in chapter 5, that uncertainty and confusion provided opportunities that merchants and refiners were happy to exploit. Conflicts over metrology were displaced onto languages of corruption, fraud, and deception. The Treasury Department was responsible for interpreting and enforcing America's growing list of protectionist rules about sugar. The government finally took a stand over 712 bags from South America, but the case, rather than clarifying official bewilderment, instead entangled techniques of production and measurement with Euro-American anxieties about racist and civilizational purity. This chapter also sets a kind of stage for the remainder of the book, as we see the metrology of commodities in action where state measures crashed against commercial values.

The largest refiners claimed that the polariscope would rationalize and revolutionize the collection of customs because, unlike the old color standard, it could exactly detect the value of sugar. But, as I argue in chapter 6, this was a sleight of hand, a misunderstanding, or a combination of the two. In its commercial usage, the trade was aware that the polariscope was susceptible to as many interpretations as it had users, if not more. But in public, refiners claimed it was a robust, scientific, and progressive tool that would help alleviate the fraud and fears of fraud around the collection of duties.

What a polariscope could tell you, under ideal circumstances, was the content of a sample, not its commercial value. Did a polariscope tariff measure sucrose, did it measure value, or did it measure neither? Further, as other refiners and witnesses warned, the polariscope was no less susceptible to corruption, and possibly more so. To many observers, the confusion over how to collect sugar tariffs facilitated the rise of the Sugar Trust.

The difficulty of grasping what sugar actually was provides in chapter 7 the background for a very Gilded Age hoax. In the 1880s, an outfit called the Electric Sugar Refining Company attracted a small fortune from investors. Its sugar impressed the best industry men and baffled the best chemists. When it was finally exposed, the company's appeal attested to the precarious state of much of the American sugar industry and the uncertain characteristics of sugar in the marketplace.

5

Acarus sacchari

The Sugar Mite

An English physician named Arthur Hassall found the bug first. This was 1855, and Hassall's microscope had recently terrified Londoners into cleaning up their water, thanks to his position at the head of the Analytical Sanitary Commission. Now, peering at sugar through the same instrument, Hassall discovered an abattoir of insects. He saw specimens of the insect "in all the stages of their growth and in every condition, some alive, others dead, some entire, and others broken into fragments, bodies here, legs there."[1] Hassall called it *Acarus sacchari*: the sugar mite.

Beetles had arrived in brown sugar before, but no critter so vicious as this.[2] The sugar mite burrowed into human skin, where its eggs irritated and scabbed. The resulting affliction was generally called the "grocer's itch," but it appeared especially on the arms of stockboys and warehousemen who worked with sugar.[3] *Acarus sacchari* was everywhere except the most refined sugar; only modern industrial practices could guarantee its absence. The normal processes of manufacture on tropical plantations would not wipe it out. The bugs were "formidably organized, exceedingly lively, and decidedly ugly."[4] Hassall and his fellow public chemists thought raw sugars were barely safe to handle, let alone eat, and nothing demonstrated this better than the sugar mite. Obviously, sugar refiners, who whitened raw sugar for final consumption, adored it.

After the mites' discovery, they reproduced. Extrapolating from his sample, Hassall had calculated that forty thousand of the mites crawled in each pound of raw sugar. By the late 1860s, the city analyst of Dublin guessed a hundred thousand per pound; a surgeon figured more like a quarter of a million. As individual specimens, they also bulked up, to the point that *The New American Cyclopaedia* reported that sugar mites were visible to the unaided

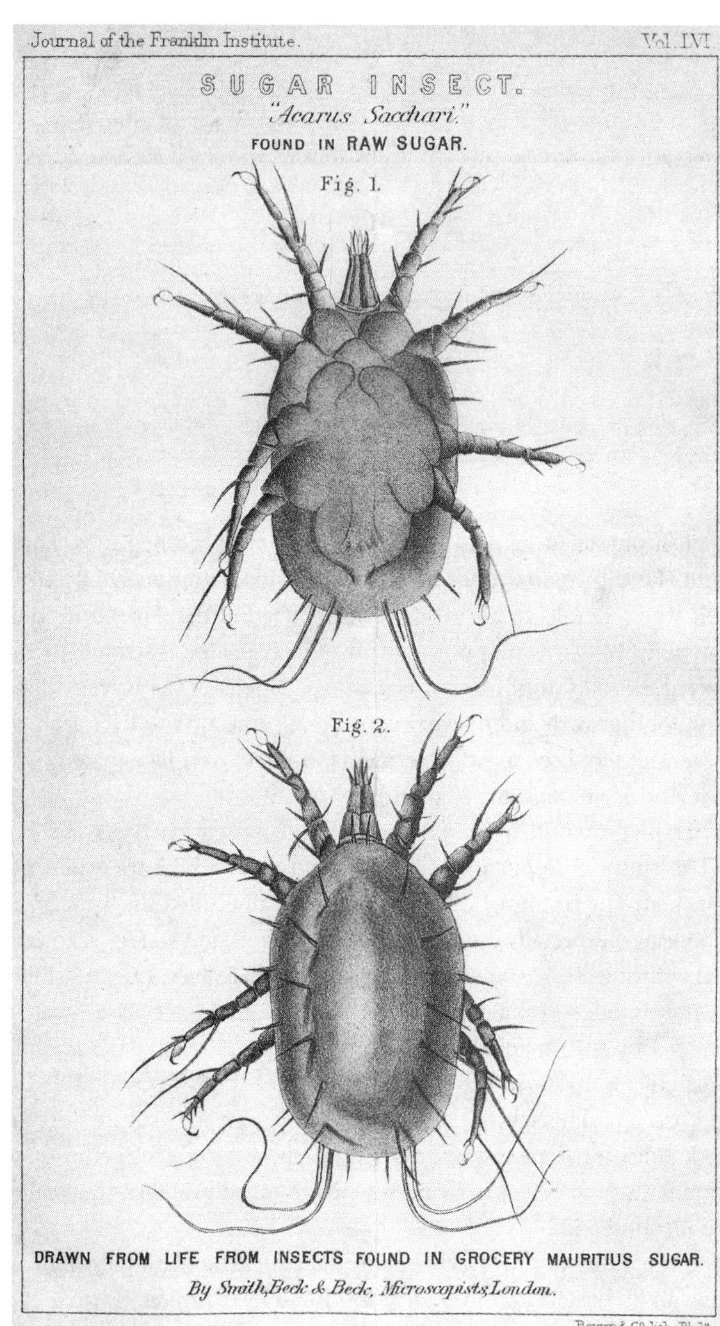

FIGURE 5.1. The sugar mite (*Acarus sacchari*). From Robert Niccol, *The Sugar Insect: "Acarus Sacchari," Found in Raw Sugar* (De Armond & Goodrich, printers, 1868).
Provided by U.S. National Library of Medicine.

eye. All you needed to do was dissolve sugar in water, let it rest an hour, and you could see the "animalcules" on the sides of the glass.[5] A terrifying Scottish pamphlet about the raw sugar mite—published in the refining town of Greenock—found readership among United States senators. And in 1879, when *Popular Science* described them for a wide audience, *Acarus sacchari* swarmed 350,000 to the pound.[6]

In the United States, an ambitious member of the House of Representatives from Ohio brought the *Popular Science* article to the floor of that chamber, and the insect wriggled into the *Congressional Record*. The mite's presence or absence in a country's sugar, according to Congressman James Garfield, indexed that country's progress toward civilization. Another index was the kind of sugar that they made. The cruder the product, the cruder the people. Inhabitants of the tropics, from Cuba to the Philippines, made sugar in a "black cheap" form, which only American skill and knowledge could render into white crystals suitable for American digestion.[7] As April Merleaux shows in her book *Sugar and Civilization*, the perceived qualities and properties of sugar were, by early twentieth-century America, inseparable from the people who made them, consumed them, and even governed them. White sugar was not only more luxurious, but better for the health of the white peoples of the world.

Previous chapters showed how new techniques and devices of sugar manufacture transformed the Caribbean factories supplying the United States and European nations with sugar. In this chapter, we will see how such sugar bewildered officials and even the sugar trade itself. The terms and concepts used to understand sugar had long since uncoupled from the processes used to extract sugar itself from the sugarcane plant, or even what the referent substance looked like. In the United States in the late 1870s, these confusions provided an opening for fraud, corruption, and adulteration—or for baseless allegations of those things, depending on whom you asked. Even the spectra of people and sugar were similar: dark to light, scorned to esteemed. But unlike people, sugars of lighter color found it harder rather than easier to cross borders. An important moment in the history of sugar in the United States, it turns out, was the late November day in 1877 when the brig *Mississippi* sailed into Baltimore's harbor carrying hundreds of bags of sugar.

Dutch Standard

The largest refineries in the United States were in New York and especially in Brooklyn. Compared with these, Baltimore's were tiny. But they were substantial enough that cargoes of sugar arrived in the port from across the

Caribbean and South America.[8] When the *Mississippi* tied up, perhaps even as she neared the dock, the captain passed the ship's papers to a messenger ashore. Long before the bags piled up on the wharf, and long before anyone in America knew how dark the sugar was, the messenger set off for the custom house, so that the owner of the ship's cargo would learn the price of its entry into the United States. The sugar she bore was bound for refining, not for direct consumption, a crucial distinction in the structure of the tariff on imported sugar.

Tariffs were one of the dominant political issues of the Gilded Age, a nationalistic reaction to the changes wrought by America's existence within a larger global economy.[9] And of all the period's protectionist tariffs, the one levied on imported sugar was among the most important and most controversial. Its effects were almost universally felt, because sugar was consumed by nearly every American in increasingly large quantities. From ten pounds per capita at the beginning of the nineteenth century, Americans ate almost thirty pounds by the Civil War, and eighty pounds by the turn of the twentieth century.[10] Sugar duties were also, in this era before the income tax, by far the largest single item of federal revenue. In 1879, for instance, the national government earned $274 million from all its sources of revenue put together. More than half of that total—over $137 million—came from customs, and of that half, about $40 million came from sugar alone. The sugar tariff, to put it differently, was as big as the US Army's budget, and perhaps more important, it was as big as the pensions owed to the veterans of the Grand Army of the Republic, on whose votes the Republican Party depended.[11]

The papers told customs officers that the *Mississippi* had come from Demerara, in British Guiana, and that the sugar belonged to one Adolphus William Perot, a Georgetown merchant. Anticipating their arrival in Baltimore, he had consigned them to another Perot, his brother and business partner William Henry. Whatever the customs appraisers decided the sugar owed the government, William Henry Perot would pay, and how much it owed depended on what sort of sugar it was.

Just as a rancher brands cattle in a herd, a sugar plantation marked its bags or barrels with a symbol denoting their place of origin and production. If a plantation shipped several sugar products of significantly different qualities, each might have its own variant of the plantation's mark. Two hundred and forty-three bags bore the mark S, conveying that they had been made by a plantation called Vergenoegen, using the relatively new centrifugal technology rather than the older muscovado process for draining molasses.[12] An E on others meant they were centrifugals from a plantation called Blenheim.[13] Both of these were among the most modern plantations in the territory, but

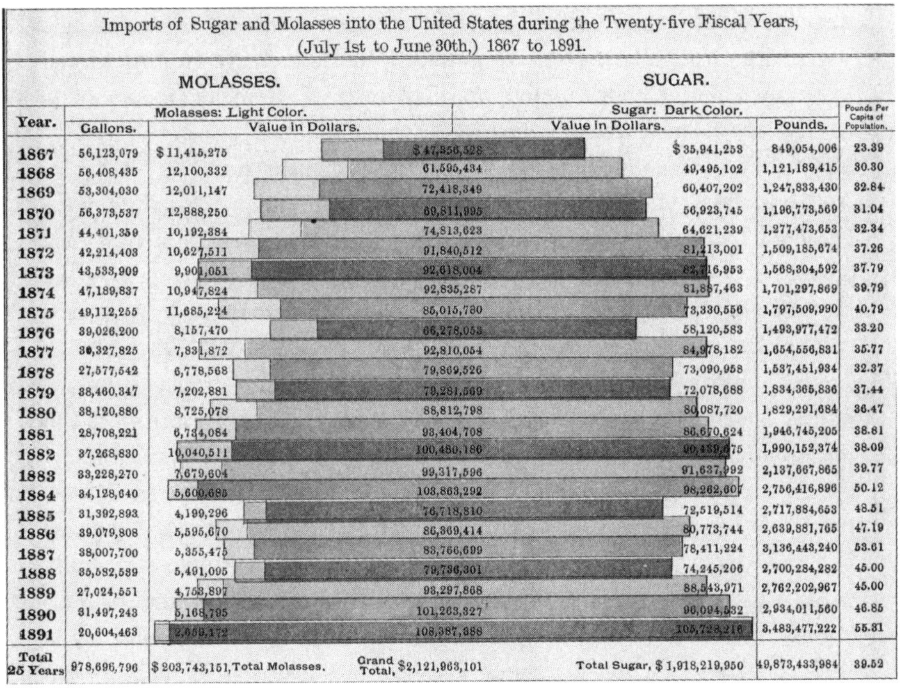

Imports of Sugar and Molasses into the United States during the Twenty-five Fiscal Years, (July 1st to June 30th,) 1867 to 1891.

Year.	MOLASSES. Molasses: Light Color. Gallons.	Molasses: Value in Dollars.		SUGAR. Sugar: Dark Color. Value in Dollars.	Pounds.	Pounds Per Capita of Population.
1867	56,123,079	$11,415,275	$47,356,528	$35,941,253	849,054,006	23.39
1868	56,408,435	12,100,332	61,595,434	49,495,102	1,121,189,415	30.30
1869	53,304,030	12,011,147	72,418,349	60,407,202	1,247,833,430	32.84
1870	56,373,537	12,888,250	69,811,995	56,923,745	1,196,773,569	31.04
1871	44,401,359	10,192,384	74,813,623	64,621,239	1,277,473,653	32.34
1872	42,214,403	10,627,511	91,840,512	81,213,001	1,509,185,674	37.26
1873	43,533,909	9,901,051	92,618,004	82,716,953	1,568,304,592	37.79
1874	47,189,837	10,947,824	92,835,287	81,887,463	1,701,297,869	39.79
1875	49,112,255	11,685,224	85,015,780	73,330,556	1,797,509,990	40.79
1876	39,026,200	8,157,470	66,278,053	58,120,583	1,493,977,472	33.20
1877	30,327,825	7,831,872	92,810,054	84,978,182	1,654,556,831	35.77
1878	27,577,542	6,778,568	79,869,526	73,090,958	1,537,451,934	32.37
1879	38,460,347	7,202,881	79,281,569	72,078,688	1,834,365,836	37.44
1880	38,120,880	8,725,078	88,812,798	80,087,720	1,829,291,684	36.47
1881	28,708,221	6,734,084	93,404,708	86,670,624	1,946,745,205	38.81
1882	37,268,830	10,040,511	100,480,186	90,439,675	1,990,152,374	38.09
1883	33,228,270	7,679,604	99,317,596	91,637,992	2,137,667,865	39.77
1884	34,128,640	5,600,685	103,863,292	98,262,607	2,756,416,896	50.12
1885	31,392,893	4,199,296	76,718,810	72,519,514	2,717,884,653	48.51
1886	39,079,808	5,595,670	86,369,414	80,773,744	2,639,881,765	47.19
1887	38,007,700	5,355,475	83,766,699	78,411,224	3,136,443,240	53.61
1888	35,582,589	5,491,095	79,736,301	74,245,206	2,700,284,282	45.00
1889	27,024,551	4,753,897	93,297,868	88,543,971	2,762,202,967	45.00
1890	31,497,243	5,168,795	101,263,327	96,094,532	2,934,011,560	46.85
1891	20,604,463	2,659,172	108,387,388	105,728,216	3,483,477,222	55.31
Total 25 Years	978,696,796	$203,743,151, Total Molasses.	Grand Total, $2,121,963,101	Total Sugar, $1,918,219,950	49,873,433,984	39.52

FIGURE 5.2. Imports of sugar and molasses to the United States, 1867–1891. In a quarter century, per capita consumption of sugar more than doubled, and Americans also ate a lot less molasses. *Rand McNally & Co.'s Indexed Atlas of the World* (Rand McNally, 1897), 326.
Courtesy David Rumsey Map Collection, David Rumsey Map Center, Stanford Libraries.

this was especially true of Vergenoegen. A newspaper survey, recovered by the Guyanese historian Walter Rodney, described Vergenoegen as having "one of the most complete plants of modern sugar machinery of the very finest kind that can be made."[14]

To make sure that no refined sugar slipped through without paying the appropriate tariff, every cargo had to be directly appraised by an officer's eye—as did every cargo of wool, tin, machinery, jewelry, or any other good whose American producers clamored for Congress's help. A system of protection this extensive not only enriched manufacturers, which was the point, but inevitably engorged the nation's customs service too. Congress kept for itself the power to tell customs officers how to appraise each good and assign it a tariff. The tariff on some goods was buckled to their value, rising or falling with the price at market. Others simply paid flat rates.

Sugar had once paid by market value too. But beginning in 1861, sugars had been judged by their color. Protecting, say, a woolen-goods manufacturer was straightforward: It took no special expertise to tell yarn from a jacket. To distinguish between raw sugars and refined ones was harder. So every single sugar cargo had to be directly appraised by an officer's eye and his special instrument after it landed on the wharf but before it was melted in a refinery.

This special instrument, known as the Dutch standard, was in fact twenty-five glass jars containing the whole spectrum of sugar, numbered in order from a molasses-dark brown (no. 1) to bright white (no. 25). It got its name because it had begun as a tool for Dutch merchants to grade sugars from their colonies in Java, and by the 1870s it came certified by the government of the Netherlands. How exactly it came to be a wider standard, in use by America's own traders, was obscure even to them.[15]

Each custom house held a set. Presented with a sample of sugar, an appraiser matched its appearance to one of his jars. It had been introduced to the American tariff system in August 1861, in a bill that raised taxes on imports of sugar, coffee, tea, and other goods, as well as introducing the nation's first income tax, all to pay for the Civil War. That March, Congress had passed a larger tariff bill that simply stated that "raw sugar, commonly called Muscovada or brown sugar, not advanced beyond the raw state by claying or other process," would pay three-quarters of one cent per pound, while "refined sugars, whether loaf, lump, crushed, or pulverized," would pay two cents per pound.[16] The supplemental August bill had doubled the rate on refined sugar and more than doubled it on raw sugar. But it also specified the particular tool as the means by which customs officers were to protect refiners' profits and distinguish between raw sugars and refined sugars suitable for direct consumption by American residents:

> On raw sugars, commonly called muscovado or brown sugar, and on all sugars *not advanced above number twelve Dutch standard* by claying, boiling, clarifying, or other process, and on sirup of sugar or of sugar-cane, and concentrated molasses . . . two cents per pound; and on white and clayed sugar *when advanced beyond the raw state above number twelve Dutch standard by clarifying or other process*, and not yet refined, two and a half cents per pound; on refined sugars whether loaf, lump, crushed, or pulverized, four cents per pound.[17]

Number 12, a medium-light brown, became the boundary between raw and refined. In doing so, Congress accidentally made a radical link between a certain value according to a particular tool of measurement and a specific history of production. To these two, it also joined a less empirical question: What was sugar meant for? There was a great deal of play in these connections of metrology and manufacture.

The addition of the Dutch standard was probably meant as a way of preventing the tariff from subversion by crafty or persuasive importers—of which the risk was higher now that the tariff was also higher. Within a few months, the government had seized a shipment of Cuban sugar that it claimed had falsely been labeled "quebrado," a type of intermediate raw from the middle of the loaf. But witnesses for the accused testified that, in Cuba, the term could mean anything from dark brown to bright white.[18] And by the 1870s, the attempt to circumvent fraud meant that the system had become more complex. To each increment of the Dutch scale, Congress had assigned a level of tariff: one rate below no. 7, a higher rate from 7 to 10, still higher from 10 to 13, and so on.[19] The whiter and more refined the sugar, the higher the cost to bring it into America.

When they received the paperwork, the appraisers at the custom house wrote their orders for the samplers—one sample from every ten packages, more if they suspected something was off about the cargo—and sent the order back to the docks. The sampler manned the front lines of America's war against dark foreign sugar. Each day he confronted containers of all sizes and shapes, none of which stayed the same from place to place or vessel to vessel. Centrifugal sugar tended to travel in sacks, but by saccharine standards this was first class. In steerage, most sugar was cooped up in colossal barrels called hogsheads. Hogsheads from Demerara came lined in blue paper.[20] At loading, stevedores piled hogsheads atop one another until they strained the limits of the ship's hold, and when it was full, they jacked in a few more to wring every last dollar from the voyage. Then they filled the remaining space with more pliable packages: bags, boxes, cases.

The sampler's only weapon was the trier, a hollow metal tube sharpened to a point with a perpendicular handle like an awl. Taking a sample was more than a matter of stabbing the closest bag. Inside the hold of each ship and even inside each package not all sugar was the same. So to establish to his satisfaction how many tries he needed to make, the sampler needed to know where the sugar had come from, how the sugar had been created, and how it had traveled.

The largest question, on whose answer a bag of sugar might radically change, was how the molasses had been drained from the almost finished product, whether by the ancient process of gravity or the new process of centrifugals. The sugar's origin played out differently depending where the sugar had been stored on the ship. As soon as the hogsheads and packages were crammed into the hold, gravity took over again, and the molasses and syrup remaining inside headed straight for the ship's keel. As much as 10 or even 20 percent of raw sugar's weight might seep out as liquid, ruining the cargo (and its buyer) and playing tricks upon the sampler.

Belowdecks on a sugar ship quickly became a filthy place. Temperatures rose as bacteria fermented the sweetened bilge water, melting even more of

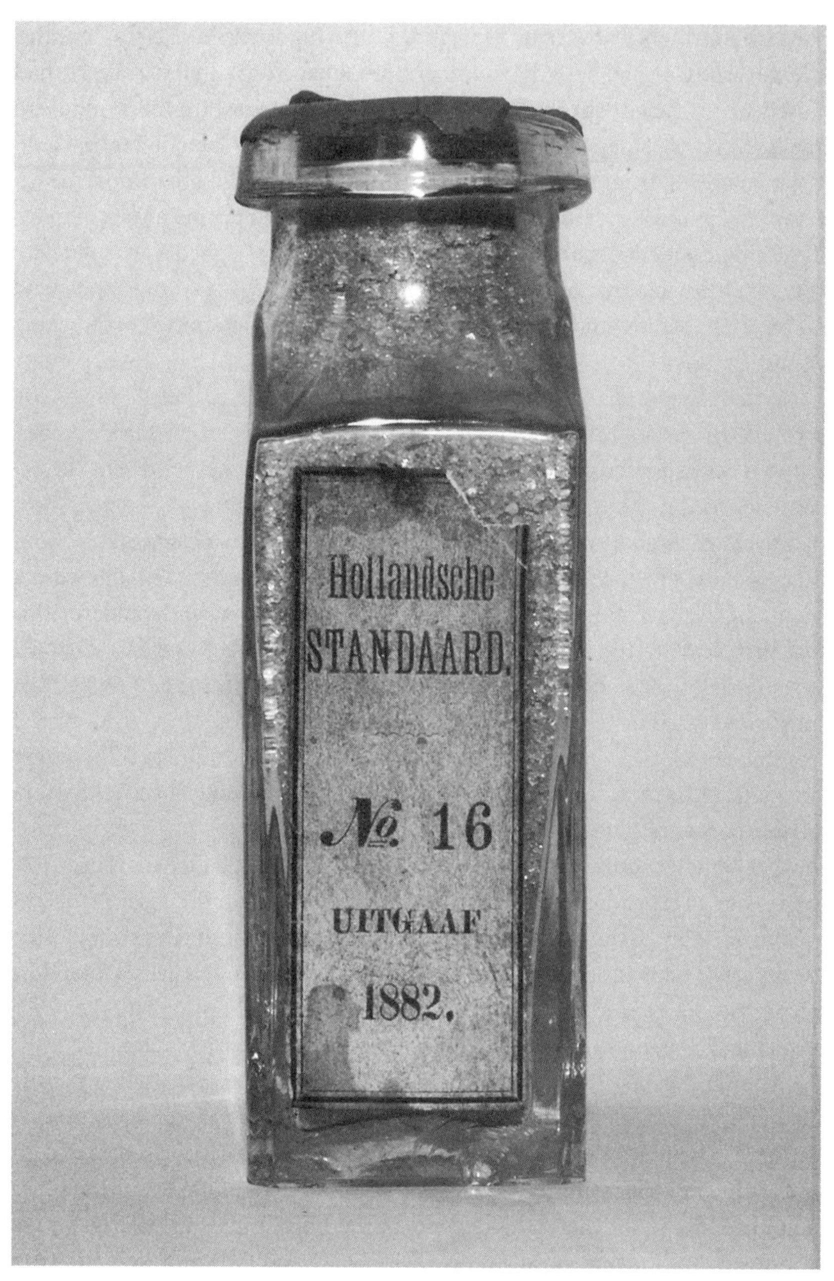

FIGURE 5.3. Jar of no. 16 sugar from the Dutch Standard set of 1882. Division of Medicine and Science, National Museum of American History, Smithsonian Institution, Washington, DC

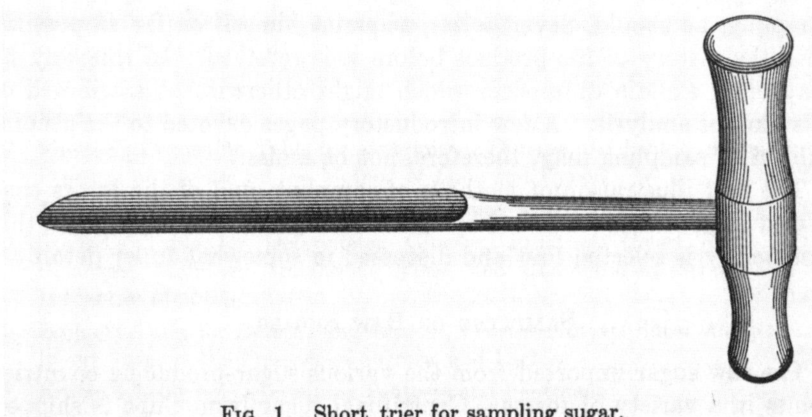

FIGURE 5.4. A testament to the importance of sampling: of the hundreds of illustrations in Charles A. Browne's 1912 *Handbook of Sugar Analysis*, as well as subsequent editions, the very first image was a drawing of the humble trier. C. A Browne and F. W. Zerban, *Physical and Chemical Methods of Sugar Analysis: A Practical and Descriptive Treatise for Use in Research, Technical, and Control Laboratories*, 3rd ed. (John Wiley & Sons, 1941), 4.

FIGURE 5.5. Samplers (lower left) extracting samples from flour barrels in New York, using the same trier as for sugar. *Harper's Weekly*, November 15, 1873, 1012. Courtesy of HathiTrust.

the cargo. Crews sometimes stuck bamboo poles deep into the hold to blow off noxious fumes, and acidic products of the fermenting bilge might even corrode newer iron hulls. A hold that contained molasses barrels was even stickier and grimier, because all the slippery muck sometimes led a barrel to overturn, dumping its whole contents out through the hole that had been used to fill it. Gradually, syrupy sugars high up in the hold could moisten dry sugars underneath them, and dark sugar above could spoil white sugar below. Inside each container the liquid pooled into a "foot" of darker and wetter sugar at the bottom. If a hogshead on the bottom lost too much of its contents, it might even crumple from the weight on its shoulders. Of course, such a sugary swamp was paradise if you happened to be a rat, a complement of which could devastate a sugar cargo.[21] And it hardly stopped enterprising sugar captains from selling their bilge water to a rum distiller.[22] Customs officials were responsible for pulling a coherent story about origins and destinations from these seasick cargoes.

But as one sampler noted, "Any one familiar with these sugars must be aware that samples can be drawn from different sections of the same cask that will vary from four to five [Dutch scale] numbers in color."[23] It took years for a sampler to learn how to read the "character" of a cargo or to judge the depth of the foot based on a few strikes of the trier.[24] Even after decades two samplers could disagree with unsettling frequency.[25] If the sugar fooled the sampler into thinking it was lighter and drier—and worth more—then the buyer would be charged too much in taxes. If the sugar's character fooled the sampler on the dark end, the buyer would owe too little.

Even before anyone sampled his bags, however, William Perot was unlikely to pay too much. If the samplers and appraisers claimed his sugar was more valuable than he thought it was, he would raise a challenge and exercise his right to demand a new sample. And if they claimed it was not as valuable as he believed it was, Perot would just keep his mouth shut. It was possible for a customs officer to request new samples if the duties seemed too low—but that was less common, not least because it meant more work for him and his colleagues.

There was another reason that importers like Perot rarely worried about overpaying. The integrity of the American fisc depended on the sampler's ability to determine the truth out of sugar like that from the *Mississippi*. But the variability of every cargo meant that, along with the drab spectrum of sugar itself, there also existed gradations of plausibility and deniability. Within unnervingly large bounds, a sampler could make a sugar cargo's character into whatever he wanted.

The Treasury was paranoid about the possibility of bribery by merchants, and sometimes regretted that of all the employees in the customs service, it paid samplers the least.[26] But samplers could just as easily strike out on their own. One importer of 350 hogsheads, having paid the government what he expected he owed, was surprised that the final appraiser's bill was much too small. The samplers had reported that the sugar was quite dark. Soon afterward, this importer received an anonymous note, claiming credit for the dark sample, and supplying an address where he might send a sample of his own cash.[27] The customs service tried to station officers to look over the shoulders of samplers, but even this guaranteed nothing. There was no way to tell whether the sampler was doing his job faithfully, from the government's point of view, or whether he was favoring some other party. "I sampled for three years myself," said one small-time refiner, "and if I were in the business now I believe I could cheat the government out of a quarter of a million dollars . . . and I believe anybody else could do it."[28] No matter the sampler's intentions, his behavior always looked the same: He sized up the container and contents, plunged the trier's point into the sugar and, with a twist, deposited the collected tubeful onto a paper. He repeated this act until he had included the foot and the head in whatever proportion seemed to capture the character of the sugar, though he took care to keep them separate on the paper. Then he folded the paper's edges, inscribed its identifying information, labeled it "To the appraiser's office," and sent it on its way.[29]

The Treasury feared a swindle beginning the moment a merchant gave sugar a name. If a Baltimore customs officer in November 1877 had consulted volume 4 of *Johnson's New Universal Cyclopaedia*—published that very year—he would have found the entry on "Sugar" to be a head-spinning centrifuge of terminology.[30] There the country's most famous food chemist, Charles Chandler at Columbia University, explained that there was clayed sugar (drained in molds and whitened by letting water from wet clay drip through it) and potted sugar (drained in molds but not clayed). There was muscovado, made by boiling sugar in open kettles and drying it in troughs. There was also centrifugal muscovado, which had nothing necessarily to do with centrifuges. Sugar boiled from diluted molasses was called molasses sugar, sugar that contained molasses was called melada, and sugar that settled at the bottom of molasses barrels was called "cistern bottoms." Concrete was just like melada, only boiled a little more. Then of course there was a set of different kinds of molasses: muscovado, clayed, and centrifugal.[31] And each loaf of clayed sugar might contain a dozen or more different grades, running from first whites to superior yellows to second browns to "cucurucho." The word signified dark sugar's place at the tip of the cone, but anglophone translators mistook it as a reference to the color of a cockroach.[32]

The Treasury suspected that it was being fooled by refined sugar on the way out, as well. Like other exports, sugar that had paid taxes on entry was entitled to be refunded some of its money on exit. This was known as the "drawback." But just as not all raw sugar was the same, not all products from the refinery were the same either. The customs officer at his *Cyclopaedia* would have learned from Chandler that, in the modern refinery, the dark liquor that ran out the bottom of molds was called "first greens," which were not green, after which "white liquor," which was not white, was poured through the molds to make "drips." That process was called claying even though it involved no clay. Then the first greens were diluted, boiled again, and drained again, producing a light-yellow sugar called "C" or "coffee sugar" plus second greens. These greens were boiled again into "X" or "yellow sugar" plus green syrup, which got diluted, filtered, and reconcentrated again as golden syrup. The sugar that had had the first greens drained out of it was the purest and driest and was called "tittlers." Refiners who processed sugar from beets, rather than cane, used their own terminology, which was similar but involved more French. Older refineries, of which a few were still around, made a moist and crumbly sugar called "bastard." Beet refiners called theirs "bastarde," and considered it "too offensive to be used" except by chocolatiers.[33] Perhaps there was something to those early nineteenth-century slanders about beet sugar's taste after all.

To calculate the drawback rates on refined sugars, the government needed to know what raw materials went into them. That equation had many variables, and each refinery solved it differently. Some were more efficient at avoiding losses, others had developed idiosyncratic ways of dealing with raw sugar's various constituents—salts, clay, plant matter, plus other sugars in various commercially useless forms—all of which could interfere with crystallization.

Refiners made their money by manipulating sugar in secret, not by revealing their magic to agents of the Treasury. So American consumers and officials could be forgiven for thinking that sugar refiners had something to hide. What was public about refining was already ghoulish. At various points in the process, the sugar was mixed with corrosive lime and buckets of blood, and it was passed through towers of bone-black—charcoal made from the skeletons of animals. It did refiners little good to insist that these were merely intermediate steps and not actually ingredients in the final product.

Chandler, the most prominent public health chemist in New York, assured citizens that their worries were groundless. In his years of experience, he had never seen refiners using toxic metals or adding marble dust to powdered sugar. Sometimes refineries used a little glucose syrup—though not the "car load after car load" that whistleblowers claimed—and sometimes they used a tin salt to improve the color, but always did so in small enough quantities that consumers had no "cause for alarm," which was not quite the same thing as saying that these additives were harmless. If pale sugars and syrup were strangely "inky" when added to tea, it was not because they were secretly adulterated but because they had carried dissolved iron from the tanks and tubes of the refinery, which was "entirely unobjectionable," he said, "perhaps useful."[34]

To set the drawback rate, the US government was forced to lean on hearsay, speculation, old foreign books, and refiners long out of the trade. Refiners still in the business either refused to speak or, when they did, were understandably suspected of being financially interested in the outcome.[35] The goal of the drawback was to give proper refunds, not to award subsidies to the makers of refined sugar. Refiners' knowledge of their process became one more invisible source of their power. It matched their apparently sorcerous ability, amply evidenced at every election, to convince unfriendly politicians to suddenly switch sides and support policies friendly to the refiners' interests. Commission after commission spent most of the nineteenth century wrestling with the problem of setting the right drawback without ever feeling they had come close to solving it.[36] Each time the Treasury set drawback rates at a level to reflect its latest understanding, the composition of refined sugar exports suddenly and suspiciously tilted in favor of those products with the highest drawback. And, of course, if the darkened sugar were also paying too little duty on import, then the sugar business as a whole was profiting off the people at both ends. The government's coffers slowly drained as more golden syrup for refiners.

Sugar, Embarrassments Attending the Collection of Duties On

The Demerara bags ought to have been easy to appraise. The name commanded the attention of those who knew their sugar. Among all the sugar regions of the Americas, this stretch of the British Guiana coast possessed an influence that belied its compactness. Lighter-colored sugar sold for more on the market, and bright yellows were Demerara's specialty. The colony's legendary sugar boilers found work across the Caribbean, their passage smoothed when employers saw their official certificates of achievement. From Scotland, Demerara imported some of the best sugar-making equipment in the world.[37] Earlier in the century, its planters had so eagerly adopted new technologies for purifying and drying that the colony had given its name to an attractive pale yellow crystal of remarkable purity, one of the most desirable sugars on the market. So when officials at the customhouse unfolded the samples from the *Mississippi*, they might have been surprised at the darkness staring back at them. Some of Perot's cargo rated between Nos. 7 and 10 on the Dutch scale, but 712 bags were declared below No. 7, making them among the darkest sugars entering America and paying the least in taxes.

Then again, sugar from Demerara had been getting darker for the past two years. In fact, sugar from everywhere had been darkening, and darkening so quickly that the appraisers could have been forgiven for checking their eyesight. Treasury officials from the docks all the way up to Washington understood why, even if they could not often prove their belief to legal satisfaction. The law by this point specified five brackets of the Dutch standard—below 7, 7–10, 10–13, 13–20, and above—each taxed at a higher rate than the last to roughly match sugar's value. Two years earlier, in 1875, Congress had raised rates on all five brackets. And ever since, the sugar arriving on America's shore had grown darker and darker.[38]

In itself, such a change wasn't troubling. Higher tariffs might eat into the profit margin of light sugar more than that of dark sugar. So though the price of dark sugar was lower, under the new tariff regime it might be relatively more profitable and thus more attractive to merchants. What was perplexing was that, despite imports of supposedly lesser quality over those couple of years, the products of American refineries stayed the same or even improved.

Importers had an explanation ready: A sugar's color no longer bore a direct relation to its quality. The manufacture of sugar had changed, not just the tariff, and the new sugars were not like the old. Ideas about color, purity, and price that animated an antique tariff no longer applied to centrifugal sugars or others made with new machinery. Far more sweetening power might hide under a darker tint. Yet even if the government's "eye test" could no longer distinguish between higher and lower sugars, buyers and sellers always seemed able to agree on how much their sugar was truly worth. Bureaucracies rarely admit their fallibility, so it is remarkable that just a week after the *Mississippi* arrived, Secretary of the Treasury John Sherman called such transparent subversion of the tariff an "embarrassment" to his customs service.[39] "Sugar, embarrassments attending the collection of duties on" even got its own entry in the report's index.

By the summer of 1877, Sherman had tired of being embarrassed. He told customs officers to seize sugars that they decided were artificially colored; these might pay even higher rates than simple refined sugars. Import firms loudly assured the government that they had specifically told their Demerara counterparties not to artificially color their sugar. Merchants wrote Washington in anguish, and major cities sent delegations to plead their case. Baltimore's emissaries—including the president of one of its refineries, the head of its custom house, and William Perot himself—swore again that there was nothing artificial about dark Demerara sugar. But the city might have already protested too much. In the spring, Treasury commissioners in New York and Boston had warned that the department was bleeding money from artificial coloration of sugars and urged punitive tariffs whenever such sugar was discovered. Not in Baltimore, whose commission assured the rest of the government that the current system was working just fine.[40] In September, the Treasury cabled the consul in Georgetown to investigate artificial coloration. Mysteriously, the very next day William Perot and his fellow Baltimore importers had not only heard the consul's instructions but also received his response as quickly as the Treasury. The consul insisted that the Demerara sugars were innocent, and then he boarded one of Perot's own ships and came back to the United States.[41]

All of which is to say that when Sherman's special agent in Baltimore, Simon Chamberlin, received a tip about 712 of Perot's bags, it was entirely in keeping with the government's suspicions about the sugar industry in the city. About two dozen agents like Chamberlin roamed the country keeping tabs on the customs service. Until a few years earlier, this force had been known for a combination of shaky legality and extraordinary powers, not least because they, like all customs officials, got a juicy cut of goods forfeited by merchants. One particularly zealous crusader against smuggling had recently extracted $66,000 from a merchant who, possibly by accident, had underpaid less than $2,000 over five years.[42]

Of that infamous $66,000 bounty, the agent had given two-thirds to his tipster. Neither role was nearly so lucrative after this system of spoils had been abolished in 1874, but whoever raised the alarm for customs fraud might still earn several thousand dollars. Plus, like almost everyone else who dealt with the Treasury in the 1870s, if things went well he might be in line for a job in the custom house—especially given the publicity of Sherman's fight with the sugar importers. So, in the days after the *Mississippi* arrived, a former sugar manufacturer named Henry Abbott snitched.[43]

Chamberlin brought the statements and evidence to the district attorney and to Sherman. By now, the bags had been hauled to a warehouse, and Perot was busy negotiating with the city's refineries for a sale. As soon as a deal was struck, the evidence would be melted. The order came from Washington: Seize the sugar. Perot got no warning. Late on a Friday afternoon, just four days after it had landed in Baltimore, Chamberlin seized a hundred thousand pounds of dark Demerara sugar, all of which Perot claimed was below No. 7, Dutch standard.

It had been years since the sugar market bubbled in Baltimore, but Chamberlin's seizure congealed it. For one thing, those 712 bags represented the majority of the stock in port at the moment. For another, Guiana was a major export market for the city, and captains would not sail there if they had no hope of coming back loaded with sugar. The planters in Demerara also had no interest in shipping sugar they could not sell. When Sherman's men seized another Demerara cargo a month later, this time in New York, the colony more or less froze its shipments to America.[44] British consumers found they could get Demerara sugar unusually cheap, since it had nowhere else to go.[45] No one intended the case to linger for a year. But the arbitrators appointed by both parties failed to agree on the value of the cargo, and so a commission traveled to Demerara to collect testimony from those involved in making it, while another went to Louisiana to interview those who made sugar from cane in the United States.

Meanwhile, Perot reached out to the country's most esteemed chemists for their analyses. He wrote to Henry Morton, a professor at the Stevens Institute in Hoboken, and to Charles Chandler, the cyclopaedist, at Columbia. Each agreed to produce a report on the chemical contents of the cargo. By the time Perot sent them samples, in early 1878, he was already paranoid that someone might try to swap out their contents. Chandler commiserated with Morton about the stream of anxious letters from their new client, who wanted both of them to appear in court as experts, and eventually Chandler just ignored him. It took until April for the Demerara commission to report, and Louisiana did not return until August. In October the trial got underway at last. Perot maintained his innocence in private as in public. "My cause is just," he promised Chandler, "and I have no fear of the result if nothing but the truth is elicited."[46]

But the court, like the samplers, found that eliciting the truth about sugar was impossible. The jury had to affirm two statements to find the sugar and its importer guilty: that the sugars were colored by artificial means after they were crystallized, and that Perot had intended to defraud the government by importing them. Perot's attorneys and witnesses didn't deny that the sugar was meant to be dark. The manufacturers of this very sugar swore that they had only run the juice through coarse filters, hadn't skimmed impurities, and hadn't even followed the common practice of washing the sugar while in the centrifuges. Nothing to affect the color had been added. Nothing had been added at all except lime to temper the solution in the pans—but then, all sugar was made with lime.

These planters recognized that their position was a little odd. For all the history of the many values and kinds of sugar, the most money had usually gone to those who took the time and effort to make sugar of the finest grain and lightest color. The profit had always in the high end of the market. But that calculus had changed when Congress had meddled with the natural order of things. The higher tariff rates meant that lighter sugar was now so costly to import as to be worthless, and the real profit awaited whoever could produce sugar that was potent for refiners, but still dark enough to slide under the Dutch standard. They were honest manufacturers who made innocent sugars, but they "did not affect to deny (what the Government seemed to regard as very wicked) that they manufactured sugar to sell."[47] In short, the dark-sugar process was so much simpler that it was actually more profitable to produce the crude dark sugar than the lighter ones for which Demerara was famous.

Witnesses for the defense accused the government of sending spies to Demerara under false pretenses. The court-appointed expedition had brought back samples of sugar, sealed in jars and kept in the customhouse. One of them, the defense insinuated, was now so much darker and "muscovadoish" than before that it must have been secretly replaced. Remarkably, these conspirators had taken care to reproduce the impression that authenticated the wax seal.[48] Or the sugar might just have been, as one analyst claimed, "changed by keeping."[49]

But the planters refused to contemplate that the sugar had been colored artificially. What was the artifice? Nothing outside the ordinary had been added to the process. Indeed, other witnesses testified that the process was normal for Demerara, and had been ever since a New York importer traveled there in early 1875, carrying dark centrifugal sugars from Cuba and seeking boilers in Demerara who would make them.[50] The two sides sparred over whether a pan-boiler could grow big beautiful crystals from even impure juices, whether the coloring matter on the surface of these sugars was clingy, whether all raw sugars or just these lightened when washed, and whether the dark color was therefore intrinsic to the crystals or an external addition to its otherwise natural surface. If sugar was taxed on entering America, then the true border of the crystal needed to be found. Where did the sugar end and the world begin?

In *United States vs. 712 Bags of Dark Demerara Centrifugal Sugars*, the jury split. They found the sugar guilty of being artificially colored, but they acquitted the importer of knowing about it. Legally, Perot was clean. The same reforms that had ended agents' ability to profit from seizures also required the government to prove intent before it could force anyone to forfeit their goods. So the merchant walked out of court as the owner of 712 bags of sugar that, being legally no lighter than No. 7 on the Dutch scale, owed not a cent more to the Treasury. In short, the jury decided that these sugars had fooled even their owner.

The Olive Man

Vindicated yet thwarted in his course, Sherman tacked, and he spent the next few years trying to persuade Congress to change the tariff law.[51] (Another entry in the index to the Treasury's 1877 report was "Sugar, the Dutch standard as a basis for the assessment of duty on, unsatisfactory.") He called on legislators to empower customs agents to use the polariscope, which he characterized as a sophisticated chemical instrument that could precisely and accurately quantify the purity and value of a sample of sugar. But in the meantime, since Caribbean sugar was still being manufactured to subvert the tariff, he wanted to know how that subversion worked.

In early 1880, he dispatched three Treasury agents back to the Caribbean. One was a chemist deputized from the Smithsonian, one was Chamberlin, and the third—a jab at his mercantile opposition—was Henry Abbott, the man who had ratted out Perot's sugar in the first place. Despite the ambivalent verdict, Abbott had gotten his reward: He was now an inspector of customs. The three set sail in February, carrying samples of the offending sugars back to the Caribbean in their baggage. They attempted to recapitulate the historical trajectory of the sugar revolution in their expedition's own path. They chose to visit Dominica first because its production seemed the least sophisticated—from the old-fashioned mills to the "natives" whom they said walked into town carrying the finished sugar on their heads. The agents witnessed the making of Dominican sugar firsthand, experimenting to see what would happen if they didn't remove impurities until the end or if they could wash it after it was finished to change its color.[52]

Next was Barbados, whose technology and methods were more advanced. The party spoke to the owner of a Barbadian plantation, who said that he boiled his sugar as high as he could without spoiling it and admitted that his American customers had complained it was too lightly colored. English customers, they learned, wanted high purity but preferred a light-yellow color, so planters added a touch of sulfuric acid to tinge their white crystals.[53] On Barbados, the agents gave a sample of sugar seized in New York to one veteran boiler, who, the agents claimed, revealed to them "the secret of making sugars of this kind so dark." Boil it for three days to make perfect crystals and only then add the molasses to darken it. Any earlier and the color could never be washed out of the sugar to whiten it for sale.[54]

At last the agents were ready for Demerara, and they soon learned why Demeraran boilers were so coveted. Dark sugars made there, dark enough to register among the muscovados below No. 7, turned out to measure 96 percent purity or more under the polariscope—as much as the sweetest and brightest centrifugals from Cuba. The boilers achieved such feats by adding far more lime during the process of clarifying the juice than what they needed to neutralize the juice's natural acids. The Americans considered this "superfluous" lime that had no business in sugar made by a "legitimate" natural process.[55]

But the true wizardry, they learned, happened inside the vacuum pan, where the boilers earned their pay by meticulously cultivating the crystals. The boilers on the La Jalousie plantation had perfected the art of breaking the vacuum at just the right moment to make molasses "adhere tenaciously" to the surfaces of the crystals, leaving their sugars far darker than any competitor for the same polariscopic measure. Perot had claimed his sugars were dark because of their impurities and inherent color. But these Demerara sugars were carefully produced to take the impurities out and guarantee their otherwise "superior" character.[56]

All along the way, Chamberlin and Abbott showed Perot's sugars to boilers and planters on the islands. These men could read a sugar sample for its manufacture the way a detective infers a supplicant's history from frayed shoelaces. Fingering a tiny amount, an experienced sugar maker could not only tell that a sugar had been made in a vacuum pan but could also elicit details about his fellow sugar boiler's technique with the pan, his method of clarification, how much sulfuric acid he added, and how he liked his molasses. After examining the samples by the eye and under the microscope, these craftsmen reported there was no chance the sugars were "natural."

By protecting American refiners from light sugars disguised as dark, the tariff also protected white American sugar workers from deceitful foreigners who would steal the rewards of their honest labor. Nearly six thousand men worked in sugar refineries in the United States, and four times as many in dependent industries: making barrels, working iron, fixing machinery.[57]

Some peoples represented a threat through the baseness of their products. Take the Philippines, the source, at the time, of some of the darkest sugar shipped to America. Filipinos, said Congressman Garfield in the same speech where he brought up *Acarus sacchari*, lacked the intelligence to make sugar that might compete with American refined products. They boiled sugar in the "crudest, rudest, simplest way, by labor the cheapest and least skillful," and the sugar they made was a "black cheap form," "the dirtiest yet known." Only in America, by American craft, could guileless sugar from simple Filipinos become free from invisible bugs and fit for the civilized table.[58] (In the early twentieth century, when the Philippines were a conquered American possession, this hierarchy would invert, and the islands became famous for factories that could produce "plantation whites," nearly chemically pure sugar without any need for mainland refineries.)

This racialized vision was the context for "Labor Side of the Great Sugar Question," the title of a pamphlet, published in 1878, by "A Workingman," under the nom de plume "Robert Howe." The entire piece was a barely pseudonymous defense of the Theodore and Henry Havemeyer, the owners of the largest and most modern refinery in Brooklyn. It described an imagined dialogue between a Cuban visitor to New York ("a sallow colored gentleman") and a longtime sugar-refinery employee named "Ludwig Kraft."[59] Refineries had long been a German-dominated industry in greater New York, so much so that the companies printed management notices in two languages, and in case the dialogue itself failed to deliver the message, the final pages of the book contained a reprinted editorial from the *New Yorker Staats-Zeitung*.[60]

Some of the Havemeyers' most vocal antagonists were owners of a competing refining firm, Booth & Edgar. All but naming them directly, Kraft dramatically reveals the existence of the conspiracy of Cuban planters to destroy the refining industry of the United States by constructing one on the island. Their scheme is to claim massive sugar frauds and adulterations by loyal (German-immigrant) refiners in order to "create a public opinion hostile to them, and by the pressure of that opinion force the hand of Congress and the Senate to frame the tariff for which they have all along been plotting."[61] The American importers who are their allies in the scheme were annoyed by the private dock system the Havemeyers had constructed alongside their waterfront property, which excluded middlemen and allowed ships to pull up directly to the refinery itself. "If the refiners had never imported a cargo of sugar direct themselves, but had taken it from the importers, there would never have been a single word about frauds."[62] At the book's climax, Kraft stares "steadily into the eyes of the olive man"—color again—and warns his visitor not to antagonize a nation that had just fought a war for the privilege of freely engaging in wage work. "You Cubans and your New York agents think Congress is going to pass a law to take the bread out of the mouths of American citizens to enrich your Cuban slaveholders, without even doing any good to the poor wretches you drive on your plantations." Cuba had just emerged from ten years of war, in part over the future of slavery, which had left its sugar industry in ruins.[63] "Your slave labor is not going to destroy our free labor."[64]

On the one hand, foreigners were mocked for their supposed stupidity: It took no ingenuity to make raw sugar the old-fashioned way. "The foreign labor of gravitation on the soil of Cuba" drained the molasses out of the sugar, said an Illinois congressman. "That is all the labor there is in it."[65] The parallel between how the Dutch organized their sugar scale—dark to light, lowest to highest—and how the West constructed its worldwide racial hierarchy was obvious enough that it became an official joke. Congressman Garfield asked his honorable friends to imagine a hogshead sitting still for a month before a sampler bored into the top and the bottom. The samples would be lighter and drier near the top, wetter and darker below. "How much foreign labor," the antislavery Benjamin Butler inquired, "is there in the settlings of that sugar?" "There certainly is a good deal of dirt in it," Garfield replied, and he and Butler shared a laugh.[66]

On the other hand, foreigners who were overly clever served as a justification for ditching the Dutch scale in favor of chemical tests. In a congressional hearing room, a former Treasury agent unwrapped two samples of imported sugar, one dark and one white—no less than 20 on the Dutch scale, he said. He then pulverized the former into the latter, right in front of the committee, to show how clever makers could disguise light sugar's true nature.[67] Without new means of detecting the true sweetness of imported sugar, America would only bring in "the sugar that has been advanced by the higher and more intelligent processes of our nearer neighbors."[68]

Imaginations raced about what else foreign sugar might contain. Conditions in Cuba, Americans read in the papers, were so miserable that enslaved people chose suicide by sugar cauldron. And Chinese workers who died in other ways might find their skeletons exhumed and burned, their char mixed with that of livestock to better filter the juice. The sugar mite and the grocer's itch were minor irritants compared to the horror of sugar adulterated with the corpses of workers who, if not slaves, were still not free.[69] It was not long before refiners in Canada learned to weaponize the sugar mite as part of even more overtly racist campaigns against Chinese sugar.[70]

Sympathetic or not, foreign workers were still dangerous. Pamphleteers for US refiners prophesied huge new refineries constructed in Havana if the tariff laws were not reformed, which would put thousands of Americans out of work.[71] Many of America's sugar workers had, not long ago, fought and bled to defeat the slave power. They were certainly not about to give up their jobs to enrich Cuban slaveholders now. There were quicker and cheaper ways to liberate Cuban slaves, as one critic put it, than to ship the whole island to America in hogsheads, its "dirt" intermixed with raw sugar at six cents per pound.[72]

Charley Ross

While they waited for their 712 bags to go on trial, the Perots played roles in a curiously parallel public drama about darkness and foreign goods.

In January 1878 America was still looking for Charley Ross. It had been three and a half years since the toddler was snatched from the yard of his affluent Philadelphia home, tricked into boarding a carriage by the original

FIGURE 5.6. Charley Ross, maybe. "Bring Back Our Darling, or, The Stolen Child" (1875). Courtesy the Lester S. Levy Collection of Sheet Music, Sheridan Libraries, Johns Hopkins University.

strangers with candy. Five months later, the suspected kidnappers died in a shootout, but still there was no child. Would-be Charleys materialized around the country, each of whom the family felt obliged to investigate. By the time a pseudo-Charley arrived in Baltimore aboard a ship from Demerara, Charley's father had noted 573 alleged sightings of his son. As one headline put it, "Charley Ross—Another of Him Found."[73]

The mystery was how Charley #574 had wound up in South America at all, let alone living, as newspapers gossiped, with a disreputable "mulatto" woman. She claimed that she had been born in Demerara, moved to the United States and married a white man who promptly disappeared. She had sought work in Boston, where rumors placed her with a known accomplice of the kidnappers, before returning to South America with the child. Some reports suggested she had abused him, and that from the moment they arrived in Demerara, the boy had insisted he'd been stolen. And then a merchant named Adolphus Perot recognized the child on the street, fed and clothed him, and "consigned" him to his brother William, just like cargo.[74]

This boy was "the most probable Charley Ross yet," said one paper. A crowd gathered at the docks just to watch him alight to buy a hat. But this cargo's papers soon proved suspect. The boy spoke crude "foreign" English and inconveniently called himself George Sylvano Signio Rio, although he supposed he might once have been called Charley Ross. His skin and hair were also darker than Charley's. After reading that the boy had "an olive complexion peculiar to those who live in a warm climate," the exhausted Ross family concluded that they had not found their son.[75]

The boy was cut adrift. Wisely the merchant Adolphus Perot had forwarded money for his care, so the upbringing and education of this false Charley were entrusted to the Home of the Friendless. Under American supervision—one might say refinement—at this wholesome institution, the dark foreign boy underwent a miraculous change to his presentation. His accent evaporated as he began to recall details of early life in Philadelphia. He had been artificially colored in Guiana, he said, and been forced to bathe apart from other children "in a peculiar kind of dirty water." Then identifying marks—from vaccinations—became visible. "The swarthiness of his complexion was wearing off," a correspondent exulted, "and he was becoming whiter."

Now the Ross parents liked what they saw in a photograph. Their doctor visited Baltimore and summoned the family. The child, he said, was looking like Charley more every day. The father arrived, but the Home for the Friendless still kept him away until Charley's appearance "improved." They wanted to wait for his "natural color" to return, and as the days passed "his skin was assuming a fairer color, the dark hue gradually disappearing." On the docks, time lightened a sample of sugar as water evaporated. The child at the home was held back so that he might make the sweetest possible impression on his appraiser.[76] Finally the home allowed Mr. Ross to interrogate the boy. After a few hours, he left without a word to the press.[77] Under all the coating, the boy was yet more contraband from Demerara. He wasn't to be trusted, wasn't fit for a white table. Perot's young charge, like his bags of sugar, had allegedly been disguised in the tropics to slip into America.

Some contaminants were more real than others. There were maggots, flies, and worms that could devour sugar in a ship's hold. The sugar mite, however, "this formerly much advertised insect," with its nasty proboscis and hooked appendages, turned out not even to exist. Sugar chemists were laughing about *Acarus sacchari* for decades, but even fictitious impurities could affect the real world.[78]

6

The Unpracticed Eye

A Little Case Which Only Begins a Big One

Especially by the standards of 1877, William H. Grace was a powerful man. He looked down on you from more than six feet in the air, five inches taller than the average Union soldier. One newspaper, which was not a fan of Grace's, said he was "as burly as an ox," while another, which was more sympathetic, said that he was "broad shouldered and muscular looking." When he died, years later, at the age of fifty-seven, his friends were surprised that he had not lived longer, because "mental and physical characteristics were so much in his favor."[1] The point is that William Grace was large enough that his size and strength were a big deal in mid-nineteenth-century America, and he was almost certainly bigger than George Sharpe, the man whom he punched so hard, and caught so unawares, that he almost knocked him into the gutter.

Sharpe, no small man himself, was a veteran of tougher fights than this one.[2] An attorney in peacetime, he had organized wartime intelligence for the Army of the Potomac until General Ulysses Grant had personally called him up to join his advisory council. At Appomattox it had been Sharpe's responsibility to parole surrendered Confederate soldiers so they could go home, and he ended the war a major general. Five years later, Sharpe's old commander, now the president, sent him to New York City to oversee the census and undermine the Tammany Hall machine of the Democratic Party. At the beginning of his second term in office, Grant rewarded Sharpe for his services by making him Surveyor of the Port of New York, the second-most powerful post in the nation's most powerful federal custom house.[3]

That was where Sharpe crossed paths with Grace, a boy from Kilkenny grown up to become an investor in Brooklyn real estate. Grace had used his knowledge of the city and his skills at appraising its properties to become a decent-sized fish in its Republican politics. In the early 1870s he, too, had

FIGURE 6.1. Barrels on the wharf of the Pennsylvania Sugar Refining Company, late 1880s. Historical Society of Pennsylvania Photograph Collection (#V59), box 115, folder 5.

been rewarded with an appointment to the custom house, though a much lowlier one. He worked as a customs inspector, and that meant he reported to the Surveyor.

Whichever political party occupied the presidency controlled the New York custom house and a thousand jobs to disburse to friends and followers. The New York custom house was the biggest plum on the tree. Being a zealous and evangelical believer in the wisdom of protection was not by any means a requirement for an appointment in the customs service. But for Grace, protecting the revenue was more than just an income: It was a calling. In the late nineteenth century it was possible to be genuinely passionate about the importance of high tariffs, and it seems that all his adult life one of Grace's animating principles was that the industries and workers of his adopted nation should be insulated from foreign competition.[4] And, Grace would later argue, it was his energetic commitment to the cause that brought him into conflict with his boss.

After years as an inspector he had been reassigned, on the first day of February 1877, to take charge of Sugar District 13, a section of Brooklyn waterfront that included the world's largest refineries. His was an independent and authoritative position. The sugar companies in District 13 had private docks. Everyone unloading the ships worked for the refiner or the ship itself. Rarely would Grace have had a fellow agent to watch his back, peer over his shoulder, or go over his head.

Not long after he took up his post, however, the companies of District 13 began complaining to Surveyor Sharpe that the new man was literally drunk with power. He was inebriated on the job, said the firm of Havemeyers & Elder, and he was a "nuisance and a loafer," but that was not the worst of it. Another company, De Castro & Donner, said he had threatened them with extortion. If the companies didn't pay him in cash—or in sugar—he swore to slander them with allegations of fraud, corruption, adulteration, and smuggling.[5] Sharpe suspended him in March, pending an investigation, and in late April he fired him. Three months after that, Grace and Sharpe found themselves on the same street corner in Lower Manhattan.

Sharpe had just come from the offices of *The New York Times*, while Grace was on his way to plead his case before an investigator from the Treasury Department. Grace demanded that Sharpe apologize. The Surveyor did not give satisfaction, so Grace took it, and clocked Sharpe under his right eye. The old spy staggered and fell, and he took a beating from a "fist like a sledge hammer" before Grace backed off.[6] Then the police dragged them both to jail.

On the basic outlines of the encounter, Grace's defenders and prosecutors did not disagree. Of course they contested small details. Did Sharpe offer a handshake when the two met on the street? Did he spit at his former subordinate? Witnesses to the fisticuffs could not settle these questions. There were more substantial disagreements too, though they also did not really matter to the issue of whether Grace was guilty of assault and battery. Sharpe was the war hero, while Grace's character had more color to it than Union blue. Was Grace the erratic drunk the Manhattan papers made him out to be, an irritant to his superiors long before he split open one's cheek? Or was he a faithful public servant whose record, as his champions at the Brooklyn *Daily Eagle* insisted from across the East River, was equally unimpeachable to Sharpe's, though it might be lowlier? The real brawl was not the fistfight itself but what it signified.

For while Sharpe had fired Grace because the sugar refiners said that he was drunk on the job, that was not the whole story as Grace told it. In his version, as soon as he assumed his post in February, he became a witness to rampant fraud and corruption by the firms he was responsible for policing. The Havemeyer and De Castro management routinely robbed the revenue through false weighings, artificial coloration, overvalued exports, and the admixture of harmful adulterants. Refinery employees smuggled ashore the sweepings of sugar that had come loose aboard ship, and cargoes were weighed on company scales rather than government-authorized ones. Grace reported all this at once to Sharpe, only to discover to his horror that the Surveyor was part of the scheme. And, according to Grace, Sharpe himself had solicited the charges from the sugar firms as a pretext for dismissing him.[7]

Grace claimed that the Havemeyer brothers, Henry and Theodore, operated a "Sugar Ring" of officials and merchants that ruled customs in New York. His allies did their best to spin Grace's trial away from assault and battery, of which he was guilty, and toward his revelation of conspiracy along the waterfront, in which he might be a hero. It seemed fanciful, and it certainly didn't keep Grace out of prison. But within a few months of the assault, De Castro & Donner's treasurer fled to Cuba, or fled as fast as he could while carrying all the company's books. The Havemeyers took the firm over, Sharpe was sacked, and Grace accepted a pardon from the governor, a Democrat all too happy to needle the Republicans in Washington.[8] He also obtained an official order from the Treasury entitling him to his back pay from the date of his dismissal through the July day he socked Sharpe. "Mr. Grace's official vindication is about as complete as it could be made," exulted the *Eagle*.[9]

In pursuit of further government employment, Grace tried to take credit for revealing the existence of sugar frauds, and, with his claims of a "Sugar Ring" masterminded by Theodore Havemeyer, invoked the ghost of the recent "Whiskey Ring" scandal that had tarnished Grant.[10] Grace insisted that there were false weighings, sugars artificially colored, sugars wrongly classified, overvalued drawbacks, and the "most criminal" use of harmful substances as adulterants in the Havemeyers' refineries. He demanded that Congress rewrite the tariff to beat back their emerging monopoly. But, attributing the demise of the Whiskey Ring to the appointment of incorruptible officials, Grace insisted that the new tariff had to be enforced by "competent" appraisers, chemists, and samplers, and "a thoroughly honest collector" at the head of the city's custom house.[11]

FIGURE 6.2. The Havemeyer & Elder sugar refinery looming over the waterfront in 1886. R. R. Bowker, "A Lump of Sugar," *Harper's New Monthly Magazine*, June 1886, 83.

FIGURE 6.3. Weighing a hogshead of sugar on the wharf. Note the inspector on the left and the man collecting the "sweepings" of sugar. R. R. Bowker, "A Lump of Sugar," *Harper's New Monthly Magazine*, June 1886, 79.

Polariscope-Struck

Grace's fight with Sharpe was, said the Brooklyn *Daily Eagle*, "a little case which only begins a big one."[12] In the decades following the Civil War, the sugar trade in the United States was plagued by charges that the largest and most powerful sugar importers and refiners exercised undue influence on how sugar was sampled, weighed, measured, and valued. The gravest charges centered on the custom house in New York, already the focus of American politics as the juiciest plum in the spoils system of political appointments, and thus the target of civil-service reformers. It was also where half the tariff revenue was collected and where the fruits of empire arrived.[13] During a time of falling refining profits, these charges were leveled primarily by smaller firms outmuscled by newer, larger, more efficient competitors, and their claims have been dismissed as resentment at being outcompeted in the marketplace.[14] The fact that sugar refining was a high-volume, low-margin business created particularly strong incentives, in the eyes of contemporaries, to manipulate the collection of tariffs: Even a small saving in the duty paid could amount to a significant proportion of the profit on a pound of sugar, which never rose above a few fractions of a cent.

In chapter 5, we saw how new sugars from the Caribbean could be a source of anxiety, envy, curiosity, and opportunity all at once. It had become clear, at least to the agents and officials of the Treasury, that the Dutch scale was no longer a tool that could measure the value of sugar with any precision, and it did not seem to be what the sugar traders themselves were using. By the late 1870s, it was an artifact in use only by Customs; no importer sold or refiner purchased sugar on the basis of what number of jar it happened to look like. In 1878, therefore, Treasury Secretary Sherman called on Congress to do away with the Dutch standard and adopt the polariscope. Color, he wrote, "bears no definite relation to the value of the sugar" and could be subverted by "foreign substances."[15] By 1879, however, with no legislation in sight, Sherman simply announced that he was reinterpreting existing tariff statutes. From that point forward, the official epistemology stated that the Dutch scale was "essentially a measure of saccharine strength and not of color as an abstract physical attribute of matter." By doing so, Sherman justified an accompanying order: that customs houses levy sugar duties using the polariscope.

The polariscope was a marvel of nineteenth-century optics and physics. Here, the polariscope's advocates promised that it would measure accurately and reliably the sucrose content of sugar—which they equated with its purity and, from there, its value. At a stroke, it would solve the problems of fraud, corruption, and mischief that were bleeding the Treasury and rotting trust among the mercantile class. The polariscope cared not at all whether sugar was muscovado or centrifugal, nor did it allow for the subjectivity that plagued the collection of duties from start to finish.

It was still a newfangled gizmo in the late 1870s. The public health chemist Harvey Wiley, who later crusaded for pure food and drug legislation, had heard of it but never used one before he visited a hospital in Vienna in 1878.[16] In principle, the way the instrument worked was simple: It allowed an observer to measure the degree to which substances rotated the direction of the polarization of light. Sucrose, the molecule at the heart of cane sugar, rotated that polarization in linear proportion to its concentration in a solution. So if you shone a beam of polarized light through a sugar solution and measured its angle of polarization when it entered and when it left, you could calculate how much sucrose the light had passed through.

When this principle took physical form it looked like a tube of brass, with lenses at both ends and levers and knobs sticking out, all mounted on a base. Set on a table, with an observer in a chair, the tube was at eye level. It was an impressive piece. When the House of Representatives Committee on Ways and Means invited two instrument dealers to show their wares, they were "polariscope-struck," one member recalled. "It seemed to us that the polariscope was a beautiful and splendid thing."[17] What government officials and energetic private citizens discovered was that the polariscope in fact made it exceedingly difficult for nonexperts to point to exactly how and where the corruption of the "proper" exercise of judgment in the assessment of duties had taken place.

Following his plea in his 1877 annual report to Congress to replace the Dutch standard, in early 1878 Sherman began a campaign to replace it with the polariscope. In 1843 a commercial polariscope laboratory had opened in Philadelphia, and at about the same time, one of that city's refineries began to rely on the instrument. By the 1870s, New York was full of commercial polariscopic laboratories, on whom sugar brokers relied for their transactions.[18] Within the United States, the instrument's advocates claimed in public and private that it was simple, reliable, and accurate. The polariscope's operation, the Smithsonian physicist Joseph Henry told Sherman in February 1878, "can readily be taught to any intelligent person of ordinary education . . . by a person thoroughly acquainted with the theory and practice of the instrument." Sherman's allies in Congress, most notably Congressman James Garfield, argued that the government could therefore safely hand over the collection of sugar duties to polariscope operators. If Henry also felt comfortable that "a layman, a man without special skill, can be taught to use this instrument accurately," Garfield said, "I have not quite the courage to say it is not so."[19] Not all were so sanguine or confident that just anyone could be taught to use the polariscope and agree with other users. When Garfield suggested that the Treasury secretary could, in any case of doubt, authorize a laborious analysis of a sample's component elements, he was challenged again by a congressman from North Carolina.

MR. ROBBINS: One word more. The chemical analysis test is too costly for general use, is it not?

MR. GARFIELD: Oh, yes; it would be too cumbrous and costly to be used ordinarily. But it can always be used to verify the polariscope test in any important case.

MR. ROBBINS: But who is to know that any correction is needed?

MR. GARFIELD: My friend in that question has taken up the conflict of ages. Who shall do anything except the men appointed to carry out the law? Who shall find out any blunder or correct any wrong unless you appoint

> somebody to do it? Congress, I take it, can hardly determine the sweetness or strength of sugar or the amount of glucose in it unless we appoint an agent. The Treasury cannot do it except by its agents.[20]

The polariscope was just one of many tools of chemical analysis brought to bear on organic substances, above all food, in the late nineteenth century. Newspapers published breathless stories of swill milk, bogus butter, and toxic meat.[21] When an individual's senses lost their power to evaluate the character of foods, or seemed to lose that power, people turned to practices that promised to transcend simple sensation.[22] Systematic institutional efforts at using chemical analysis to standardize American economic products had begun—of all places—in the world of fertilizer, carrying out the Department of Agriculture's mission to boost American food output. By the 1870s and early 1880s, states across the Union were empowering official chemists to patrol consumer goods too.[23]

But even if the polariscope promised to escape the need for a person's senses, it could not escape the need for a person. It was the continued dependence on agents that made the Columbia chemistry professor Charles Chandler another skeptic of the polariscope's government use. "The ordinary testing of sugars by the polariscope is not accurate within 1%," Chandler wrote to an import firm from Washington, where he was lobbying Congress against the instrument. "Differences of 2 or 3% or even more may occur when different samples are taken from the same lot of sugars and tested by different chemists."[24] Judgments of color, texture, and grain using the Dutch standard were based on what "sugars show to the unaided eye,"[25] and "a great many more men are judges of color than are judges of the test by the polariscope."[26] The new instrument, wrote one former Treasury agent, would "furnish a vehicle for the safe practice of fraud [at] the option of operators and their instigators." He enumerated some ways polariscope tests might be imperceptibly modified: "The operator adds a few drops more of water than the proper quantity, and the grade is lowered; or a few grammes of sugar more than the proper quantity, and the grade is raised."[27]

Moreover, the ways the polariscope was used commercially indicated its potential unsuitability for government work. In private exchanges, buyers and sellers each hired their own samplers and chemists, who delivered a certificate of the resulting test. Predictably, negotiations began with buyer and seller quoting widely different values for the same sugar. One refiner might find that his own values differed from another's while both disagreed with the seller's chemist. Parties frequently called for resamples and retests, and the sale might even be aborted if no agreement could be hashed out.[28]

The manipulability of polariscope tests was thus accepted as fact among its commercial users, and especially among those who used it regularly. Effectively deploying that manipulability was part of the skill that went into making use of the instrument in transactional settings. "Some will go on color, some on classification, and the seller will have his goods sampled and tested, and will make them test 95 [polarization], and then the buyer will have them sampled and tested and they will test perhaps 92," complained the refiner Lawson Fuller, "and then both parties knowing how much they are cheating each other, will come to a square understanding and strike a balance just [as] though they had not sampled or tested the goods at all."[29] Compare this with what Jean-Baptiste Biot had said in 1840 of the utility of his analysis to the sugar industry: "Thus sugar merchants, as well as refiners, would know what they are selling as well as what they are buying." The sentiment was the same, even if its polarity had been rotated to the opposite direction.[30] To polariscope opponents, this proved that chemists were not to be trusted with the administration of customs. For while an importer might challenge a high assessment by a customs chemist, there would be no adversarial challenge to a value that was too low. The apparent accuracy of polariscope tests was inseparable from their commercial and negotiated context.

In fact, the polariscope's opponents, which included most of the smaller refiners and importers, argued that it was supported by the large refiners precisely because of, and not despite, the fact that it would make customs manipulation easier to commit and harder to detect. "I think Havemeyer is trying to persuade the Government to use the Polariscope in connection with the duty & we are against him & don't think it wise as it will leave such an open door for fraud," wrote the refiner William Booth to his friend Chandler.[31] If Havemeyer and his fellow large refiners were capable of bribing samplers on their docks, or of punishing Grace, nothing would seem to stop them from equally colluding with polariscope operators, whose work would be even more difficult if not impossible to hold to account.

In the context of civil service "reform," it was impossible to disentangle questions about the unreliability of the polariscope as an instrument from claims about the unreliability and incompetence of customs employees. The claimed goal of these reforms was to depoliticize the custom house by routinizing the appointment procedure and eliminating the practice of "assessment," in which the beneficiaries of patronage were canvassed for donations to the political party through which they had landed their posts. But trustworthiness and expertise were not separable qualities, and neither were political means of appointment and perceived reliability distinct. Hence, Sherman placed great emphasis on matters of trust when instructing the Collector

Arthur precisely which of his custom house employees to release following the commissioners' reports. At the highest levels, Sherman wrote, Arthur was to pick "the more efficient and trustworthy," while for lower-ranking officers, he should "employ only those who are competent for the special work assigned them, whose industry, integrity, and good habits give guarantees for faithful services, honestly rendered."[32] The Collector of Customs in New York, in charge of all tariffs at the port, was in 1879 a man named Chester Arthur. He testified that in his opinion many of the men overseeing "the delicate duties of the appraiser's office, requiring the special qualities of an expert, were better fitted to hoe and plow."[33] The concern of the polariscope's opponents was that if its operators were on the payroll of one of the large refiners, no one would ever be able to tell. One refiner put it best: "Men are generally honest," he said, but "some men's sight through the polariscope would be very defective if a fifty or a hundred dollar bill got into one end of the instrument."[34]

In February 1878, as the Perot case rumbled along, and Congress considered various proposals for revisions to the tariff, New York's importers and smaller refiners huddled to appoint a committee that would propose an alternative system for the assessment of sugar.[35] To circumvent the vicissitudes of sampling, they endorsed a "uniform" tariff of a single rate below 13 Dutch standard as Sherman himself had proposed the previous year when he admitted his embarrassment. Such a tariff, one refiner suggested, would specifically exclude "the class of sugars which trouble the Appraisers."[36] The idea was to discourage the importation of all sugars except the most valuable ones that bumped up against the 13 Dutch standard ceiling, and rendering it pointless to make sugars darker than they already were.

Everyone in the sugar trade seemed to view their opponents as part of a vast conspiracy. The uniform tariff's proponents were no different, to the point that two large public meetings on the question, in New York City in 1879 and 1880, ended in brawls. The 1879 meeting proposed "a revision of the tariff as will provide for the honest collection of duties on sugars, and also to protest against adulterations." The organizers invited notables like Columbia's Chandler, whose attendance would bolster the meeting's credibility.[37] The meeting was significant enough that even President Rutherford B. Hayes and several cabinet members sent telegrams regretting their inability to attend, but they were applauded in absentia.

Statements against the polariscope were at the top of the agenda, so pro-polariscope forces had attempted to pack the hall. A squad of "rough-looking" Havemeyer employees had crossed the river from Williamsburg, in an attempt to "capture" the meeting, but they were blocked by an augmented police guard arranged at the last minute. Theodore's brother Henry Havemeyer himself appeared at the door, theatrically dismissed his men, and then made an entrance himself. When one of his punchiest antagonists, the refiner Lawson Fuller, accused him from the dais of having "on the proceeds of fraud and adulteration, built up eight new refineries," Havemeyer and David Wells—a free trade economist hired to report on Cuban refinery conspiracies—declared their eagerness to settle the matter with their fists. "The callers of the meeting had prepared a series of resolutions," the *Times* reported, but given the tense atmosphere, "did not deem it advisable to even read them."[38]

By April 1880, when a second meeting was called, no one in the cabinet was getting applause. The meeting produced resolutions expressing "public opinion against the unjust private dock and tariff systems." Sherman was pilloried for not releasing the many affidavits about fraudulent importing practices that it was rumored he had collected over the last fifteen months. According to the speakers at the meeting this was because he wanted the sugar refiners as backers for the Republican presidential campaign in 1880, in which he would be a candidate. He had, speakers alleged, promised a uniform tariff one morning, submitted testimony to the Ways and Means Committee that afternoon in favor of the polariscope, privately recommitted to the uniform tariff in the evening, and then been spotted playing polo with Theodore Havemeyer in Newport the next day.[39]

In spite of such opposition Sherman pressed ahead with plans for the polariscope. He circulated an order in 1879 asserting his administrative authority to implement the device. The existing statutory language, he claimed, implied that the Dutch scale was intended as "essentially a measure of saccharine strength and not of color as an abstract physical attribute of matter."[40] Several importers sued the federal government, claiming that they had been overcharged on a cargo as a result. The case worked its way to the Supreme Court, which in 1881 ruled in favor of the plaintiff: If the department was to use the polariscope, then Congress had to authorize the specific use of that instrument. As a result, the Treasury was forced to refund two million dollars in sugar duties, and telegraph custom houses that the "apparent color" was to be the basis of future levies.[41] Only in 1882 did Congress eventually mandate the polariscope's use, as a result of Sherman's own legislative maneuvering. Having lost his campaign for the Republican nomination to his polariscopic ally Garfield, he had resumed his seat in the Senate in 1881. From there, he watched Chester Arthur—whom he had eventually dismissed as Collector—ascend to the White House after now-President Garfield was fatally shot.[42]

Polariscope opponents continued to lodge allegations of fraud just as Henry Havemeyer began consolidating the refining industry along the East Coast. Several attempts to form price cartels among refiners in the early 1880s failed, especially when the Havemeyer refinery burned to the ground in January 1882.[43] It would be set ablaze again in June 1887, allegedly in retaliation for a recently crushed strike. But by then Havemeyer had consolidated the East Coast's refiners into the Sugar Trust, which controlled 84 percent of the refining capacity east of the continental divide. What had finally given his firm the power to compel others into cooperating was that his new refinery was the largest and most efficient in the world. It could melt three million pounds of sugar a day, twice as much as the nearest competitor. Its refining costs were 0.44¢ per pound of product, but the price margin between raw and refined sugar had fallen to just over 0.7¢ by 1885. Havemeyer's firm may in fact have been the only one in the country to earn a profit that year. In 1888, the year after he formed the Trust, the margin jumped to 1.258¢.[44]

The United States' economy at the turn of the twentieth century was reshaped by "the great merger movement in American business." The horizontal combination of producers in so many different kinds of output—everything from petroleum to whiskey to tin-plate to bicycles to salt—was not, the historian Naomi Lamoreaux concluded, "an inevitable component in the rise of modern industry." Instead, it was a confluence of two factors: The development of new and powerful but expensive techniques raised the fixed costs of manufacturing, and the explosive growth of the American economy at the same moment left industries unstable and their leaders uncertain.[45] The depression of 1893 made clear that the consolidated industries survived best in bad times.[46]

Sugar refiners had been combining to force down the prices of their raw materials and force up the prices consumers paid for their refined sugar since at least the early eighteenth century in England.[47] By the 1870s, even Americans' ballooning consumption of sugar could not make refining a broadly profitable business. The same sorts of machines that had reshaped raw sugar production in the tropics had had an even more dramatic effect on metropolitan sugar refining.[48]

Refining was a concentrated and enhanced version of the crystallization and drying process from the plantation. In the newer refineries, sugar boats were unloaded by hand and the sugar carted into the refineries. But once inside, it was washed, remelted, filtered, cleaned, crystallized, dried, and screened, and then packed, cubed, or powdered. By 1890, in the latest refineries, sugar did not have to be handled on its way to the melters at all.[49] Ideally this was a continuous-flow process of pipes and pressure chambers and automatic machines, but in reality, the work was almost as physical and as taxing as in a raw factory. The capital required to build and run a refinery was immense, and the difference between profit and loss was very fine. Between the 1880s and 1910s, the American sugar industry's consolidation was one of the most visible and despised of all monopolizations. Sugar refiners' cartels became a textbook case of what happens in a viciously competitive and ruthless industry.[50]

The spiral was vicious. As the competition in refining became increasingly brutal, margins fell.[51] By 1887 the difference between the price of refined sugar and the price of raw sugar was barely half what it had been a decade earlier. The only way out was volume. Refiners built ever larger, more sophisticated, and more efficient plants, but these cost more money too and left their owners only more hostage to those same collapsing margins. Some refiners even shut down their plants altogether rather than lose money on every barrel. After prolonged and delicate negotiations, eighteen of the twenty-three sugar refining companies in America agreed to swap a controlling share of their own stock for certificates of the new Sugar Trust. The trust form kept its internal workings secret, unlike a holding corporation.[52]

The point of a combination was to idle plants and thus constrain supply. Total revenue and profits would increase overall, and then it was simply a matter of dividing up the larger pie. The size of each slice was determined by the capacity of each refinery, which was subject to daily reports and routine audits. A few refiners who remained outside of the Trust got the best of both worlds: They did not have to restrict their production, but they enjoyed the benefits of high monopoly prices. Most were small enough, however, that the leadership of the Trust, and later the American Sugar Refining Company, chose to ignore them.[53] Everyone was making plenty of money; any time a pound cake went into an oven, anywhere in the country, a refiner earned an extra penny.[54] When the muckraking Ida Tarbell pitched a study of how the Sugar Trust affected the prices women paid for groceries, her editors dismissed it as unimportant.[55]

In 1890, the Trust reorganized itself as the American Sugar Refining Company, using a friendly corporate environment in New Jersey to hide, temporarily from the new Anti-Trust Act sponsored by now-Senator John Sherman.[56] Over the next four years, over 280 trusts reorganized themselves in New Jersey.[57] In 1895, it survived a legal challenge from the federal government, when the Supreme Court ruled in *United States vs. E. C. Knight* that manufacturing combinations did not restrain trade.

Two well-capitalized rivals also emerged over the next decade.[58] One was Claus Spreckels, a German immigrant who had pioneered the tight vertical integration of Hawaiian raw sugar production with his California refineries.[59] Spreckels had refused to combine with the Trust, and when the Trust launched a price war against Spreckels, he retaliated by building the world's most advanced refinery in Philadelphia to conquer the eastern market.[60] Eventually, the American Sugar Refining Company and Spreckels negotiated a profitable ceasefire. The other was John Arbuckle, a coffee-roasting tycoon, who pushed his new packaging technology into sugar. By the early 1900s, Arbuckle, Spreckels, and a few others existed alongside the American, but this small oligopoly still kept the price of sugar high—unjustifiably high, to progressive and populist critics.[61] In his dissent in *E. C. Knight*, Justice John Marshall Harlan shook his head at "this stupendous combination."[62]

Sugar refiners were famously ruthless about making money at the expense of the public, which included writing tariff laws to its own benefit. Fiddling with polariscopists allowed sugar companies to pilfer from the federal treasury. But ultimately, public corruption was not the point. The true motivation for a refiner manipulating the price he paid on imported sugar was to gain an edge on his competitors. This was as true when they were part of the same arrangement as when they were outside it.

A Doubt Is at Best an Unsafe Standard

It was in this tense environment of the mid-1880s, and this uncertainty over the value of sugar, that Boston refiners complained that the New York Custom House was charging a lower duty than its Boston counterpart. Identical samples of sugar, they alleged, were declared as much as 3 percent less pure by the New York chemist than by the chemist in Boston. In October of 1886, a journalist for the Boston *Herald* named T. Aubrey Byrne began to look into their claims, and shortly thereafter he was commissioned by the Treasury as a special investigative agent.[63]

The Boston merchants claimed that this discrimination was sufficient to render them uncompetitive. Conditions were so bad that sugar cargoes were being diverted to New York and then reloaded onto ships to be sent up the coast to Boston refiners, in order to save $150,000 in duties each year. Relatively small amounts of money could affect trade substantially, as Byrne pointed out at the beginning of his final report. "Inasmuch as 1/16 cent per pound is considered a margin of profit in handling and refining sugar, the net cost of that commodity, including customs duties, must be very carefully considered for one degree by polariscope for duty is nearly equivalent in value to 1/16 cent per pound." The Boston importers wanted Dr. J. T. Leary, the chemist in their city's custom house, fired, and "a more liberal man substituted" who would round down fractions as they claimed New York's chemist did.[64]

The Custom House sugar laboratory in New York had been headed, since Sherman's 1879 order to institute the polariscope, by a chemist named Edward

Sherer. With his brother John, he had run a polariscopic laboratory on Front Street in Manhattan from the early 1870s. They became the most prominent private chemists in the city, and before the Custom House had acquired its own polariscope it had sent the Sherer brothers samples of sugar that needed testing.[65]

Byrne's investigation and subsequent reports to the Treasury and to Congress exhaustively documented all sorts of malfeasance on the part of both brothers, supported by hundreds of dense pages of statements from weighers, samplers, examiners, laboratory assistants, brokers, refiners, and other custom house officials. The Sherers would finish their duties at the government laboratory and then head downtown to their private one, working all night for the same importers and refiners whose cargoes they were meant to be evaluating for the Treasury. Witnesses also testified to the persistence of a ring of customs chemists and Treasury agents in New York that conspired to lower sugar duties by hundreds of thousands or even millions of dollars by manipulating samples and especially by shaving fractions or whole degrees off polariscope readings. Among them was Ira Ayer, a longtime Treasury special agent who had been responsible for drafting much of the sugar tariff legislation since 1880.[66] The ringleaders, however, seemed to be not the Sherers but the broker for the Havemeyer refinery, who moved with ease among the appraiser's stores, the refinery docks in Brooklyn, the Sherer firm's laboratories, and Havemeyer offices around the corner at 120 Wall Street. It even emerged that Edward Sherer was an honorary consul for the Ottoman Empire. While it seemed strange that a federal official should also represent a foreign power, odder still was that Byrne was unable to ascertain whether the consul's office was in the waiting room of the Sherer brothers' laboratory, or the laboratory in the waiting room of the consulate.[67]

Under pressure, the Sherers resigned from their posts in 1888, only to be reinstated in September 1889 when reports surfaced that the New York Custom House's polariscope had somehow been tampered with.[68] They were fired again in 1890, and the circumstances of their last sacking testify better than anything else to the impossibility of determining whether polariscope operators were acting improperly. In December 1889, a mere three months after the Sherers were reinstated, enough doubts had been raised about Byrne's report that three special agents, led by one O. Z. Spaulding, were asked to either corroborate or condemn his findings. Spaulding's report, received in June of 1890, rendered a harsh verdict on Byrne himself. Having spoken to those in the Boston sugar business who had complained of the lower New York tests, the agents found that nearly all "denied any conversation with [Byrne]" or else "accused him of a perversion of their language." Moreover they claimed to have always "complained of Dr. Leary's rigid tests at Boston, not of Dr. Sherer's liberal tests at New York."[69] They concluded there was no Havemeyer sugar ring "or any other sugar ring" in existence, at any level of kinship, friendship, or political appointments.

The trouble was that the discrepancy between the New York tests and the Boston tests refused to disappear, and Spaulding chose to turn his attention to the polariscopes themselves. In particular, much of his report was occupied with puzzling out the fate of a quartz plate mounted in a brass tube. Quartz, like sucrose in solution, rotates the polarization of light; thus a slice of quartz crystal of established rotatory power could be used to calibrate polariscopes. In 1885, on Treasury orders to rectify their laboratories' differences, Sherer had met with Leary, but "neither was willing to accept the other standard to the discredit of his own." So Sherer had purchased, and sent to Leary, a quartz plate, on the side of whose mounting tube was inscribed "99½," indicating the number of "sugar degrees" that a properly configured polariscope should indicate when the plate was inserted. But this seemingly straightforward bit of notation proved just the opposite. Between the Customs chemists, the National Academy of Sciences, and the Massachusetts Institute of Technology, different chemists disagreed on the simple matter of what a fraction meant. Did "99½" mean 99 plus "between one-tenth and two-tenths" or 99 plus one-half? Finally the exasperated agents dispatched the plate to its maker, the German instrument firm of Schmidt & Haensch, who, after several months, replied that their engraved value was correct. The only trouble was that the firm interpreted their own engraving as "99.12." "Several other eminent professors came near it or had exactly the same," the firm wrote. "Only one gentleman who is in the sugar trade found it one-tenth less," they added, "but then these gentlemen find always a little lower than others."[70] This was a dizzying dissensus on a matter as simple and fundamental as calibrating an instrument.

Yet even that didn't explain pervasive discrepancy between the ports. As the Boston *Daily Globe* wrote, "the controversy reverted to the main question, the honesty of the chemists at Boston and New York."[71] After an elaborate series of over 250 blind tests of sugar samples exchanged among the Boston, Philadelphia, and New York custom houses, it became clear that the Manhattan laboratory consistently tested sugars as containing over half a percent less sucrose, on average, than the others. "New York tests are on a plane below that of all the other ports," shrugged the agents, "and these planes never intersect." The same was concluded by the Agriculture Department's chief chemist, Harvey Wiley, after testing samples from the three custom houses. Overwork was no excuse: The fact that New York handled a hundred samples a day would not, Wiley argued, by itself produce systematic errors in one direction or another. The "rapid work" at a customs laboratory should not, he thought, mean that "the instruments used should not read practically the same numbers with the same solution, why the flasks should not be accurately calibrated, the tubes of proper length, and the operations of weighing, dissolving, and reading made with a fair degree of accuracy." Wiley visited each city's laboratory, but even such personal inspection could not trace the source of the variation. Slightly over half of it he attributed to "the tendency of the New York assayers to read too low, combined with a tendency to fill the flasks too full," yet that still left a 0.2 percent difference.

To Spaulding, Sherer offered a startling explanation: Different people read polariscopes in different ways. This was just what critics of the polariscope had always claimed and what its advocates had insisted could be eliminated through training. "He believes that the Boston chemists are high readers," noted the agents of Sherer, but they were not convinced by his practice of "giv[ing] the doubt to the importer" and rounding down his decimals. "A doubt at best is an unsafe standard," they cautioned. Yet without any evidence implicating Sherer in any illicit practice or against his "personal integrity," all they could do was recommend that the New York laboratory's tests "be 'toned up' by an average of nearly .5 of a degree."[72] They deemed such an adjustment sufficient as a compensation for Sherer's tendency to measure low—what was known as his "personal equation"—since they could not point to any way in which he was being dishonest in his use of the instrument.[73]

In late June 1890, Sherer was again dismissed from government service.[74] That left the Custom House in desperate need of good publicity. A few days after Sherer's firing, a reporter for the *Times* was invited to accompany visiting customs appraisers from across the country on a tour of its sugar procedures.[75] They began at the dock, where the reporter described in exquisite detail the samplers' rigor, down to the positions of their fingers. The visit to the laboratory was meant to emphasize that both polariscope and polariscopists could make reliable and consistent measurements.

The results did not go exactly as planned. The correspondent was invited to look through the eyepiece, but to him the two halves of the view "looked enough alike in color to seem to the unpracticed eye just alike," so he struggled to make his measurement. When he gave the polariscope over to the supervising chemist for confirmation, the chemist replied that the reporter was wrong by two-tenths of a degree. But the second chemist in the room disagreed with the first, by three-tenths.[76]

Most of all, the customs service wanted to advertise the secure procedures they had built into the architecture of the laboratory itself. The confidence of Spaulding and his colleagues that no sugar ring existed rested on the fact that the chemist "could not discriminate in favor of particular persons unless he knew whose sugars he was testing," and they felt that no conspiracy of the required magnitude to inform the chemists could long remain invisible. The *Times* reporter agreed: "To complete the safeguards for a purely scientific test, the chemists and polariscope readers are not permitted to know anything whatever about the ownership of the sugar to be graded, its place of production, the steamer by which it was brought to port or the date of its arrival." The only thing the polariscope operator knew was a serial number assigned by a clerk in a separate room. The reporter lauded the physical isolation of the laboratory, which was connected to the clerks' room only by the dumbwaiter used to deliver samples. "To reach the laboratory from the classification or examiner's room, [I] had to descend two flights of stairs, pass through the anteroom to the series of laboratories . . . through a small hall and then up two other flights of stairs." One message of the demonstration to the reporter was that the polariscope users were the most important men in the New York sugar trade, but another was that the polariscopists were also the men most subject to suspicion. The constant inability to determine just what a polariscope chemist was doing—even when the observer was a sugar chemist himself—was exactly the problem of unaccountability to which critics of the polariscope had pointed ten years earlier.

The fundamental change, however, was that the reason to discriminate among sugar samples from different cargoes or bound for different refiners had vanished. There was only one refiner of sugar, one ultimate buyer, not just in New York but up and down the East Coast, and "Cuban sugar growers found that they either had to deal with the trust or forgo selling their product in America."[77] The beneficiary of a low polariscope reading would always be the Sugar Trust. Later, in fact, one of Henry Havemeyer's employees testified that Havemeyer himself had ordered him to report raw sugar by the polariscope "half a degree lower than it really was," but by that point Havemeyer was dead and could neither deny the charge nor implicate any other executives of the company.[78]

The formation of the Trust might also have explained the sudden disappearance of Byrne's Boston witnesses. Whereas it had, in the mid-1880s, been in the interests of New England refiners to complain about Sherer's liberality in New York, it was now decidedly in their collective profit-sharing interest to complain of Leary's stinginess in Boston and to deny that they had ever told Byrne what he reported they had told him. Spaulding concluded his report on the government's laboratory with a double entendre that may or may not have been an accident. "Its employés," he wrote, "are as faithful to their trusts as are the employés of private parties, or the employés of the great corporations."[79]

Infamous Practices

As the largest manufacturing employer in the North's urban centers, the refining industry was politically powerful and consistently pushed for a greater differential between raw and refined to protect its profits at consumers' expense. These efforts infamously culminated with the greatest political triumph of the Gilded Age monopolists: the rewriting of Congressman William L. Wilson's 1894 free-trade bill by Senator Arthur P. Gorman into the protectionist Wilson-Gorman Tariff, which raised rates on refined sugar and protected the Trust's own profits instead.[80] Such a blatant demonstration of political "friendship," including supposed secret agreements between the Secretary of the Treasury and the Trust's executives, alarmed the reformist editors of *Harper's Weekly*. "The tariff question is not a mere question of schedules and rates," they wrote in September 1894. "It does indeed involve the integrity of republican government."[81]

Stories about corruption have framed our understandings of the Gilded Age for more than a century. Over the last few decades, however, historians have questioned the justifications for that frame. They have pointed, rightly, to the origins of corruption rhetoric in anti-Reconstruction revanchism and its rejection of federal power to enforce civil rights. "Progressive reformers targeted corruption in all its forms," Jackson Lears writes, "hoping to cleanse individual and society alike." But reformers' emphasis on purification could fit almost any political movement, from progressive crusaders for food safety to Confederate lost-causers.[82] Moreover, scholars have suggested that the Gilded Age was unlikely to have been more profoundly corrupt than earlier periods. Its chief distinguishing feature, on this account, is the "discovery" by newspapers and political entrepreneurs that corruption and reform were powerful electoral issues.[83]

But there also really was plenty of corruption, in the sense of public officers benefiting by granting favors to particular citizens or corporations while trying to keep the whole thing quiet.[84] "Although historians have tried to diminish the corruption of the Gilded Age," writes Richard White in *Railroaded*, "it is hard to study the period without being aware of how corrupt the normal procedures of business and governance became."[85] Economic historians have suggested that the very scale of the late nineteenth century's new industrial systems increased the incentives for corruption while simultaneously reducing the ability of democratic institutions to cope, as larger

FIGURE 6.4. When *Harper's Magazine* illustrated the Sugar Trust's victory in the 1894 tariff battle, it represented the Trust as a hogshead. "The True Issue," *Harper's Weekly*, September 8, 1894. Courtesy of HathiTrust.

and better-financed corporations saw enormous gains to manipulating the processes of a national government that itself wielded greater authority.[86] Much of that authority and wealth, argue Peter Andreas and Andrew Wender Cohen, derived from the illicit movement of goods. Smuggling was a foundation of US economic might, a justification for growing state power, and a vehicle for expressing racialized anxieties about citizenship, nationalism, and empire.[87]

Historians, however, reflexively defer to the analytical tools of modern science when those tools are counterposed to corrupt practices.[88] And they, too, have been polariscope-struck, arguing that concerns over the rectitude of the sugar tariff ended when the Treasury began employing chemists in its customhouses. In doing so they have concurred with the Havemeyers, who, though they denied any misconduct of their own, nonetheless insisted that the Treasury could avert fraud only by adopting new instruments and techniques.[89] Scientific practices were hardly inevitable techniques of accountability and good governance. They could also be put to work to obfuscate and confuse.[90] The process by which certain forms of behavior became defined as corrupt and unscientific, and others as proper and progressive (and even Progressive), becomes more visible when we leave the level of legislation and descend to the docks, customhouses, and laboratories where state authority met private power.

The month the Wilson-Gorman tariff was enacted, a dark joke in Brooklyn's *Daily Eagle* tied power over politics to power over measurement:

> "Name some of the qualifications for a United States senator," said a professor to a young man who was being examined for admission to college.
>
> "He must be 30 years of age, be above sixteen, Dutch standard, and be able to stand the polariscope test," replied the applicant.
>
> He got marked 100.[91]

At the best of times, sugar refining was a high-volume, low-margin business. By influencing customs chemists to raise certain refiners' tariff burden, even by fractions of a cent per pound, the more powerful refiners might entirely wipe out the profits of their rivals. These, wrote the *Daily Globe* in 1888, were the "infamous practices which made it so easy to organize the Sugar Trust."[92] When the Treasury turned from the Dutch scale to the polariscope, it did not substitute a modern and objective form of chemical measurement for one that was based on color and thus subjective and corruptible. Instead, it helped to delegitimize one variety of human expertise in favor of another.

There is no better evidence of this than the way the polariscope actually worked. Light entered the instrument from the far end, then was split so that one beam passed through a sugar sample while the other did not. This created different colors or shades in each half of the eyepiece. By turning a dial linked to a prism, the operator manipulated the colors; when both halves in the eyepiece appeared the same color or shade, a scale on the dial indicated the angle of rotation. By combining that angle with the known rotatory power of sucrose and the weight of the sample in the polariscope tube, the purity of the sample could be calculated. In other words, for all its elaborations of brass and glass, the polariscope's operation ultimately also consisted of the comparison of two tints of light. As one of the instrument's critics had written in 1879, "only the most accurate eye and skillful adjustment can exactly determine the perfect color blending which marks the record of true grade upon the scale." This was precisely the same charge that its advocates had leveled at the users of the Dutch standard, and before that, at the embodied knowledge of artisans on sugar plantations.[93]

7

The Electric Apartment

Anything by Electricity

They all agreed, even after everything unraveled, that they had never seen sugar like it. In New York the crystals were praised as "defined and glitteringly beautiful."[1] The Havemeyers tried to recreate them and failed. In Chicago skeptics said that there was "no discounting" them, "no gainsaying." In England it was "the finest sugar the dealers had ever seen."[2] Months afterward, a journalist still admitted to seeing "sugar that is unlike and superior to any other sugar known in the market." You could not buy it, not for ten dollars a pound.[3]

That was a lot of money for sugar. For ten dollars you could refine an entire ton of raw sugar from Cuba or Puerto Rico. An ordinary pound of refined sugar, in the middle of the 1880s, cost about six and a half cents at the grocer. For each pound of refined sugar sold at retail, a little more than one-half of one cent of its price was absorbed by the refining process, soaked up in the filters and bone-black. After covering their costs, refiners pocketed about three-quarters of a cent more as profit.

But when the towering new Havemeyer plant, the most efficient in the world, came on line in the summer of 1883, it could produce twice as much sugar as the nearest competitor for a much lower cost, just over four-tenths of a cent per pound. On American shelves, the price margin between raw sugar and refined sugar fell that year, and fell the next, and the next again. At those prices, Henry Havemeyer could make money, but, as he happily acknowledged, no one else had a hope.[4]

No hope, that is, until December 1885, when a chemistry professor from Germany disembarked in New York City and made struggling refiners a dazzling offer. He would teach them how to turn raw cane sugar into refined sugar not for ten bucks a ton, but eighty cents. All they had to do was pay him a million dollars.

Could he really cut the cost of sugar refining by 90 percent? His offer seemed too good to be true, yet also too crazy to fake and so specific, some refiners figured, that it must have been the result of a calculation. Why say eighty cents, one investor asked, not unreasonably, when saying a dollar would be much simpler?[5] The audacity of his precision inspired confidence, and so did the way he promised to make refined sugar. He would use electricity.[6]

The professor's name was Henry Friend. His victims later described him as an unsurpassed confidence man, which was transparently a way to justify their own error. He really had chosen a scheme that perfectly fit its moment. The professor alighted in New York the same year that Thomas Edison built the city's first direct-current power station and made its night incandesce.[7] The electricians—people who could harness lightning—seemed poised to transform society.[8] But the nature of electricity remained a mystery to the brightest physicists. Not even Edison could tell you what an electric current really was. And yet the public believed electricity's potential applications to be infinite. There was no easy way to tell where plausibility ended and credulity began. "In America, as everybody knows," sniped one English killjoy, "anything and everything may be done by electricity."[9] When acquaintances and investors described Friend's presence as "electric," "magnetic," even "chain lightning," they were speaking to audiences that devoured stories of miraculous electric treatments for incurable ailments and applauded inventors who zapped huge voltages through their fingertips.[10]

Friend took advantage of the mania for electrification. He also followed a well-lit Gilded Age path for a swindler. Plenty of con men before him, such as the cosmopolitan French oleomargarine adulterator Alfred Paraf, had found "manufactured food" to be easy pickings. The industry's innovations and novelties had made American consumers uneasy about mapping ideas of natural and artificial onto those of real and fake.[11] Long before the Treasury Department figured it out from Byrne's investigations, Friend intuited that the sugar industry was not so precise as the polariscope made it seem.

Prowling for investors, Friend began inviting refiners to demonstrations at his East Side home.[12] On the parlor table, a seductive cloth was draped over unseen shapes. He declared that the fabric traced the outlines of his unique machine, but he refused to let anyone touch it or even approach too closely. Instead Friend made them wait in the next room while they listened to crunching noises and other familiar sounds of the refinery. Later, one newspaper quipped that the whole affair seemed like a "sugar seance."[13] But after they were invited back in, the guests saw samples of Friend's famous sugar, and plenty of them were convinced.

His secret, like those of an alchemist, would not come cheap.[14] He offered it to the Havemeyers first, who were persuaded by his wares, and prepared to offer $2 million if he would reveal to them his method, which he would not. So they walked away from this "pig in a poke." Eventually, he recruited two respectable sugar figures, the refiner and Trust opponent Lawson Fuller, and the sugar broker and newsletter publisher Wallace Willett. They could serve as his front men. The company issued stock at a valuation of a million dollars, and people lined up to buy it.

New Yorkers, even then, had been trained to approach spectacular displays with a sense of irony and detachment. But this was no Barnumesque artful deception.[15] Spectators were not drawn to electric sugar by the sense of excitement they got from P. T. Barnum's shows, which came from trying to determine whether and how they were being fooled. The attendees at Friend's demonstrations wanted to believe, in part because he was good at identifying marks. He needed gullible sugar men who saw their money slipping away but still had enough to invest in a hopeful venture.[16]

Even a magnetic con man occasionally slips. At one demonstration, Friend accidentally turned 250 pounds of sugar that he said was "full of impurities" into 249 pounds of sugar he claimed was refined. Unfortunately, a chemical analyst working for some of the visitors calculated that there was more "saccharine matter" in that output than in the input. Not even Friend had claimed to find the philosopher's stone that could make sugar out of not-sugar. The visitors decided that Friend was ignorant of chemistry, and to be safe they hired a detective who looked fruitlessly into his background.[17]

Friend normally refused to let samples leave his person, but he took another risk in 1885, sending a sample of the Electric Sugar Refining Company's raw sugar supply to the public health chemist Charles Chandler at Columbia.[18] The company deployed the chemical reputation of the Sherer brothers, even though their laboratory and their ethic of public service were just then in question as part of the Sugar Ring.[19] The Sherers were sufficiently well known to provide legitimacy, but they could not match Chandler's imprimatur. The results of Chandler's own analysis, however, made him more suspicious rather than less. Searching for the organic components he expected, Chandler found that they added up to significantly more than 100 percent. Whatever he was analyzing, it wasn't raw sugar.

Chandler didn't say what he thought had been done to the sample. But whether he understood the electric sugar refining process itself, he understood enough about the Electric Sugar Refining Company to want to distance

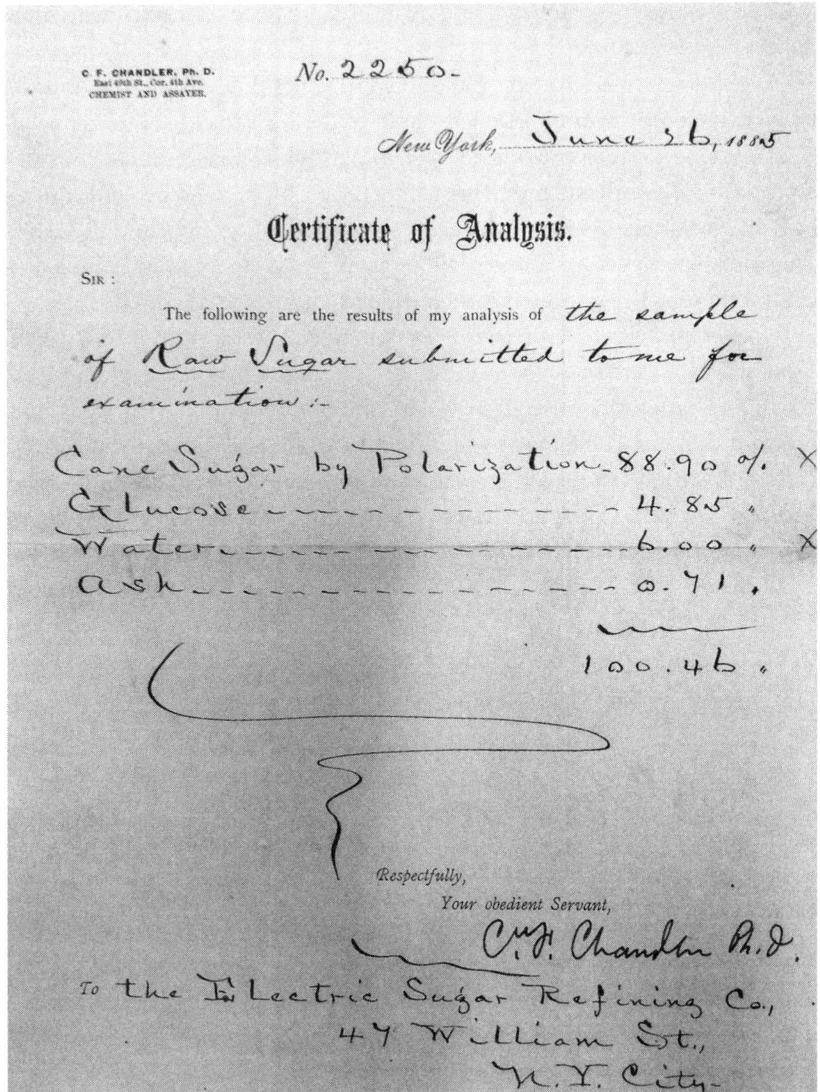

C. F. CHANDLER, Ph. D.
East 49th St., Cor. 4th Ave.
CHEMIST AND ASSAYER.

No. 2250.

New York, June 26, 1885

Certificate of Analysis.

SIR:

The following are the results of my analysis of the sample of Raw Sugar submitted to me for examination:-

Cane Sugar by Polarization 88.90 %. X
Glucose 4.85 "
Water 6.00 " X
Ash 0.71 "

100.46 "

Respectfully,
Your obedient Servant,
C. F. Chandler Ph.D.

To the Electric Sugar Refining Co.,
47 William St.,
N.Y. City.

FIGURE 7.1. When Charles Chandler tested the Electric Sugar Refining Company's samples, the components didn't add up, and he refused to be associated with the scheme. In Electric Sugar Refining Company, June–July 1885, box 260, folder 10, Charles Chandler Papers, Rare Book and Manuscript Library, Columbia University Library.

himself from it. He warned his office that the company wanted a certificate on Chandler letterhead "as a sort of endorsement, to make people think I am working on the enterprise," and told them to keep the results of his analysis private, to refuse any payment, and to return the sample. "I do not care to be connected in any way with your sugar enterprise," he told Lawson Fuller, and the prickly Fuller replied that they didn't care to be connected with him, either.

As the stock price rose, people wanted in, and hundreds of thousands of dollars arrived. A disproportionate share of the investment came from England, Friend's last stop before New York. A few Britons became converts there or in New York and began preaching the gospel back home. In Birmingham in November 1886, two shares were advertised for £250, and the seller promised that they would double before the year was out and quadruple again by the end of 1887. "The affair is thoroughly genuine," he reassured readers.[20]

Flush with investors' funds, Friend purchased a defunct flour mill near a Brooklyn wharf and set it up as his refinery. It was also his laboratory and even his living quarters. He ordered special machinery, of proprietary designs, and placed the orders with faraway firms in Philadelphia and Pittsburgh, because, he said, his process was not patented and he wanted to make sure no rival engineer saw all his designs at once. The machinery arrived at groggy hours of the night, in containers of odd shapes, and was taken directly to "burglar proof" rooms. In fact, Friend insisted on paying for the machinery out of his own funds, so paranoid was he about leaks. The Electric Sugar Refining Company dutifully reimbursed him $150,000 of the cash it had received from stockholders. In the postmortem period, this bit of credulity amazed observers the most.

At his electric refinery, Friend choreographed even larger and more spectacular demonstrations. Investors demanded to know when the refinery would begin commercial operations, so he told them that these were test runs of the process on the factory scale. Just as in a real refinery, raw sugar was hauled to the top floor and gained in purity as it descended. In locked rooms on the middle floors, his "mysterious electric apartments," Friend performed his magic alone. On the ground floor, through chutes in the ceiling, witnesses "saw the beautiful crystals that they could not buy in the market tumbling down," wrote *Harper's Weekly*, "and they were content."[21] Nine different kinds of his electric sugar circulated among the sugar men of New York, from lumps to a granulated sugar of a finer character than anyone had ever seen.

By late 1888, word of the marvel had even reached Kansas. In the name of national sugar self-sufficiency, the federal government was lavishly financing experiments with the goal of producing sugar from sorghum. Kansas's committed representatives in Washington ensured that the state received a disproportionate share of these funds, and other states and territories across the temperate zone were also convinced that sorghum would turn them into America's next sugar bowl. But while sorghum was cheap and easy to grow, its yield of crystalline sugar proved stubbornly difficult to increase.[22]

The Sterling Syrup Works in Kansas wondered what the most famous electrician in America thought about Friend's new process. Might it be applied to the sorghum grass? It was easier to generate sorghum molasses than crystalline sugar, and if someone figured out how to remove the slightly unappealing color, demand for Sterling's product would surely rise. This syrup company had heard that electricity could bleach paper pulp. If it could refine sugar, perhaps it would also bleach sorghum, and help this wholesome Midwestern industry supplant molasses imports. If true, "it will be a great thing for the country" and, they added, "also for the manufacturers."[23] They mailed their letter to Thomas Edison just before Christmas. By the time it arrived, electric sugar was over.

Peddling Rainbows

By all accounts, Friend enjoyed his time at the head of the Electric Sugar Refining Company. He surrounded himself with horses, jewelry, big houses in Manhattan and the countryside, and plenty of brandy. By December 1887, all the brandy had gotten to him. Rumors of his ill health prompted one of the English promoters to call on the unwell professor at home in New York. Writing home, he assured his fellow investors that the man who held the company's entire value in his head would recover to deliver the profits they had been promised. Doubters were to remember that "he impresses a visitor as being himself an electric man."[24] If Friend recharged, it wasn't for long. By March, he was dead. ("At least it is supposed he died," *Harper's* pointed out later. "Somebody died and was buried, and the weight of testimony is that it was Friend.")[25]

Despite this obvious setback, investors didn't desert the Electric Company, for Henry had left both his secret and the business in care of his wife, Olive. Along with her mother, stepfather, and two other relatives, Olive Friend stuck with the program. The oddly shaped machinery kept arriving, and the refinery kept up its schedule of tests and demonstrations. The last display for investors took place in December, and Olive had contracted with the company's president and secretary to sell them her secret on January 1, 1889, for $75,000 in cash.

In mid-December, the company secretary sent a letter to shareholders. The factory was almost ready, he reported, but there was even better news. As well as being on the verge of revolutionizing the business by producing electric cane sugar quickly and cheaply, the company had also recently reproduced an experiment made by the late Professor Friend and confirmed that the process also worked to remove any unpleasant odors left in sugar from beets. If they could really make refined beet sugar for retail sale, the real value of the company was double the investors' already high expectations. And then, on the verge of their success, Olive Friend and her relations disappeared.

The president found them in early January in their rural Michigan hometown, where they lived in mansions and drove carriages all paid for by electric sugar. Olive Friend demanded more cash in exchange for the secret, and, amazingly, the company president was even willing to pay her. He asked only that she affirm that the secret process did, in fact, refine cane sugar from its raw state. She hesitated, conferred with her attorneys. Would the president accept a recipe for making already refined sugar look even better? He would not, and her attorneys suggested he might want to go back to New York.

In Brooklyn, the company officers hired some muscle and broke down the doors of the refinery. Its electric apartments and secret floors were almost devoid of machinery except for the same crushers and sifters that you would find in any refinery. The rooms were, however, crammed with sugar—both the raw sugar that Friend had brought in to allegedly refine, and the refined sugar that he had smuggled indoors within the "queer shapes" of the machinery boxes. There were barrels and barrels of the refined stuff, ice axes to break the barrels, and mallets to pound the sugar before it was dumped into the crusher. The company officers found they needed those mallets. Something strange—something "chemical," they supposed—had been done to the refined sugar in those barrels, "for when the staves of all were ripped off it stood solid and alone."[26]

Back in Michigan, the conspirators were arrested and extradited to New York. The discovery of the swindle made front pages from London's *Pall Mall Gazette* to Joseph Pulitzer's *New York World*. On Saturday, January 5, 1889, the day the story broke, men with money on both sides of the Atlantic were all talking about the same thing.[27]

Plus some men without money and far from the Atlantic. The Electric Sugar "fiasco" was the first item of the next Wednesday's *Semi-Weekly Miner* in Butte, Montana.[28] Every day, miners in Butte crawled hundreds of feet into the earth in search of copper for electricity's new wires. Electric dazzle hid this deadly work from middle-class audiences happy not to know where their copper came from.[29] The paper in Butte covered the story all year, including the convictions of the fraudsters.[30] Friend's method, had it been real, would have made the world a little hungrier for copper. But as a "gigantic swindle" it lifted the edge of the cloth and revealed a bit of the truth behind the supposed magic of electricity.

A dozen years after the sugar from Demerara sailed into Baltimore harbor, the collapse of the Electric Sugar Refining Company revealed that the polariscope could only measure so much about sugar. Sucrose content and polariscopic purity mattered—but they were not everything. The value of sugar—as measured by the amount of money that even the savviest sugar men, like the Havemeyers, were willing to pay for Friend's process—still came down to questions of character: the character of the people who made it, the character of the people who assessed it, and, above all, to the character of the sugar itself. Friend made beautiful crystals, and even Chandler's analyses could not shake that.

For all that people agreed about the quality of his sugar, nobody ever bothered to pin down where Henry Friend was supposed to have come from. Was he a German immigrant, as he claimed, or was he native-born American descended from French stock, as one English visitor said? If so, why did the same visitor say he spoke English with a non-native accent? At least one newspaper wrote his name as Henri. Others wrote it as Freund. Certainly, nobody knew where the professor had allegedly professed before he got to New York.

But he could talk about every subject, especially chemistry. For his company officers, few of whom knew anything about sugar, he ably played the character of obsessive inventor. He once berated a carpenter building a hopper in his refinery, who had hidden a knot in the wood by facing it inward. Friend loudly insisted, making sure that an audience saw him, that what the audience saw was less important than a perfect surface for his refining process. Of course, the wonder afterward was that anyone had been seduced by this disagreeable and even "wicked" man at all, let alone by his ridiculous plan.[31] After seeing how much money had been entrusted to Friend, the New York *World* proposed "to organize a company to chop up rainbows and peddle them out as a substitute for aniline dyes," themselves another miracle of late nineteenth-century chemical engineering.[32]

The knowable facts of Henry Friend's life are less than exotic. He was born in Manhattan in 1842. He was drafted into service by the Union Army and after the war, lived in Brooklyn, where he claimed to work as a wheelwright. Neither the detective nor anyone else could figure out what he had done during the 1870s, but on the day he married in 1883, he told the judge that he was a chemist.[33] The witnesses were Olive's mother and stepfather, who both soon became part of Friend's operation if they weren't already.

By then Friend had already been running a con in Chicago for two years.[34] As the historian Amy Reading observes, and as Herman Melville suggested, the history of the "big con" itself started in the booming Western states before taking on established cities in the East.[35] This one was called the Grape & Cane Sugar Refining Company, claiming to turn grape sugar into cane sugar, and "into a saccharine product never before equaled for purity and cheapness." The script in nature's metropolis prefigured the one he used in Gotham. He supplied samples of better quality than anyone had ever seen. In Chicago, though, he described his invention as a chemical process. He hadn't yet discovered the au courant idea of electrification.

For the purposes of general credibility, Friend recruited two local businessmen—outside the sugar industry—as secretary and treasurer. The largest sugar refiners in Chicago were the Matthiessens, whom Friend approached just as he later would the Havemeyers. They were initially intrigued but, following the advice of their superintendent, they turned him down, and they managed to intervene when one of the city's richest families was about to fall for Friend's appeal. But once again small-time investors, including a man who came all the way from St. Louis with $2,500, had lined up to fund Friend. With their money, he bought an old building for his refinery and began shipping machinery from out of state. But he does not seem to have hit on the idea of funneling company money through his own pocket. Instead he only came away with living quarters in the refinery, and the secretary wondered later whether all that work was worth it just to live rent-free.

Friend also learned how to prolong his schemes. In Chicago, he promised too much too quickly. The refinery would start in March 1882, he claimed. That date came and went, then another. Sometime after his wedding, in late 1883 or 1884, Friend took the only signed copy of the contract, "shook the dust of the town from his foot," and skedaddled.[36]

Henry and Olive spent some time in New York setting up the Electric Sugar Refining Company, recruited a few English connections, and then sailed for Europe. By the time they returned, late in December 1885, Henry Friend was styling himself professor.

All Hope Abandoned

The details that Friend chose to invent for his new persona suggest that wherever he got his knowledge of sugar, he knew enough about refining to calibrate his image for the industry insiders whose allegiance he needed. The Havemeyers and Matthiessens might have blown him off, but at least they had been initially interested, and it had been worth a try: They had so much money that they might solve all Friend's worries at once.

To raise money from small investors Friend would have to appear a credible sugar refiner. Eventually, he managed to recruit someone in sugar who also needed him. That was why Lawson Fuller was useful. For almost a decade, Fuller had been complaining about and lobbying against the Havemeyer practices that had squeezed his own refining business dry. The story Friend invented was designed to appeal to someone like Fuller, who at that stage of his career was far from the cutting edge of sugar refining, either as a business matter or as an area of knowledge.

Germany and France were traditional centers of refining expertise, so it made sense for Friend to cultivate the impression that he had come from one of those countries. Both of them was even bolder. The evolution of his self-described métier from chemist to professor was a signal, too, that he knew sugar-making was moving from an art to a science, or at least that he knew his contemporaries believed so. And he too understood how sugar could dazzle and deceive—the difference between ordinary refined white sugar and beautiful crystals that you could not buy in the market. In the end, all Friend probably did was crush refined sugar through a handful of fine sieves, but it fooled almost everybody.

The peaks of Gilded Age fortunes could be steep and high and the valleys abyssal. Hundreds of thousands of real dollars had been entrusted to the Friends and their co-conspirators. For investors, $1.5 million in paper wealth crumbled away, most of it through frantic British fingers.[37] The company collapsed onto the middling professional classes of Birmingham and Liverpool. Six hundred and fifty of the cities' doctors and lawyers, "and not a few women" too, had purchased upward of £15,000 of the company's stock.[38] In Huddersfield, a dressmaker had paid a draper's assistant £42 in September for a mere half share in the company.[39] They watched as Electric Sugar shares, the ones that had promised to reach a thousand pounds, fell to £10, then £2, then 30s.[40]

Each new cable from New York to England carried more damning details. A man in Birmingham lost £2,000 alone; one in Liverpool, £7,000.[41] One unfortunate, who had bought in late in 1888, not only chased one of the company's chief English promoters to the dock from which he was sailing to America, but climbed aboard for the journey so that the promoter couldn't escape his demands for repayment.[42] Another bankrupt threw himself into a furnace, and after rescuers pulled him out, he tried to cut his own throat[43].

"All hope abandoned in England," said one headline, and the editors were not exaggerating, because many holders of Electric Sugar Refining Company stock had invested more than just their earthly savings. Company secretary James Robertson was a Christadelphian, a member of a sect which believed that the end of the world was near and that Christians needed a spiritual restoration to the practices of the church's earliest days. He and Robert Roberts, the Christadelphian leader, planned to use the riches from the electric sugar process to bring about the return of God's kingdom on Earth.

The millennium was approaching, and the Christadelphians had much to do. They would publish pamphlets and give lectures, distribute charity to the poor, and, most important of all, settle Palestine full of Jews. Roberts described Friend not just in electric terms but in messianic ones. He was pleased that Friend dismissed doubters in his process as "unbelievers." Even the unbelievers, though, could be saved and delivered into wealth. What mattered was that they believed the process was real and bought shares.

"Every form of prosperity for the work of God upon earth," Roberts lamented after he returned from New York, "was bound up in the success of that enterprise."[44] When Friend's prospectus had dropped into the brotherhood's lap, they had allowed themselves to think it a sign, "the providential form of the initial beginning of God's returning goodness in the latter days." Most of the stockholders in Britain got in thanks to Roberts's evangelism, and he himself was buying shares at inflated prices almost until the collapse.

To these Christadelphians at least, something certainly seemed electric about Friend. They had heard the questions about his impossible demonstrations and suspicious demands, but they had invested anyway. The complete collapse of the scheme did not shake their belief. Some of them held fast even after the police had smashed the crates of the pseudo-refinery and the trick was revealed. Roberts maintained his "faith" that Olive Friend possessed two valuable processes, though he no longer thought either was electric. And even after she was arrested, a group of his comrades were still prepared to offer her $75,000 for her secret.[45]

The lost savings vaporized by electric sugar were the inverse of the profits earned by the new refining monopoly. Even Lawson Fuller maintained, for a while, that there must be something real about the process, if could produce such unmistakably good sugar. You did not, in 1888, have to be a Christadelphian to believe that the world was ending or that its fate rested in Friend's hands. You could just be an aging sugar refiner, desperate not to see your business pulverized between bigger and more sophisticated rivals. The fraud "had its foundation," wrote *Harper's*, "in the keen competition among sugar refiners."[46]

Friend knew his marks and tried to tickle refiners as far away as San Francisco. An agent for the Electric Sugar Refining Company opened an office at the Board of Trade, showed samples, and sold some shares. He even set up a demonstration room of his own—hidden, of course. Soon enough, the California-Hawaii tycoon Claus Spreckels came to see what all the commotion was about. The agent promised that by joining forces they could defeat the Sugar Trust. But when he found out there was a secret, Spreckels wanted nothing more to do with it.[47]

By the middle of January, the tragedy had already turned into a joke. On the 17th, Brooklyn's high society threw its biggest gala. In attendance were city's very finest, wrote the *Daily Eagle*, "the creme de la creme, the doubly refined, like Professor Friend's electric sugar."[48] That same day, Thomas Edison finally picked up the letter from the hopeful sorghum company in Kansas, and scribbled some notes for his private secretary. Mr. Edison knew that electricity could bleach sugar, the secretary wrote to the Kansans, but didn't know anything about electrical refining. Edison also advised checking the papers for the Friend story. The secretary left that part out of his letter.[49]

PART THREE

What did it take to create a modern commodity market in an uncooperative substance like sugar? By the turn of the twentieth century, the raw sugar business had become standardized around a product that had not existed even a quarter-century earlier: centrifugal sugar measuring 96 percent by the polariscope. Its stability and constancy, the usual story goes, changed the economics of the business and laid a path toward futures contracts in sugar. But establishing a liquid market in sugar meant overcoming the profound ways in which such sugar and its measurers were both unreliable. The consequences of sugar's instability are explored in chapter 8 through the controversial question of whether polarization fell as the temperature rose. When the US government changed its customs procedures to reflect the mounting evidence that polarizations did change with temperature, the Sugar Trust sued, and the case eventually came before the Supreme Court.

In chapter 9 we'll see how metrological power began to be consolidated in the sugar business. Beginning in 1907, a small laboratory in New York acquired the authority to set rules for the trade's chemists. The story of this laboratory is very different from the ordinary idea of how standards are enforced. Its authority derived not from perfecting technical methods, but from forcing biased private chemists to adjust their own readings out of economic self-interest. This laboratory's existence was crucial to the liquidity of the sugar market.

There was another problem with the idea that all raw sugar was standardized at 96: It wasn't true, at least not in the way historians have understood it to be true. Raw sugar still varied wildly enough that producers, merchants, and refiners found that they had to develop uniform terms for pricing those variations. The "allowances" for sugar that was not 96 were made consistent, long before the sugar itself, and then those pricing allowances were embedded in the contracts that were traded on the new Sugar Exchange. But as chapter 10 explains, the notion that sugar was now uniform was a powerful illusion—so powerful, in fact, that in the late 1920s, the refiners' cartel used it to disguise a conspiracy in restraint of trade so thoroughly that not even the Justice Department could find it.

The final chapter tells four stories. A memo from an engineer, a microbe, a broken instrument, and a contractual innovation: Each of these helps us explore the relationship among sugar's value, its qualities, the instruments designed to detect those qualities, and the people meant to use them.

8

Instructions Relative to the Use of the Polariscope

Starchamber

The United States swung its tariffs around in the early 1890s, lifting the Cuban sugar economy to unprecedented heights and then smashing it into the ground.[1] The tariff of 1890 was championed by the arch-protectionist William McKinley, but it put sugar on the duty-free list. This was a gift to refiners but delivered a beating to cane and beet growers, who were granted a bounty instead—that is, they were subsidized by the Treasury to help compete with the newly cheap foreign sugar. The sugar bounty plus the elimination of the revenue from tariffs cost about $60 million a year, letting McKinley proclaim he had "given the people free and cheap sugar, and at the same time we have given to our producers, with their invested capital, absolute and complete protection against the cheaper sugar produced by the cheaper labor of other countries."[2] When the economy turned sour in 1893, the Treasury was thoroughly punished, and in the meantime the bounty generated a mind-boggling amount of paperwork.[3]

When the McKinley tariff was replaced by the Wilson-Gorman act in 1894, sugar was returned to the duty list. But for a brief period, the polariscope was left out. Now, sugar was taxed purely ad valorem, based on its market price. There were different rates for sugars above or below No. 16 on the Dutch standard.[4] Antimonopoly critics called this "the secret of the Sugar Trust's power," because it served to keep foreign sugar too dark to sell straight to consumers, no matter its polarization.[5] Customs offices were still equipped with polariscopes, which mostly gathered dust. Because the number that mattered was the price, not the purity, the controlling test for duty classification purposes was actually the "settlement test" between the buyer and the seller, and the price was determined by the purity they agreed on. What the customs officers said was irrelevant. It took the Treasury some time to admit that it could not directly calculate the market value from its own instruments.[6]

In the 1897 Dingley tariff bill, Congress reinstated the polariscope test for the duty in such a way that it maintained the refiners' protective margin. Now the tariff was 0.95¢ per pound up to 75 degrees plus 0.035¢ more for every additional degree, but a flat 1.95¢ per pound on sugar "which has gone through a process of refining." In other words, a pound of raw sugar that tested 100 sugar degrees would pay 1.825¢ in duty, while a pound of sugar of equal purity that had gone through a refinery would pay one-eighth of a cent more. That eighth was pure protection.[7] But the tariff bill still maintained the Dutch standard, at the no. 16 mark, as the distinguishing tool between raw and refined. "The injection of the Dutch color-standard into the Tariff Bill was either a case of heredity," Harvey Wiley wrote in an article for *The Forum*, "or else it was due to some influence brought to bear upon our legislators. It is not probable that a company consisting mostly of lawyers, sitting in a starchamber to determine the methods of levying duties, would think of the abuses to which the Dutch standard could be subjected."[8]

But for all the secrets of its power, not long after the Dingley Act went into effect, the American Sugar Refining Company filed a legal challenge against the way the Treasury had instructed its agents to use the polariscope. The case hinged on a technical dispute over how sugar behaved in the instrument, and yet it found its way to the Supreme Court of the United States, which in November 1908 rejected the industrialists' final appeal. The case revealed an epistemological question that had never been explicitly asked and whose very existence distressed the judiciary. The polariscope, judges learned, was simply a tool that could serve many different ends. Sometimes its methods were intended to produce values of the market, and at other times and with other methods it was asked to produce values closer to natural truth. The market value, legal classification, and true worth of sugar turned out to all be slightly different. On which sort of value did a scientific tariff rest? If chemical constituents did not translate directly to commercial worth, then which mattered more, and which did the polariscope actually measure? The cases also revealed a geography of precision in the sugar trade. Where could proper scientific measurements be taken, with what tools, and by whom?

Teasing out those differences among human and natural values pulled the executive and judicial branches of the federal government up to the bench of sugar chemistry and optical physics alongside the most skilled practitioners of the sugar industry itself. At the same time, the industry's own practices to standardize and evaluate its goods were shaped by the desire to find accurate values and simultaneously by the need to portray themselves as fair and accurate to the government, which were not always the same goals. The place to begin is with the obscure technical question that made it to the United States Supreme Court. That question was: What happened to sugar when it got hot?

Gyrodynat

More specifically, the question was: Did heat change how sucrose rotated polarized light? The European chemists who toiled away during the middle decades of the nineteenth century perfecting polariscopes thought, generally, that it did not, that sugar's optical activity was impervious to temperature.[9] In 1836, in one of the earliest papers on sugar polarization, Jean-Baptiste Biot had admitted that for "complete rigor in calculation," the experimenter needed to take temperature into account, but he worried about changes in the volume of solution, not about the solute itself to be analyzed.[10] Sixty years later, it did not bother Harvey Wiley, who published an authoritative volume on the chemical analysis of agricultural products in the late 1890s. He leaned on the work of several of the most widely known Central European sugar chemists, and proclaimed that "the influence of temperature on the gyrodynat of common sugars is not of great importance."[11] His proposed replacement for "specific rotation"—the term denoting the power of a given substance to change the angle of polarized light—never caught on.

Almost the instant that the book was published, Wiley considered it outdated. Temperature had become a broader concern among American and European sugar chemists since the end of the 1880s and had been the subject of many meticulous studies and at least one international convention all its own.[12] The magnitude of the phenomenon was tiny but still noticeable: For each degree Celsius that the temperature increased, rotation fell by a little more than one one-hundredth of an angular degree. If a sample of sugar were tested at 15°C and the same sample were also tested at 25°C, the second test would show the sugar to be about 0.15 percent less pure.[13] In a $200 million annual business, this was a substantial sum, and given how fine refiners' and producers' margins were, it mattered even more.[14]

"My conviction has undergone a change," Wiley testified a few years after his book was published, "due to my own researches."[15] In practice, it was fiendishly difficult to nail down how much, if anything, happened to the optical properties of sugar as its temperature changed. In trying to perform such fine experiments, the experimenters often just discovered new ways to emphasize the artificiality of their own laboratories. The idea of a single experimental temperature was itself a fiction that fell apart as chemists pursued ever more taxing degrees of precision. Each paper seemed to discover a creative new way of undermining something that had seemed like a stable thermal unit. The sugar sample might be at one temperature, but that temperature was not necessarily the temperature of the water into which it was dissolved, which itself was not necessarily the temperature of the instrument, which itself was not necessarily the temperature of the room. As rooms got warmer, flasks expanded. Tubes expanded. Quartz wedges expanded. As the wedges expanded, perhaps their own optically rotatory powers changed too.

Then there was the water. Sugar was measured in grams, but the water into which it dissolved was measured in cubic centimeters (and not all cc were even created equal, depending on the temperature and pressure at which the flasks were graduated).[16] Water became less dense as it warmed, so if twenty grams of sugar were dissolved in 100cc of water, the warmer the water, the fewer water molecules in those 100cc, and the more concentrated the solution. Even in his soon discarded textbook chapter, Wiley worried, and had gone to the effort to determine for himself experimentally, that the number of molecules of sugar floating in solution might affect how much each of those sugar molecules rotated light. Not "practically," he concluded.[17] If experimenters mixed more solution than the polariscope tube could hold, then they had to worry too that the solution's temperature would change while, and because, their hands delivered it from preparation to pour.

There were yet more insidious sources of error. Even a single scientific instrument on a tabletop, it turned out, was too large a system to coherently say that it had a single temperature. The end of the polariscope that was closer to the light source would be hotter than the end into which the user peered.[18] The polariscope's sturdy appearance belied inner thermodynamic chaos. It could take a whole morning for the innards of a polariscope to come to temperature. To prepare for some experiments at 40°C, Wiley preheated his polariscopes in the room for half a day.[19]

And just by breathing and working, all experimenters would change the ambient temperature. The most celebrated thermodynamic experiment in history, James Joule's isolation of the mechanical equivalent of heat during the 1840s, had been persuasive only because Joule had avoided mentioning how hard he and his assistants had sweated.[20] As we have seen, chemical control of sugar factories consumed an enormous amount of physical labor. Likewise it was easy to underestimate sheer physiological demands of running a chemistry laboratory, especially one in a high-volume sugar port.

If instruments and materials were not expanding from the warmth of all the work, they were shrinking in the chill of its absence. It was just as important to know how polariscopes performed closer to freezing in northern ports and beet factories as to know how they did in the hellish conditions of tropical cane mills. Drawing the lower slopes of temperature curves was tricky for practical reasons. Four degrees Celsius, for instance, was not a temperature that even the Department of Agriculture's laboratories were equipped to reach, so Wiley and a colleague had to take their sugar and their polariscopes to the cold storage room of Washington DC's Central Market, where they "cheerfully endured the fatigue and discomfort" before surrendering to the cold, retreating after taking just two of their planned three sets of readings.[21]

Heat of the Hands

Faced with the new rules in the 1897 Dingley tariff, the assistant secretary of the Treasury ordered his special agents to oversee their enforcement. That assistant secretary was a young attorney named William Howell. He had joined the department as a clerk during the original polariscope controversies of the early 1880s, and had been present for the Byrne and Sherer investigations later in that decade. In 1899, President William McKinley would appoint Howell to the Board of General Appraisers when a resignation suddenly opened up a seat. The appraiser who resigned was George H. Sharpe, whom William Grace had punched on a New York street and who had been a major general before Howell was even born.[22]

Howell was determined not to repeat the mistakes of the 1880s. He empowered an experienced agent named Chance to form a three-man committee of Wiley from Agriculture, Charles Crampton, the chief chemist of Internal Revenue, and Andrew Braid, the officer of the Coast and Geodetic Survey in charge of its Office of Weights and Measures.[23] "It was decided that the regulations should embrace the improvements of science in chemistry," Chance recalled, and as might have been expected from three such professionals. They were not to be rewritten from scratch but rather "modeled upon those that had prevailed under the act of 1883, and those in force under the Internal Revenue Bureau for the bounty provisional law" of 1890. From the perspective of their authors, the addition of new temperature corrections was evolutionary rather than radical, and in their testimony at trial they professed surprise that the commercial interests had taken such an objection to it, especially since they had met with most of the practicing private chemists in the Northeast.[24]

The Treasury Department's initial 1880 instructions for the polariscope had been written by Edward Sherer, the private chemist hired as the first polariscopist for the customs service, and who was later sacked and then rehired as part of the Byrne investigation. He only mentioned temperature to warn

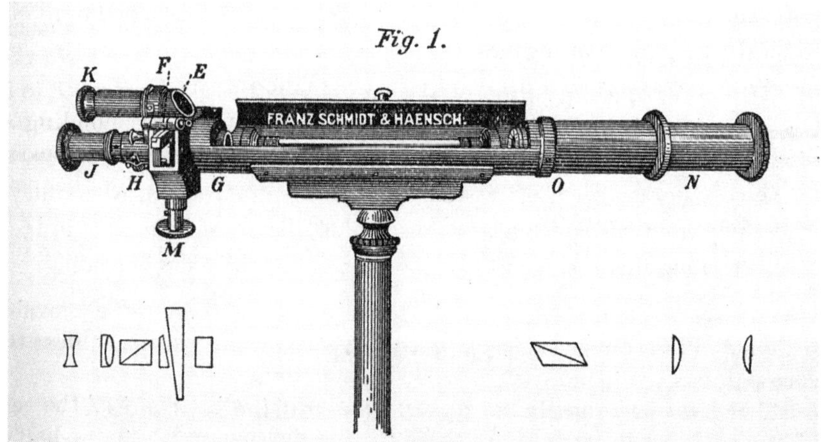

FIGURE 8.1. A detail of the Schmidt & Haensch half-shadow polariscope authorized for use by the Customs Service. The shapes underneath the body of the instrument show its arrangement of lenses and prisms. The operator looked through the tube at *J* and turned *M* to move the large wedge against the smaller one. One scale was on the moving wedge, and the other reference scale, known as the vernier, was on the smaller fixed wedge. When the two halves of his field of view were identically shadowed (hence the name), he read the result off the tube at *K*. This illustration was included in the instructions to customs officers, and was itself a frequently reproduced image, found in catalogues, advertisements, and chemical handbooks. US Department of the Treasury, *Synopsis of the Decisions of the Treasury Department on the Construction of the Tariff, Navigation, and Other Laws for the Year Ended December 31, 1897* (Government Printing Office, 1898), 991. Courtesy of HathiTrust.

appraisers to keep the temperature of the solution constant: "Take the flask by the neck, between the thumb and forefinger, and shake gently in one direction . . . the body of the flask should not be handled, as the heat of the hands will increase the temperature." These instructions were meant for someone familiar enough with a chemical laboratory to aspirate bone-black over sulfuric acid, but also unfamiliar enough with polarization of light to be told that the optical activity of glass would change under pressure from the liquid inside the tube.[25] Sherer's entire instructions ran to just four printed pages, including a translated portion from the German notes written by this particular model's inventor.[26] But in 1890 Sherer was in no position to write the new polariscope rules. Instead, they were assigned to Crampton as the chief chemist of the Internal Revenue division, and he, "having knowledge of the difficulties experienced in the customs branch of the service in obtaining accurate and concordant results of polarization in different laboratories,"[27] had enlisted Wiley, along with Braid's predecessor at the Coast Survey.

By 1898 the Treasury's new "Instructions Relative to the Use of the Polariscope" ran to twenty-eight pages, an intricate choreography of how sugar should be sampled, labeled, transferred, weighed, dissolved, polarized, and recorded, and how the instrument should be handled and used. The rules specified exactly what kind of polariscope the Customs would employ, a half-shadow device from the Berlin firm of Schmidt & Haensch (although a larger and more sophisticated triple-field was allowed and even encouraged), and how often observers should take a break. French and German polariscopes, until the end of the First World War, were widely judged to be better made and more accurate than anything produced in the United States. Europe benefited from an enormous pool of skilled instrument technicians. Despite enormous tariffs on optical devices, American instrument makers judged that they could not make them for a competitive price.[28] The rules began by defining the very word "degree" as "the percentage of pure sucrose contained in the sugar, as ascertained by polarimetric estimation." What the department provided was in essence a patient user's manual, guiding "the observer" through the consequences of setting up and adjusting the preferred model of polariscope, and what the observer would see when he looked through its two eyepieces. The instrument was called "half-shadow," because the prisms rotated to darken or lighten the left or right halves of the view through the eyepiece. But these were not monochrome views, and the Treasury went to the expense, in its annual synopsis of customs decisions, to print glossy color plates showing the shadow patterns of the different instruments.

The rules also enumerated in meticulous detail all of the work a polariscopist had to do to set up for even one polarization. Every model of polariscope was different. This one allowed the user to exactly set the zero point by making tiny adjustments to the position of one of the quartz wedges. The alternative to this high-stakes maneuver was to note the deviation of the zero point and add or subtract it at the end, but this wasn't simple either. The Treasury instructed its polariscopists to take the mean of half a dozen readings of the zero point, throwing out any that exceeded the average by 0.2 or more. Then the other end of the range could be calibrated using quartz control plates, meaning another half-dozen observations at least.

This rigmarole was ordered to be carried through at least once an hour, said the Treasury, "and oftener when the operator has reason to think that any of the factors indicated above have been altered." Once a sample of actual sugar was in the tube, it got another six looks. The department offered a gentle reminder that after each reading, "the eye should be allowed a moment of rest." The owner of the eye seemed to be allowed none.

In all twenty-eight pages of instruction, the detail that became controversial was the one in which the department also now took temperature into account. The Schmidt & Haensch polariscopes issued to the sugar testers each

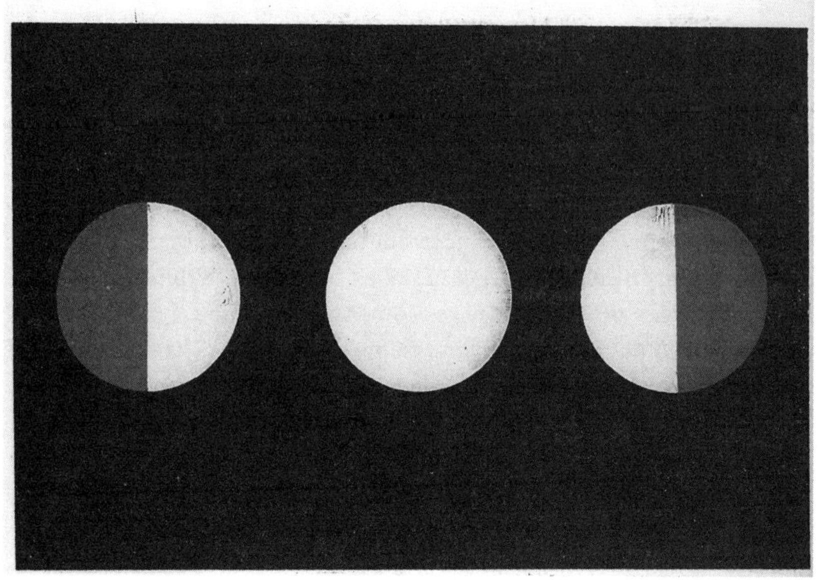

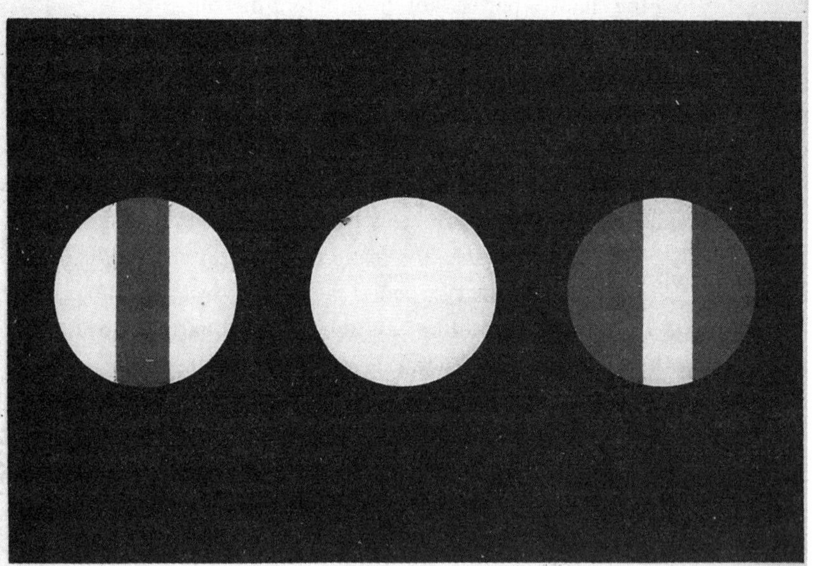

FIGURES 8.2 AND 8.3. The patterns of the polariscope, printed on special glossy full-color pages by the Treasury Department. US Department of the Treasury, *Synopsis of the Decisions of the Treasury Department on the Construction of the Tariff, Navigation, and Other Laws for the Year Ended December 31, 1897* (Government Printing Office, 1898), 991 and 993. Courtesy of HathiTrust.

came with at least one quartz "control plate." Each plate came with a printed table from the Coast and Geodetic Survey listing temperatures between 10° and 35° Celsius, and for each temperature, that plate's corrected value on the sugar scale, "as their sugar values vary with changes in temperature." The sugar examiner was to take his reading, then read the temperature off a thermometer placed "close to the instrument and in such a position as to be subject to the same conditions." Then he looked at the printed table for the corrected sugar value.

The difference between the visual reading on the quartz plate, taken directly off the scale on the polariscope, and the listed sugar value at that temperature, read off the printed table, was the factor that the sugar examiner was to add or subtract from the polarization of the sample of actual sugar. For instance, a control plate might be listed as polarizing 91.7. If the plate actually tested 91.4 in an environment at 25° Celsius, then it meant that, at that temperature, 0.3 needed to be added to whatever the sugar itself measured.[29] In general, what the Treasury's tables said was that, for every 1° Celsius the temperature went up, the sugar examiner was to add three-hundredths of a percent purity to the polariscopic observation. This table, explained Braid, would "save the operators at custom houses the time necessary to make computations which they would have to make without them. . . . That prevented their making mistakes and saved time."[30] All these procedures were a reminder that, in order to give a polariscope observation meaning, there was already a whole apparatus of calculation and adjustment that took place both before and after the observer looked down an eyepiece.

Later, at trial, the counsel for the Sugar Trust pressed every government witness on this point. The legal strategy was to differentiate work done before the observer read a number off the scale, which the Trust wanted to claim was legitimate polariscope use, from the allegedly illegitimate work done afterward to manipulate that number. On the witness stand, Braid pushed back. Every decision the experimenter made required compensation and adjustment. "It is often customary in custom houses and in the trade also to use a tube of half length in polarizing sugar," he pointed out. "Now, a solution of pure sugar polarized under such circumstances will read 50 on the instrument; would you claim that 50 was the true polarization of that sugar?"[31]

Called to testify several years later, Wiley, Braid, and Chance all talked up how much work and care had gone into the revision of the tariff. The Coast Survey's report for fiscal year 1898 complained that "much of Mr. Braid's time was devoted to the consideration of questions referred to the commission appointed by the Secretary of the Treasury," because he had had to travel up and down the Eastern Seaboard, and because it obliged him "to act in an advisory capacity to the Treasury Department in its controversies with outside sugar chemists." Meanwhile, the Treasury had also tried to preempt further embarrassment by ordering that all of its polariscopes and sugar flasks be rechecked. Braid's office grumbled about its inadequate facilities and staffing: "In consequence of the urgency of this work many other important branches are made to suffer."

And indeed the overwhelming majority of the office's work that year was for the sugar service. Between July 1897 and June 1898, Braid's assistants graduated over four hundred flasks and almost two hundred polariscope tubes for the Customs, plus plenty of polariscopes for official use themselves, while managing only to verify the odd measuring device of other types for government and private clients. At the very end of the fiscal year, they did verify one quartz plate for John Sherer and another quartz plate for Ferdinand Wiechmann, chief chemist of the American Sugar Refining Company.[32] Each plate was accompanied by a table of its sugar values. Sherer's had an asymmetric defect, so the Survey told him it was only to be used when the mounting screw pointed up and away.[33]

Ultimately, of course, all this effort was about saving the government money by charging the proper rate of tariff. If higher temperatures meant lower polariscope readings, then the federal government was losing out on its rightful tariff revenue every spring and summer, when most of the sugar arrived in its ports. When the tariff of 1897 went into effect, therefore, every sugar sample that entered the United States was scrutinized against one of these tables for temperature correction. Claiming that this scrutiny robbed them of more than $500 per ship, on average, the American Sugar Refining Company and two sugar brokers formally petitioned for reappraisement.[34] Their protest went first to the Board of General Appraisers, the special court that had been set up in 1890 to end a catastrophic backlog of customs cases.[35] The general appraisers heard the case in March 1899.

Commercial Method

Adversarial courts, as the scholar Michael Lynch observed, can become "sociology of knowledge machines," because every supposedly accepted scientific fact presents an attractive target for a skilled lawyer to undermine.[36] In the temperature corrections case, federal judges at several ranks found themselves wrestling with awkward questions about scientific instruments, their users, and their environments. In their March 1899 hearing before the Board of General Appraisers, the American Sugar Refining Company made two arguments.

First, the Customs Service tests were too high, and they were charging too much, because temperature correction was unnecessary. Sugar did not change its rotatory power. And second, more fundamentally, such a correction violated the meaning of Congress's phrase "testing by the polariscope." Their case rested on the idea that Congress meant to refer to "the commercial method," the use of the instrument as it was conducted in private laboratories, by producers, and by refiners. The General Appraisers were unpersuaded by the testimony that such a single method existed. And, they noted, the trade tests were done for the purpose of fixing a price for a transaction, not for determining truth. A trade chemist's polarization was a negotiating tool rather than a statement about nature.[37] But the Treasury's sense of the polariscope had diverged from this commercial sense. The authors of its regulations had interpreted the legislative phrase "testing . . . by the polariscope" as "the *percentage* of pure sucrose contained in the sugar, as ascertained by polarimetric estimation."[38] The appraisers noted a subtle difference. What mattered now was not that the test outcome and resulting duty aligned with "market values of sugars dependent on actual sales by merchants." What Congress had ordered was an assessment, as absolutely correct as possible, of the sugar's chemical purity in the form of sucrose: "a true polariscopic test by improved instruments and advanced scientific methods designed to determine classification rather than market value." On the matter of whether the temperature corrections were scientifically necessary at all, the Appraisers also sided with the Treasury, and they were swayed in particular by Wiley's arguments. They seemed impressed that the 1,200 polarizations he had made in pursuit of the temperature effect, or lack thereof, had led him to change his mind in the few short years since he had published *Principles and Practice of Agricultural Analysis*.[39]

There was one serious difficulty for the government. As the counsel for the Trust relished pointing out, sometimes the temperature "correction" produced impossible results. There were plenty of cases in which a tester, by adding the plate factor, arrived at a purity of a high-quality sample that was greater than 100 percent. By the Trust's argument, such impossibilities implied malfeasance, incompetence, or that the whole approach was bunk. Of course, the customs officer never entered more than 100 as the value. "I didn't think it would look well to report them higher," one sugar tester confessed.[40] But the Appraisers took a relaxed view of this as well. "It is admitted on all sides that mathematical accuracy is not attainable by any such process." No knowledge was error-free. Perhaps there was a mistake in making up the solution, or too much of a substance called raffinose, which strongly rotated light. It did not nullify the whole effort.[41]

The Trust appealed its defeat. As the case moved through the courts, the legal battle with the Treasury was avidly followed by what, in its coverage of the case, a trade journal called "the commercial body at large." For there was nothing trivial, one contributor (and polariscope-maker) argued, "with regard to 'a few tenths,'" and the case "supplies a fine object-lesson in regard to the importance of exact polarization."[42]

The district court, at the next stage, held the polariscope to a different standard. A formula that produced more accurate results in some cases but ridiculous ones in others was not justifiable. Here, the judge placed great weight on the impossible tests. He worried about the samples that, strictly applying the government's adjustment, should have been 100.2 or 100.3, "the absurd and impossible result under said test of being more than chemically pure." And against simply reporting the readings of the eye, adding 0.3 percent for each 10 degrees "violently changed" the meaning of "testing by the polariscope." To him these "so-called corrections" were in fact "arbitrary additions."[43] While the Treasury's rules brought the final inscribed value closer to the imagined truth, it had still not come up with a method for producing "actually accurate results," and thus had no basis to swap out the commercial method for its own.

One of the judges who heard the case on its way up the judiciary was the same William B. Howell who, as assistant secretary of the Treasury, had overseen 1897's exhaustive revision of polariscope rules precisely to stave off such challenges. By the time the case reached a panel of the appeals court, the judges were struggling to keep up with the chemistry, and "apparently" and "supposedly" appeared more often in the text. But they returned to a more flexible view of knowledge. What mattered was not sugar's market value but its inherent nature. No test of this could be "absolutely accurate," so the Treasury was within its rights to apply any rules to make it close in on that absolute. In fact, the court noted, the process of "striking averages" under the commercial system (and previous tariff system) was itself a kind of adjustment. As to whether it was a scientific question or a commercial one, the judges finally noted that none of the appellants themselves knew how to polarize anything. A polariscope "has never been used by the traders themselves, but only by expert chemists whom they employ."[44] In June 1904, they turned the Trust out again.

After licking their wounds for a few years, the American Sugar Refining Company and the importers who were its co-complainants mounted a last push directly to the Supreme Court.[45] They had a new argument: that temperature corrections, since they were not part of Congress's instructions to test "by the polariscope," constituted an entirely new and unconstitutional tax. The majority was unimpressed: "The present direct appeal to this court is a mere attempt to obtain a reconsideration of questions arising under the revenue laws and already determined by the Circuit Court of Appeals in due course." Appeal dismissed. But the legacy of this obscure case was more lasting than the summary rejection might suggest. As the next chapter shows, it had a profound effect on the institutions that created and enforced metrologies in the commercial side of the sugar market as a whole.

9

The Sum of the Errors

Seventeen Holes

In the summer of 1907, a man approached the Treasury Department with a story. For ten years, Richard Whalley said, he had been employed by the American Sugar Refining Company on its wharves, the same wharves that William Grace had prowled three decades earlier. His story implied that nothing had really changed since the Grant administration.[1]

Raw sugar was bought by weight, but not all sugar was bought by weight in the same way. Sugar that had traveled on a relatively short journey was purchased on the basis of its invoice weight: how much it had weighed when it had embarked from a port or even when it had left the factory. But more distant sugar (from Java, for instance) might lose some amount of water and molasses during its journey. That kind of sugar was purchased and taxed on the basis of its weight upon landing in New York.[2] To determine the landed weight of sugar, the Havemeyers' refinery, like many others, had embedded enormous scales into the wharf. A cartload of sugar could be wheeled over from the ship, weighed, and wheeled away to the melters. From inside a little cabin nearby, an officer of the Customs read the scale and recorded the weight for the tariff duty, and to his left sat a man employed by the refinery to keep an eye on the officer. This fellow was known as a company checker. Whalley had been a company checker, and for much of his time there, he told his Treasury interviewers, he had been under direct orders to make sugar weigh less than it really did.

Whalley, having turned double agent, was deputized to find work on the Havemeyer docks again, and one day late in November a few other Treasury agents joined him there for a sting. Whalley tipped them off—literally, he was to tip his hat when he saw the checker make his move—and the agents barged through the doors of one scale-house, catching the checker "monkey-

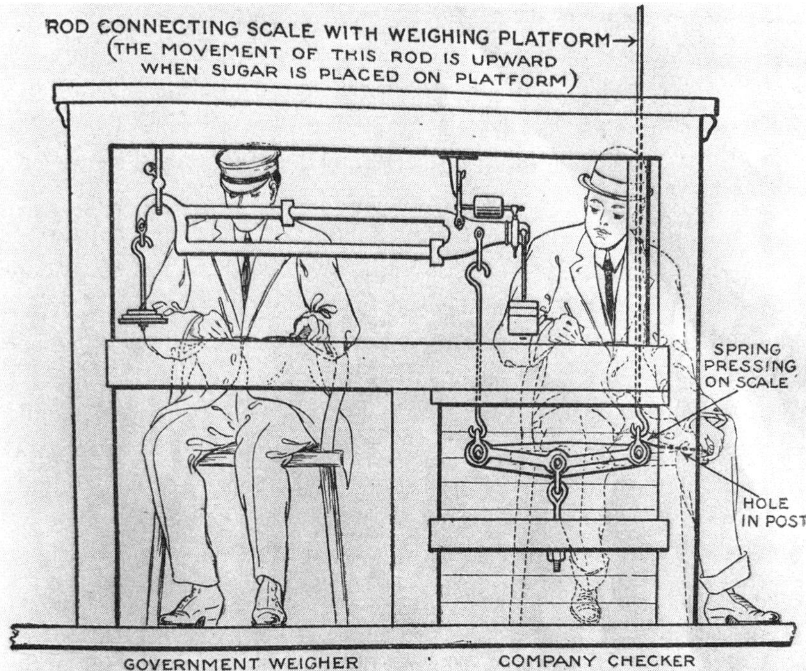

FIGURE 9.1. Diagram of the mechanism for reducing the recorded weight of a cargo of sugar without alerting the government checker. Harold J. Howland, "The Case of the Seventeen Holes," *The Outlook*, May 1, 1909.

doodling" with something under the table. It turned out that he was tugging on a curl-handled spring that ran through a hole in the wall and attached to the mechanism of the scale. By pulling the handle, the agents discovered, a checker could drag the scale's rods down ever so slightly, which would lower the value that the government weigher would see and record. The agents rolled the most recent loads back onto the scale and weighed them a second time, this time preventing the checker from pulling the spring. Lo, the bags had put on weight. Sugar might change by keeping, but it didn't put on fourteen pounds in three minutes.

One of the federal agents testified later that the refinery superintendent took him aside and asked him to name the price of his silence. When the government charged the company with fraud, it called as a witness a former customs weigher. This man swore that, on his first day at the Havemeyer docks, the same superintendent had told him to walk up to the firm's cashier every month for his disbursements. The company checkers who most routinely worked in the cabins with the government weighers also received an extra stipend, but their official pay envelopes reported the same salary as everyone else.[3] Over the course of six years, the American Sugar Refining Company's checkers had made seventy-five million pounds of sugar fall from the scales in front of the weighers' eyes. That was just about the same amount as a year's production from a modern factory in Cuba or Puerto Rico, as though the company had bought the entire annual output of a giant central factory and smuggled it into Brooklyn in broad daylight.

Over the next few months federal auditors compared the company accounts with declared weights and shipping invoices. On almost all types of sugar cargoes, the company paid buyers for several tons more sugar than it reported to the government. The exceptions were cargoes delivered in hogsheads: Because every hogshead was branded with its weight as it left the scales, the difference between the official and commercial weights would have been too obvious to ignore. No one higher than the superintendent was prosecuted, though prosecutors argued that "this company, down to minute details, was virtually run by one man."[4] Just a few weeks after the raid on his docks, Henry Havemeyer died at his estate on Long Island.[5]

Once again, the public finances were collateral damage of internal rivalries. Testifying a few years later, the longtime bookkeeper and secretary of the American Sugar Refining Company swore that he knew nothing of the weighing fraud, since he worked from the company's offices over on Wall Street, and never set foot in the refinery itself. But he offered his opinion that the real reason to drill the holes and underweight the bags was not to steal the people's money, but to make the refinery look more efficient in internal accounts.[6] The Brooklyn facility, a dreadnought when it was renovated in the 1880s, by 1907 lagged behind newer plants. Within the structure of the Trust, profits were doled out depending on the comparative capacity and efficiency of each refinery. Not for the first time, and not for the last time either, robbing the government could just be a secondary consequence of squeezing one's own partners.[7]

This scandal made the fall of 1907 an unpromising time to open a laboratory devoted to the probity of the sugar industry. Yet just a few weeks after the weigh-in on the Havemeyer docks, on Monday, December 2, 1907, an outfit called the New York Sugar Trade Laboratory opened its doors.[8] The laboratory had begun as a plan by Wiley and by John Arbuckle, a coffee tycoon turned sugar refiner whose firm, while independent, was part of the controlling oligopoly. Their idea was to create what was called a "referee" organization as a way of ending the tedious standoffs between chemists working for buyers and sellers. (In many cases, the buyer and seller often wound up having to send a sample off to a third chemist anyway for arbitration.) The term "referee" was exact: The laboratory's purpose was not to develop new rules, and not really to enforce the rules either, but to discipline the behavior of others by its very presence. This new laboratory would test every cargo of sugar transacted by any buyer or sugar who subscribed to it. Supposedly Arbuckle was so naive that when he got into sugar, he was so unfamiliar with the business and so shocked to discover that its chemists shaded their results that he had apprenticed in his own refinery and learned to polarize himself.[9]

Although never occupying more than a few small rooms in lower Manhattan, the New York Sugar Trade Laboratory came to govern sugar measurement within the Atlantic sugar trade and beyond. The self-interested version of the story claimed that its power derived from more careful methods, superior techniques, and greater probity and incorruptibility. "Sugar boasts of its orderly methods," Fernando Ortiz wrote, "and accuses tobacco of carelessness."[10] Creating a standardized commodity out of sugar took something more, and something different, than orderly methods. It required reckoning with the ways in which sugar was not all the same and could not be made to seem the same, the ways in which it was difficult to compare the qualities of one sugar to another, or even to compare the same cargo of sugar with itself across time and space. Once again, suppressing the variability of a skilled art (and reading the polariscope, as we have seen, took aesthetic as well as technical proficiency) revealed new and more intractable human and organic sources of variation. The story of this little laboratory helps us understand why certain sources of metrology become powerful in the history of capitalism. As we will see, the influence came first and the superiority of its methods second.

Referee

After decades of an adversarial system, it is not entirely clear why the largest refiners and merchants of New York chose this moment to organize the referee laboratory. London's Beetroot Sugar Association, formed in the 1880s while New Yorkers attempted to organize a sugar exchange, already had a system in place for a third laboratory. But the Londoners involved their third only in the same extremis as New Yorkers: when the buyer and seller chemists couldn't agree, or, more precisely, when the buyer and seller could not agree on a price that equally disadvantaged the other.[11]

The previous few years had seen a further consolidation by Henry Havemeyer of his direct and indirect authority over the sugar refining industry of the eastern United States.[12] Given that he already set the retail price of sugar by posting it outside his door at ten each morning, perhaps it was simply time to consolidate the determinations of purchase price as well. The available evidence suggests that the decision was about more than just private interest. Wiley and Arbuckle held their initial conversations in 1905. It took two years to persuade Havemeyer, representing the vast majority of purchases, and to persuade the largest seller, the firm of Czarnikow, Macdougall & Co.[13] All that time its organizers knew they were being watched from the bench. The litigation against the government on the temperature question was going poorly, although refiners were mustering forces for that last effort at the Supreme Court.

The new laboratory would be governed by two trustees, one representing the refiners and the other the raw-sugar sellers, and they were acutely aware of how the procedures and public image of this new laboratory might strengthen or weaken their case before the justices. Thus, the trustees first asked Wiley himself to direct, as he had a reputation for incorruptibility, and though he saw shadows moving against him in Washington, he declined. But he agreed to choose the director, someone, the trustees asked, "who will command the respect and confidence of buyers and sellers," with "no other relation to the sugar trade whatever." Wiley chose Charles Albert Browne, his deputy at the Agriculture Department, who had worked in academia and at the Department of Agriculture's Louisiana experiment station.[14] The trustees made sure to remind Browne of the legal and public context when considering his proposals for the laboratory's procedures. "We are not for the present prepared to adopt any method of analysis except what is regarded as the commercial method," they insisted. "Differences of opinion appear to exist among chemists as to methods and some of them have been, *and still are, the subject of litigation and the Trustees do not favor any change that would lessen the authority of the commercial method in use.*"[15]

They were also conscious of the broader public and political perception of the immorality of the sugar business and of how the critiques of testing methods and the critiques of monopoly might reinforce each other. Even the name of the laboratory was carefully chosen. "United States Sugar Trade Laboratory" was an initial favorite, and accurately suggested the minimal geography over which the trustees intended their values to rule: "The patrons of the laboratory would be scattered all the way from the Gulf to the St. Lawrence." But they feared that the phrase suggested affiliation with the government, something they certainly did not wish to imply. The "American Sugar Trade Laboratory," their second favorite, sounded much too much like the American Sugar Refining Company and could imply yet another chemist on direct payroll, or, as Browne put it gently, "the similarity might create mistaken ideas." They settled on the eventual name since New York City, prosaically, was where everyone had offices and, ultimately, it was where the trade in sugar actually took place. It was pleasingly anodyne and factually indisputable.[16]

What sort of laboratory was this New York Sugar Trade Laboratory? Testing laboratories like this one inhabit an odd place in the history of science.[17] Their assessment of qualities is not nearly as glamorous an activity as the work often called "pure" research. Nor are they grittily embedded in economic history like the laboratories we have encountered attached to sugar factories themselves, or the early modern apothecaries and marketplaces, or even the laboratories of agricultural experiment stations. They are also not obviously spaces for teaching, demonstration laboratories that might hold hundreds of students, where a historian might instinctively turn to study pedagogy and see the transmission of chemical knowledge and know-how in action. These latter functions, teaching and research, were often combined in dedicated institutions, as in the Audubon Sugar School in Louisiana. Conversely, to capture the awkwardly liminal nature of a sugar-testing laboratory, recall the confusion about whether the Sherers' laboratory was in the anteroom of the Ottoman consulate or vice versa. Browne selected rooms at Eighty South Street, less than a five-minute walk from the headquarters of most of America's sugar companies.

The credibility of the new laboratory rested on its procedures, but its procedures for controlling other chemists, not for controlling what happened to the sugar. The nearly fifty signatories devised a theatrical exercise in impartiality. The stage was so managed that once the laboratory opened for business none of the interested parties were allowed inside, except for its two trustees. Dress rehearsal consisted of comparative tests of six samples of various sugar cargoes, sent to five laboratories across the city as well as the Trade Laboratory, in the weeks leading up to the laboratory's opening. The results were sometimes more than half a percentage point apart. The exercise not only tested the new laboratory's capability, but more importantly, it also provided a baseline to see just how much gravity the new institution would exert on these existing centers of calculation. December began on a Sunday. On Monday, the laboratory opened. The whole sugar business district was curious to see what would happen.

Only two samples showed up on the first day. December was, after all, the slow season in the Caribbean. These sugars were from Java, sold to the National Sugar Refining Company by Czarnikow, MacDougall & Co. Soon daily arrivals picked up to an average of about ten. By the new year, that difference of half a percentage point had fallen to less than one-third. Already this small laboratory had cramped the bounds of plausible partiality, but not out of any superior reputation for integrity. In fact, Browne was apparently so worried about the honor of his own new staff that he had asked the trustees, in the planning process, to permit sugar samples be anonymized and numbered as was done in the Customs laboratories, in order to prevent his own chemists from privileging some clients over others.[18] In general, the money men who oversaw the laboratory gave Browne liberty to direct, but anonymized sampling was another method upon which the trustees had firm opinions. In refusing, they pointed out that this would just shift the risk of corruption onto whomever was entrusted with the bookkeeping, and they would rather just hang it on Browne: "We prefer to rely upon the honor and integrity of the assistants you select and upon the exactness of your own supervision." It turned out Browne was right to worry, as he caught Wiley's own nephew, one of his first hires, pocketing some lab funds.[19]

The rules Browne and the trustees worked out tested a chemist's stamina as much as any sugar. A sample arrived at the Trade Laboratory only after both buyer's and seller's samplers had witnessed its extraction and attested to its fairness. When the seal was broken at the laboratory, half was locked in a case for a week, the other half tested immediately. Two different laboratory assistants, using two different polariscopes, each tested the sugar. If these tests were within two-tenths of a degree, they were averaged and booked; if not, the assistants did the test again. The Trade Laboratory reported its results to the twentieth of a degree; commercial chemists took only one test and reported theirs to the tenth. Once the laboratory was content with its own work, its result was compared with those of the buyer and of the seller. Subscribers agreed that transactions would be settled by averaging the closest pair of the three polarizations, unless the laboratory came in "exactly midway," in which case the average of all three was taken, which would, of course, coincide with the Trade Laboratory's value.

"The effort of the buyer's or seller's chemist while still in the direction of giving his customer the benefit of the errors of experiment or observation, was now to approach the laboratory polarization a little closer than the contrasting party," Browne explained. Browne and the trustees understood perfectly well what laboratory polarizations were for. A number was still an argument. Chemists looked around, saw that all of their colleagues were shading results toward their patrons, and thus they had to do the same, Browne sympathized, "by necessity and not because they wished to do so."[20] Money was another argument. Chemists needed to keep hold of their clients.

The point, in other words, was to get the city's chemists to change their behavior, to adjust their personal equations just enough that their value would get counted, rather than the other fellow's. "Otherwise his polarization being the most distant would be discarded and the customer would lose the advantage of his partiality." The entire metrological enterprise explicitly depended

KOW, MACDOUGALL & CO., NEW YORK
LIMITED. & LONDON.
(ADDRESS CABLES CZARNIKOW, NEW YORK.)

112 WALL STREET.

CZARNIKOW, LONDON.
ZARNIKOW & CO., GLASGOW.
ZARNIKOW & CO., GREENOCK.
ZARNIKOW & CO., LIVERPOOL.

NEW YORK. November 26, 1909

M

TESTS OF SUGAR EX. S/S "BENDU"

MARKS	BAGS	Buyers	Sellers	Third (N.Y.S.T.L.)	Mean
◇	2000	~~96.5~~	96.7	96.65	96.675
	2000	96.6	~~96.8~~	96.65	96.625
	2000	96.5	96.7	96.6	96.60
	2023	~~96.6~~	96.8	96.95	96.875
	8023				96.6943
G V	1800	97.4	~~97.3~~	97.45	97.425
	1877	~~97.2~~	97.4	97.35	97.375
	3677				97.3995
PL	576	97.0	97.0	~~96.85~~	97.00
JJ	200	96.3	96.3	~~96.35~~	96.30
V D	280	97.2	97.2	~~97.1~~	97.20
G G	189	~~95.9~~	96.0	96.05	96.025
E P P	810	95.1	95.1	~~95.05~~	95.10
C M S	360	~~95.4~~	~~95.8~~	95.55	95.475
D B K	882	~~95.9~~	96.0	96.05	96.025
M X R	550	95.7	95.7	~~95.55~~	95.70
I C G	965	95.4	95.6	95.5	95.50
S W R	542	~~96.4~~	96.6	96.55	96.575
T O V	200	96.4	96.4	~~96.45~~	96.40
C J	76	96.6	96.6	96.55	96.60
		96.34	96.44	96.40	96.41

17,330 bags ----------Avge-------- 96.6109

FIGURE 9.2. A statement of polariscopic tests of a large cargo of sugars. For each mark, the importers recorded the values reported by buyers, sellers, and the New York Sugar Trade Laboratory, and which one had been rejected. From New York Sugar Trade Laboratory folder, box 30, Papers of Charles Albert Browne, Library of Congress.

on a kind of competitive pressure upon inaccuracy rather than an active pursuit of accuracy.

The presumption was that the buyer's chemist would polarize low, the seller's high, and the laboratory would come somewhere in the middle. During those first weeks of December and January it sometimes came in lowest of all. Browne attributed this to the fact that many sources of error in sugar measurements tended, in the hands of "more or less negligent" chemists, to push results higher. One refiner who had "taunted" importers that their chemists could no longer play favorites now found himself berating his hired chemist for polarizing above the Trade Laboratory. Not infrequently an embarrassed buyer would dispatch his own chemist to the laboratory's rooms on South Street in Manhattan to demand that Browne retest some sample, though it rarely worked out in the plaintiff's favor.[21]

Exactly the Right Degree

By the time the laboratory was founded, calm had settled on the legal front, but the refiners had not conceded the point. In 1906, both Wiley and Wiechmann reviewed a book called *The Polariscope in the Chemical Laboratory*, by the MIT chemist George Rolfe, who spent half his year supervising production at the largest sugar factory in Puerto Rico. Both reviewers praised the work on most counts, but each took the opportunity to lob ordnance at the other's position, and devoted disproportionate space in their limited lines to the temperature question. Writing in *Science*, Wiechmann disliked the book's endorsement of what he referred to as "the alleged influence of temperature on the specific rotation of sucrose."[22] Had Rolfe somehow not heard, Wiechmann asked, of how all those careful investigations had been punctured in federal court? Or of the scandalous finding that the "so-called corrections made to counteract the alleged influence of temperature" had resulted in so many adjusted readings over 100? Taken aback by the vitriol, Rolfe belatedly replied from Puerto Rico that a less angry reviewer might have noted his belief that temperature's net effect was negligible.[23] Meanwhile that belief was exactly what Wiley, in a different journal, chastised him for his "opinion that the sum of the errors of ordinary commercial work is practically zero. . . . This observation could hardly be applied with justice to the ordinary polarization of sugars for dutiable purposes."[24]

Between May and September of Browne's first year in Manhattan, he recorded that the average daily temperatures ranged from a low of just over 22° Celsius to a high just shy of 28°. He measured variations during cooler and hotter times of the day and in different parts of the building, and told the trustees that temperature errors could be cut in half by testing in the basement.[25] Almost offhandedly, and while noting that neither attorney had actually suggested this solution, the district court judge had sortied into the temperature debate itself: "No reason is perceived why," he wrote, "if the government desires to secure uniformity and accuracy, it should not take its tests,

FIGURE 9.3. The polariscope cabinet in the constant-temperature room of the New York Sugar Trade Laboratory. Charles Albert Browne, *A Handbook of Sugar Analysis: A Practical and Descriptive Treatise for Use in Research, Technical and Control Laboratories* (John Wiley and Sons, 1912). Courtesy of HathiTrust.

or provide that they should be taken, at the temperature at which its instruments are standardized."[26] This was not an original idea.

Relief came once the Supreme Court had rejected the last appeal, almost a year to the day after the opening of the laboratory. Browne no longer had to publicly disclaim for the benefit of his employers that that the effect of temperature was uncertain. News of the court's decision had barely reached New York City before Browne went to his trustees and requested the installation of cooling equipment, at more than twice the price of the initial laboratory's "fitting up" costs. The result was a "constant-temperature room," maintained at 20° Celsius throughout the working day. It could hold three specialized sugar polariscopes called "saccharimeters" along with five chemists, and it had enough space for "polarization in duplicate" of 150 samples a day.[27]

The indefatigable Browne then began a campaign to make his own elaborate solution appear both indisputable and indispensable. In the summer of 1909 he published a paper in the *Journal of Industrial and Engineering Chemistry* titled "The Use of Temperature Corrections in the Polarization of Raw Sugars and other Products upon Quartz Wedge Saccharimeters."[28] It related the whole history of temperature correction debates at international congresses as well as the course of the lawsuit against the government. He invariably described the work of Wiley and his allies as meticulous, and while acknowledging Wiechmann's critique, he snuck in a jab at the legal and scientific counsel given to the American Sugar Refining Company. The companies had flatly argued that sucrose's rotation did not alter with temperature and thus that the government had been wrong to issue its corrective formula. The absurdity of the more-than-pure sugars matched this critique. But they had lost.

The real issue, Browne now claimed, was that both sides had treated pure sucrose as the relevant object. What they ought to have talked about was not sucrose but sugar. Yes, sucrose's specific rotation altered with temperature, but so did that of its invert products, glucose and fructose, as well as other substances that were present in worldly sugar, and each of these affected light to a different extent and even in different directions. Browne was baffled that neither the government nor the firms seeking relief had submitted any evidence about what happened to raw sugar under changing temperatures. And "unfortunately," he wrote in his 1912 handbook, the outcome of the court case "seemed to many chemists sufficient authorization to use such corrections indiscriminately in the polarization of any and every kind of sugar-containing material."[29]

He reminded his readers that on a saccharimeter, the scale should read 100° when the sample was pure sucrose. But raw sugar was not just less pure. It was also more impure. It contained other substances that produced their own effects on the rotation of light. "The polarization of a raw sugar upon this scale no more indicates the actual percentage of sucrose than it does in the case of molasses, honey, commercial glucose, condensed milk, or any other of the numerous products which are examined by means of a saccharimeter."[30] No single mathematical correction would suffice, he argued, and any corrective formula would need itself to be corrected, depending on the idiosyncratic constituents of the sample in the tube. This was a poison pill. Browne's goal was to show that no correction formula was realistic. The superior answer, and the only one not subject to later criticism, was to work at a standard temperature, as he was now able to do in his chilled room.

Within a few years the Treasury would begin installing equipment for temperature control. The Bureau of Standards had such a room.[31] But this was beyond the ability of most producers of beet and cane.[32] The blueprints for the laboratory building at Central Guanica in Puerto Rico, one of the largest and newest in the world at the time, show no sign of any cooling equipment.[33] So "the majority of sugar polarizations in sugarhouse and trade

FIGURE 9.4. The machinery Browne ordered installed to maintain its constant temperature. Charles Albert Browne, *A Handbook of Sugar Analysis: A Practical and Descriptive Treatise for Use in Research, Technical and Control Laboratories* (John Wiley and Sons, 1912). Courtesy of HathiTrust.

laboratory will have to be made at room temperature, whatever that may be," said Browne, and thus it was important to understand better how much tropical tests would vary in practice.

Historians of chemistry remind us that laboratories are "complex artefacts created for utility, [not] stage-props in front of which well-known chemists perform."[34] It took a lot of skilled human work to make it seem as though Browne had installed a self-correcting mechanism for the difficult ontological and epistemological question of temperature.[35] For instance, although Browne reported that the machinery could reduce the temperature in the air from 26° to 20° Celsius, he found "considerable lag" among the instruments and laboratory materials, which took several hours longer to cool each morning. The solution he arrived at was to turn the refrigeration equipment several degrees below 20°, and then keep an eye on the polariscopes. Once their internal temperature reached 20°, he would kill the circulating fans "so that the temperature thereafter is kept at exactly the right degree." During the two or three hours it took the polariscopes to come to temperature, he

FIGURE 9.5. The author's saccharimeter, the 1907 model designed by the Bureau of Standards for the Treasury.
Author's photograph.

and his assistants would weigh samples and mix solutions, presumably on the assumption that these far less massive items would arrive at the set temperature sooner. And the cooling apparatus needed to be in a separate space and behind an insulated partition or else the machinery itself would overheat the rest of the laboratory. Like the supposedly self-sampling sugar factory, the constant-temperature room was an automaton that hid its human governor.

Of course, the thermal state of an instrument's guts was not easy to observe. Happily, the New York Sugar Trade Laboratory possessed the same model of saccharimeter as had been recently developed for and adopted by the customs service. The Bureau of Standards's chief sugar scientist, Frederick Bates, had worked out a gear arrangement to satisfy several once incommensurable demands: It permitted enough bright light to penetrate darker sugar solutions and it allowed natural rather than monochromatic light, while also maintaining enough sensibility for fine measurements, and without constantly resetting the zero point after each use. One of its less celebrated but still innovative features was a thermometer mounted directly to the metal case holding the quartz wedges. Only thanks to that thermometer was Browne able to switch off his refrigeration fans at the right moment. But these polariscopes were far from standard issue. Bates had arranged them to be made exclusively by the Frič Brothers firm in Prague, and they cost $900 each, about four times the price of a standard German instrument and as much as ten times the price of an entry-level model.[36]

The exclusivity of these instruments made them yet another device by which North American chemists in general, and especially Browne from his central position, could claim superiority for numbers backed by money. It was one small but important part in the decisive shift to consolidate metrological power by the backers of the New York Sugar Trade Laboratory, at the expense of a diffuse competitive model of rival analysts operating and measuring for their patrons.

Standardized by Myself

The district court judge had been impressed that, according to the Trust's counsel, no other government in the world had incorporated mathematical corrections for temperature into its use of the polariscope. "Except," he noted, "in British Guiana." There was a human and climatic "geography of precision" that itself shaped perceptions of how that climate and geography affected the science of polarization.[37] Northern skeptics could point to various ways in which the tropics degraded instruments designed in and for temperate climates, and as always, criticisms of an instrument were always criticisms of the people who claimed to use it. The most pointed of these were the exchanges between Ferdinand Wiechmann, the chief chemist of the American Sugar Refining Company, and John Harrison, the government analyst and head of the sugar laboratory in British Guiana, which took place in the years surrounding the customs case.

Harrison was called as a government witness in the sugar adjustments trial, and in June 1900 the *International Sugar Journal* had reported that Harrison had supplied testimony for the United States that was compelling, even decisive. "We expect shortly to have more to communicate on this question," the journal said mildly, "regarding which some divergence of opinion has existed."[38] Much of that "divergence of opinion" had appeared, in original or republished form, in the widely read pages of the same *Journal*, where Harrison had argued, based on decades of his own experiments in Guiana, that sugar did indeed polarize slightly lower at higher temperatures. Meanwhile, Wiechmann had reported back to American chemists from the Vienna conference in 1898 where Wiley had first raised the issue that polariscopes might encounter difficulties in the tropics, difficulties not foreseen by their makers in continental Europe.[39]

On the one hand, Wiechmann charged that Harrison had not adopted his experimental procedure for the fact that he was in the tropics, as if he had forgotten he was no longer at university in England. His sugar samples would have been wetted by the air to different amounts in each test, which Wiechmann called a "disturbing factor" caused by "the damp climate of the colony where Professor Harrison's work was done." His instruments had betrayed him too.[40] "The wedges or the quartz parts of some of his instruments may have suffered stress," Wiechmann sympathized, "owing to the unequal rates of expansion of quartz and of the metal in which the quartz was set." Perhaps Harrison and others had also missed some solution expanding and escaping through the rubber seals of the polariscope tubes? "It is not surprising that under such methods of procedure . . . the readings at the higher temperatures should always be lower than the readings obtained at inferior temperatures."

Yet on the other hand, Wiechmann also charged that Harrison had adjusted his experimental procedure too much—that he had arrogated to himself the power to calibrate instruments. (In his missives to the journal the author styled himself "Ferdinand Wiechmann, Ph.D., Member, The International Commission for Uniform Methods of Sugar Analysis," which was a still-small group of mostly European beet-sugar chemists.) In particular, Harrison had tested each of his saccharimeters and found them wanting. "I have not found it possible to take readings with the makers' normal weights of chemically pure sugar at the makers' normal temperatures in this part of the tropics," Harrison had alleged.

This is where the distinction between polariscopes and saccharimeters becomes important. All saccharimeters were polariscopes, but not all polariscopes were saccharimeters. A general-purpose polariscope measured the change in the plane of polarization in pure angular degrees. That provided useful flexibility for chemists who might have to deal with a wide variety of organic and inorganic substances, each with its own rotatory power. But that same generality also meant extra work for sugar laboratories, which conducted essentially the same test hundreds of times a day. What they needed to know, and what a specialized instrument told them, was the change in the plane of polarization measured not from zero to 360, but on a scale of "sugar degrees," zero to 100, where 100 meant purity.[41]

Regraduating in sugar degrees rather than angular degrees offered several advantages. It saved sugar analysts from having to squint at tables to calculate the sucrose percentage equivalent of an angle every time they used their instrument. It spared their brains from arithmetic, their scrivening hands from cramp, and their ocular muscles from switching from paper to scope and back again, and it avoided errors from fatigue that would have led to disputes and appeals.[42] The saccharimeter applied scientific management and movement analysis to the sugar chemist. And when a laboratory performed hundreds of tests a day, tiny efficiencies added up.

In metrology, however, local convenience demands global annoyance. If the change in the orientation of polarized light were measured in angular degrees, the observer could read off the degrees, divide them by an agreed factor of degrees per gram, and thereby compute the grams of sucrose in the sample. But what he usually wanted was not the absolute number of grams of sucrose in the sample, but the grams of sucrose as a percentage of the total grams in the sample—that is, the purity. And the whole point of the regraduation was for this percentage to leap straight off the engraved scale of the saccharimeter. So each saccharimeter needed to know, in a manner of speaking, the mass of the sample that had been put into it. Such a mass, if composed entirely of pure sucrose, should theoretically read 100 on the saccharimeter scale. This mass was known as the "normal weight."

For three-quarters of a century, French, German, Dutch, and eventually American chemists argued in favor of their various national champions for what the normal weight ought to be. The choice of mass was in some sense arbitrary—any value could work just as well as any other, as long as each was paired with the correct 100° scale. Nonetheless, each country's sugar chemists claimed that their preference was justified on the basis of some combination of convenience and natural constants, and when every instrument was built, the sugar scale engraved or etched into it anticipated a particular normal weight.

This meant that saccharimeters around the world could all at once be rendered, or judged, inaccurate. The French scale had originally been defined by the optical rotation of a 1mm-thick quartz plate, 21°40′ from which the normal weight of 16.29g was derived in reverse. That derivation left the French Empire's sugar at the mercy of a mineral. "Instruments were constructed on this basis, and when it was later found that the rotation for 1 mm of quartz was somewhat higher, the original value [was] retained in order to avoid confusion of scales." The French scale stayed in place, but eventually more than twenty different versions of its normal weight existed, ranging from 16g to 16.51g. So a chemist's instrument stayed the same, but the saccharimeter had merely displaced his metrological task, since to stay accurate according to his peers, a sugar chemist now needed to keep up with the latest in normal weight research. The accuracy of a simple polariscope test rested on the individual experimental technique of its user. The technique of a saccharimeter user mattered less, and his deference to authority mattered more.[43]

The logistical task of distributing the physical weights themselves was not a trivial problem. These so-called normal weights came in highly abnormal figures. The original German weight was 26.048g, and neither that value nor 16.29g were standard issue in a set of laboratory weights. So saccharimeter users had to use special sugar weights for their saccharimeters, manufactured in these "normal" figures, or else laboriously combine analytical weights of various small sizes. But whenever normal weights changed, new sets of weights had to be ordered. Prominent sugar chemists almost made a sport of publishing results every few years in which they announced the latest normal weight was fractionally too high or low. A substantial amount of the critical response from colleagues was, essentially, that too many people would get confused and use the obsolete weights. New versions were sometimes issued in distinct shapes, round replacing hexagonal or vice versa, to avoid this mix-up. But of course even special sugar weights, too, began to degrade. Wiechmann, in his handbook for sugar chemists, urged that weights "should be verified from time to time, as they will, in daily use, unavoidably suffer some wear and tear."[44]

If metrology is "the creation of universality by the circulation of particulars," the saccharimeter only "worked" if it was surrounded by extremely particular particulars.[45] Harrison's point was that in the tropics, where these instruments needed to be used, this circulation broke down. "If we make up the makers' normal weight to the normal volume," he had testified, "sugars which should polarize 96 or would polarize 96 at 17.5°C, polarize from 95.65 to 95.7″ at the more typical temperature of 28°C.[46] So he had weighed his flasks carefully, then "ascertained by experiment for each instrument" what amount of sucrose gave a polarization of 100, and substituted this weight for the maker's own instructions—a new value that he then called the "true normal weight."[47] If it was irreverent to suggest that the instruments needed adapting to the tropics, then it was regicidal to claim that it needed to be done "for each instrument"—that different specimens of the same model, calibrated and certified by the same reputable maker, possessed idiosyncratic normal weights, and further that the differences between them were detectable with the tools available to an ordinary if highly esteemed chemistry professor in his Guianan laboratory. Yet he and his fellow West Indian chemists had been doing so for twenty years.

"It should be borne in mind," Wiechmann had said, implying that Harrison had not done so, "that the instruments used were all built and graduated in a climate much cooler than the one in which they were used by the experimentator [*sic*]."[48] This still left open the question of whether to defer to the cooler climate of the maker, or to the environment of the experimentator. Deference was meant to extend only one other way. At the July 1900 meeting of the International Commission in Paris, capital of metrology, the temperate-country sugar chemists had asserted that saccharimeters ought to be calibrated only with "chemically pure sugar," and further that "central stations shall be designated in each country which are to be charged with the preparation and the distribution" of the calibrating substance. "Wherever this arrangement is not feasible," however, a rule-abiding sugar chemist was to fall back on quartz plates.[49]

In what sorts of places might the arrangement prove not feasible? Wiechmann summarized the sentiment of the assembly: "The preparation of chemically pure sugar is not an easy task, and that in countries having hot climates, sugar is dried with difficulty and hence is not stable, and hardly available for transportation." Guiana was the sort of place where raw sugar might be made in a factory, where it was dried and rendered stable for transportation, and counterintuitively therefore it was precisely not the sort of place where chemically pure sucrose could be prepared in a laboratory. In making his own calibrations with what he claimed to be pure sucrose, Harrison thumbed his nose at the makers of his instruments and at the alleged hierarchy of who was properly situated to work with which sorts of substances. He even recalled being told, when he first arrived in Barbados, that "C.P. [chemically pure] beet sugar and cane-sugar did not polarise alike."[50]

Harrison did not dispute that he was challenging the ability and credibility of metropolitan makers to create saccharimeters robust enough for use in hot weather. What he claimed was actually more radical: that the tropics were not merely adequate places to calibrate sugar instruments but the ideal places to do so, and ought to be the true reference point of global sugar metrology. In a letter to the *Sugar Journal*, he observed drily that his laboratory was "situated in the course of the Trade Wind, and in which conditions of atmospheric humidity are remarkably uniform over considerable periods, far more so than they ever are in laboratories situated in temperate climates." Back in New York, questioned on this point by the counsel for the sugar business, Harrison had said, "As a matter of fact our temperature is practically stationary, *so it makes things simple for us*." Not only was it never 17.5°, it was "very seldom" below 28° or above 29°. Trying to lower the temperature more than a few degrees only condensed the humidity and left everything soaking.[51]

To those who took Harrison's side, the "true normal weight" was virtuous autonomy rather than vicious independence. Just as central factories needed to be able to cobble together enough parts to keep their mills grinding, thousands of miles from the original foundry, laboratory equipment had to be repaired if it broke lest the factory spin out of chemical control. Outfits like Harrison's were unusually self-reliant, able to repair their own instruments as well as those of others. Browne, for instance, who corresponded with Harrison and visited him at his laboratory in Guiana, complained often that polariscope makers ignored the needs of their users, especially "colonial" ones.[52] And when they did try to accommodate the needs of working outside the temperate zone, their efforts showed how little they knew. At least one German manufacturer, for instance, sold instruments with "specially designed tropical protective equipment of great complexity," but what the polariscopist actually needed wasn't fortification against the environment, but ease of cleaning out the instrument once the environment got inside. The protective cases around the polariscopic prisms "became in reality inaccessible moisture chambers, especially suited for the development of fungi." Mold could strike instruments as well as sugar itself.[53]

The sharpness of the "Trade Wind" jab was that there was no substitute for experimentation and practice in the tropics because that was where sugar was from and that was where the sugar factories were. As Harrison pointed out, he was simply "standardizing my instruments on chemically [pure] saccharose *at the temperature and under the conditions in which they have to be used*."[54] He could have mentioned a comment made by Wiechmann at the trial. Under questioning at the hearing, pressed about the reliability of a quartz plate, Wiechmann had defended his procedures. "It has been standardized by myself," he explained.[55] Only chemists in certain places were allowed to make such claims.

Instruments in Good Order

Measuring the temperature effect at all could only be achieved with high levels of human and technical surveillance. The horizontal consolidation of factories, centrales, and even refineries into ever larger conglomerates made this kind of knowledge about sugar possible. The sprawling Cuban-American Sugar Company, for instance, was controlled by some of the largest refiners in New York.[56] Its chief chemist was Guilford Spencer, who had been Wiley's assistant during the tests in cold storage. Spencer made an annual practice of calibrating the laboratories—that is to say, checking the work of his subordinates in each of the company's factories—by exchanging samples, similar to the practice in the customs service. In 1909, he sent a sample of the same sugar to six centrales, to his company's Cuban refinery, and to a "commercial chemist" in Havana, in addition to the Bureau of Chemistry in Washington, DC, and to Browne in New York City.[57] The sugar in the sample measured anywhere from 99.21 at Central Constancia in Cienfuegos, where it was 26.5°C, to 99.80 in the room at the bureau, where temperature was kept steady at 20°C.

Then Spencer asked each laboratory to adjust for temperature using the formulas Browne had recently published, which although imperfect were still superior to the Treasury's crude tables. The Cuban sites, at one of which the temperature had spiked as high as 33°C, wound up with corrected values about a third of a degree higher. But the annual troublemaker of Central Chaparra came in at a perfect 100.00, whose implausibility Spencer annotated "!". He was disappointed in his laboratories, his factories, his product, or really all three: "This sugar is so dry and mixed so readily that I hoped for closer results." In fact, as the high polarizations indicated, it was not normal raw sugar for American sale, but "a grade that we make especially for melting in our Cardenas refinery. You have doubtless seen similar sugar from the Hawaiian Islands," where factories made a higher grade for the Spreckels refineries in California.

The point is that only an octopus like the Cuban-American could corral the laboratories of six distinct centrales under the same procedures, let alone to test them by the same hand.[58] "The factories all have standardized quartz plates and flasks," Spencer complained. "In three instances I have put the polariscopes in order and tested them [myself]." And yet it was not enough. "At these places I am confident that the instruments are in good order, yet there is a difference of 0.2 between Luisa and Tinguaro and a greater one between the Refinery and Tinguaro." Browne agreed that these variances displayed the urgent need for temperature correction, first by fixing the formula and then by fixing the situation with a controlled-temperature room.[59]

The New York Sugar Trade Laboratory itself created new conditions for what counted as valid studies. Another of Browne's correspondents was George Rolfe, the longtime MIT chemistry instructor who spent half the year supervising production at Central Aguirre in Puerto Rico. There, he commanded "the manufacture, and incident chemical control, of nearly 30,000 tons of raw sugar under conditions exceptionally favorable for comparing polarizations."

In a paper published in the widely read *Louisiana Planter*, a little more than a year after the Supreme Court decision, he essayed such comparisons between Aguirre and New York. Studies in metropolitan laboratories, he agreed, were lovely specimens, but he agreed with Browne that the focus on pure sucrose was misguided: "No results of these investigations yet published can be formulated into any correction applying to raw sugars in general." What mattered to a factory was factory sugar, and factory sugar could only be studied in a sugar factory. As Rolfe put it, the New York Sugar Trade Laboratory had cut a Gordian knot with its temperature room, even as the customs chemists kept applying arithmetical corrections.[60] "Naturally it is of exceeding interest to know how polarizations will agree with those made at standard temperature, especially where, as in the tropics, the temperature is notably higher than standard," Rolfe wrote. While Wiley shivered in cold storage, Rolfe sweated on the south coast of Puerto Rico alongside workers and machines. And he made his results public, unlike chemists for other companies who kept information private as valuable corporate intelligence.

Rolfe had long argued that what mattered was not temperature's effect on pure sugar but the "total errors" of the instrument, and "the most consistent and fairest way to estimate the sugar value of a commercial product."[61] Now Rolfe worked hard to make his situation at Aguirre seem ideal for testing whether he and his colleagues in the tropics could produce numbers that ought to be trusted by "northern chemists," of whom he was one for half the year. So he boasted that his sugars "had a fine even grain and were classed as first," thus unlikely to decay, and that because these went straight from central to ship to refinery, they had no time to deteriorate. He noted how rigorously he disciplined his samplers and his method for tracking the numbers of bags. And most importantly, he fortified himself against accusations of instrumental malpractice, explaining how often he zeroed his saccharimeter, "a Schmidt & Haensch half-shade instrument which had been carefully compared at the time of its purchase in 1900 with the Institute of Technology saccharimetery no. 2880."[62] The result of all this care was that his polarizations matched remarkably closely to those of the Sugar Trade Laboratory. Even Browne was impressed, but he maintained that the only permanent solution was to take all readings at one temperature.[63] For the most diligent chemists in hot climates, the New York Sugar Trade Laboratory provided a fixed and authoritative point to which they might calibrate their own little kingdoms of chemical control. But they were only as reliable and accurate as Browne said.

There are more indices of the New York Sugar Trade Laboratory's influence. Browne's 1912 *Handbook of Sugar Analysis* went through many printings and thousands of copies, and a new edition in the 1940s was issued with contributions by his successor, F. W. Zerban. The book appeared on shelves of sugar centrals across the Caribbean.[64] In 1916, Guilford Spencer's manual for chemical control in the Cuban-American Sugar Company instructed its chemists to use "Browne's formulae" and gave a quick synopsis of the method, but included in the text the page number to direct readers to their own copies of the *Handbook*, which would have been within easy reach.[65]

And even though only the trustees were initially allowed in, that prohibition either loosened, or else fellow chemists were exempted, or else it was simply ignored. Browne had plenty of visitors. The chief chemists of all the major sugar companies, and many of their subordinates, passed through the doors, as did official chemists from the US government. This open door was good advertising. The laboratory was only a few rooms, and it would

FIGURE 9.6. One of Browne's subordinates mixing samples at the New York Sugar Trade Laboratory. Note the map of Cuba overhead. From New York Sugar Trade Laboratory Photographs folder, box 30, Papers of Charles Albert Browne, Library of Congress.

have been impossible to keep the visitors separate from the work. So when the chief chemist of a subscribing firm dropped by the laboratory to meet Browne, they would have watched his assistants bustle about polarizing sample after sample, would have felt the constancy of the temperature. Browne printed photographs of the laboratory in his *Handbook* and described its methods, but visitors could witness everything. He also used his position at the center of the sugar world to hand out positions on the periphery, building what Bruno Latour called "landing strips" for the arrival of his ideas.[66] Sugar chemistry was a transient business, as "sugar tramps" worked *zafras* in the Caribbean and then packed their bags for continental beet harvests during the back half of the year.[67] As people circulated through his rooms and his graphomanic correspondence, he found places for former laboratory assistants, or the students of professors at technical institutes, or simply correspondents. In filling these positions, he built up a reservoir of indebtedness.

In 1949, *The Sugar Molecule*, a promotional publication of the industry's trade group, ran a retrospective on the New York Sugar Trade Laboratory. "The serenity of the laboratory is in marked contrast," it said, "with the remembered bickerings of the years just before and after the turn of the century." The article is a textbook example of metrological myth. The word "integrity" appears three times, twice in combination with "unquestioned."[68] Precision, accuracy, and incorruptibility triumphed over a "colorful era," and "the original aim of eliminating disputes has been abundantly realized."[69] That was true, but not because of "a quiet laboratory in which scientists measure the quality of incoming raw sugar cargoes with meticulous care." Instead, as we have seen, it was because of the laboratory's ability, through its connection with the big money in sugar, to induce changes in the immeasurable and uncontrollable elements of polarization. Rather than propagating a greater degree of accuracy and precision, the laboratory, and the large sugar interests it represented, pursued a strategy of stamping out competition among self-evidently biased and inaccurate private chemists. In the next chapter, we will see how the existence of such a laboratory made possible a commodity exchange in raw sugar, and also how the laboratory's data facilitated one of the sugar refining industry's most subtle conspiracies in restraint of trade.

10

Ups and Downs

Standards Universally Adopted

Nature does not distinguish between raw sugar and refined sugar. In the twenty-first-century United States, for legal purposes, raw sugar is more or less defined as sugar meant for refining rather than direct consumption, and refined sugar is sugar meant for uses other than refining.[1] Part of the problem, as the historian and chemist Noël Deerr observed, is that in European languages the prefix re-, when attached to the past participle, is ambiguous: It suggests both "having been made so a second time" and also "especially so."[2] "Suffice it to say that with sugar, certain phraseology has become universally understood, and grades and standards universally adopted," wrote an economist in 1925. "The word 'raw' in the sugar trade of America is known as 96° Centrifugal. . . . This makes possible the transaction of sugar business by wire and in code. The buyer knows what he is to receive and the final consumer never needs to inspect his purchase, knowing he can depend upon trade names."[3]

There was also nothing natural about raw sugar being ninety-six degrees of purity. The origins of this number as a boundary remain unclear. It had no significance in the original proposals of 1878 for the Treasury's use of the polariscope.[4] But by the mid-1880s, Edwin Atkins was writing about "96 test," and in 1889 he had instructed a chemist that "95 ½ avg test is as high as I wish it is useless to make it higher."[5] It is possible that 96° sugar was optimally dry and relatively stable, but not so pure and white as to challenge refiners' control over the consumer market. In any case, as 96° became the default basis for contracts in trade, it spread upstream to production and then to agriculture. Centrales in Cuba pegged the payments in their colono contracts to the latest price of standard 96° sugar on the Havana market.[6]

FIGURE 10.1. A plan section of an idealized modern refinery, 1917: sugar arrives by raw steamer on the left, is crushed, melted, filtered, boiled, crystallized, centrifuged, granulated, packed, and shipped out on a steamer on the right. George M. Rolph, *Something About Sugar: Its History, Growth, Manufacture and Distribution* (J. J. Newbegin, 1917).

Sometimes things are "standardized" by making them the same. So to say that sugar was "a standardized product" at 96° could mean that all of it was really the same, or at least that it all read 96° on a properly calibrated polariscope. That is how most historians have understood the 96° number. "In the early years of the twentieth century a sugar purity of Pol 96° became the standard," wrote Manuel Moreno Fraginals, "whose origin (cane or beet) or the region it came from (Cuba, Puerto Rico, Java, Australia, Mauritius, Brazil) was impossible to determine."[7] Others have gone much further to suggest, incorrectly, that in factories, "the goal was to reach the industry standard" of sugar at 96 percent purity.[8] And when sugar people spoke of buying sugar "at 96," that is probably how most non-historians would understand they meant: Sugar had become a homogeneous product, it was all the same, and it was all 96 percent pure.

But some things are standardized in other ways—for instance, by placing them in the same frame of reference, or in commodity terms, a "basis." The standardization of sugar is this kind of standardization. "As regards sugar in the Caribbean, in the nineties everything was completely different from what existed in the sixties," wrote Moreno Fraginals.[9] But what had changed was not that sugar was all the same. What had changed was that, for the first time, many of the parties in the sugar trade had agreed on how to account for, and compensate each other for, the ways in which each cargo of sugar was different.

In a world where sugar was bought and sold "at 96" degrees, very little sugar was actually bought at 96 degrees. Consider the aggregated data from all 9,215 samples tested by the New York Sugar Trade Laboratory in 1908, its first full year of operation.[10] Only in the last three months of the year did the average polarization of arriving sugar meet or exceed 96 degrees. Every year exhibited an annual cycle during which the colder months brought the highest-polarizing sugars; the lowest sugars arrived in the heat of the summer. (As we saw in the previous chapter, it wasn't clear whether that represented a real change in the sugar or simply the temperature's effect on the polariscope.) In June, sugar couldn't even clear 93, and not even an eighth of the sugar that arrived that month was above 96. The average of all polarizations for the whole year was less than 95 degrees, and barely 44 percent came in between 95 and 97.

The sugar market dealt in small differences—refiners, just a few decades earlier, had been accused of bribing official chemists to undervalue sugars by just a few tenths of a percent, and the US Treasury and Congress had spent years investigating discrepancies of the same caliber. At the boundary of raw sugar and refined sugar, two degrees was an abyss. For the purposes and people that mattered, sugar in New York in 1908 was definitely not all the same.

Sugar remained an unreliable substance. The centrifugals imported to the United States by the beginning of the twentieth century clustered their qualities more tightly than ever. And yet, as the whole industry came to expect more sophisticated sugar, these small variations retained a large importance.

DISTRIBUTION OF SUGARS BY GRADE FOR EACH MONTH OF 1908.
Polarized at The New York Sugar Trade Laboratory.
(In Percentage of Total Number for the Year.)

Polarization.	Jan. Per cent.	Feb. Per cent.	March. Per cent.	April. Per cent.	May. Per cent.	June. Per cent.	July. Per cent.	Aug. Per cent.	Sept. Per cent.	Oct. Per cent.	Nov. Per cent.	Dec. Per cent.	Total. Per cent.
97–100	0.37	0.59	0.15	0.38	0.37	0.15	0.75	0.64	3.28	5.55	4.41	2.75	19.39
96–97	0.76	2.54	3.16	2.47	1.44	0.89	0.89	0.47	0.43	0.09	1.00	0.82	14.96
95–96	1.00	5.22	5.96	6.64	3.47	2.35	1.49	0.99	0.73	0.36	0.58	0.55	29.34
94–95	0.34	2.07	2.46	2.84	3.30	1.27	1.39	0.79	0.63	0.35	0.14	0.08	15.66
90–94	0.62	0.47	0.72	0.60	1.87	1.29	1.47	0.58	0.30	0.89	0.21	0.13	9.10
85–90	0.06	0.02	0.36	0.42	1.41	2.19	1.31	1.23	0.95	0.13	0.26	0.14	8.48
70–85	0.03	0.02	0.03	0.20	0.33	0.27	0.46	0.84	0.45	0.11	0.33	0.00	3.07
Total.......	3.18	10.93	12.84	13.55	12.19	8.41	7.71	5.54	6.77	7.48	6.93	4.47	100.00
Mean polar ...	95.00	95.52	95.21	95.04	93.81	92.78	93.21	91.59	94.28	97.13	96.87	96.55	......
Mean temp. C°..	20.37	20.50	20.42	20.46	22.14	25.68	27.93	25.38	22.38	20.22	20.59	20.25	......
Max temp. C°..	21.5	21.0	21.0	22.0	27.0	29.0	31.0	30.0	24.50	22.0	21.5	21.5	......
Min. temp. C°..	18.5	20.0	20.0	19.0	20.0	23.0	26.0	20.0	20.0	18.0	19.0	19.5	......

Total samples analyzed for 1908, 9,215.
Average of all polarizations for 1908, 94.68.

FIGURE 10.2. The polarization of raw sugar cargoes in New York, 1908, as recorded by Browne's laboratory. The average quality of sugar declined in the summer, but how much that was due to temperature and how much to other factors (such as the source of the sugar) was a difficult question to answer. C. A. Browne, "The Use of Temperature Corrections in the Polarization of Raw Sugars and Other Products Upon Quartz Wedge Saccharimeters," *Journal of Industrial & Engineering Chemistry* 1, no. 8 (1909): 567–80.

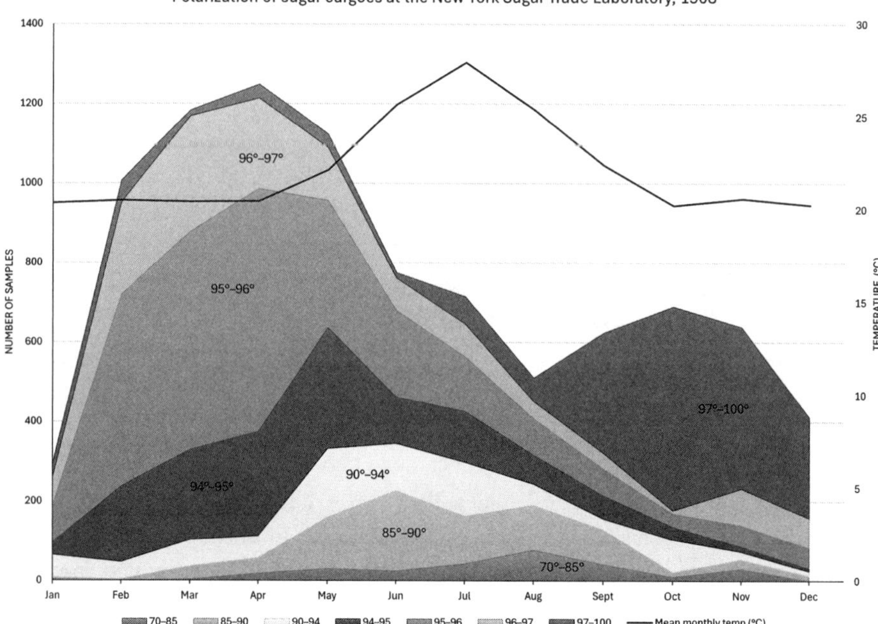

FIGURE 10.3. This graphical representation of the data in Figure 10.2 shows just how unstandardized sugar was. Only some of the sugar cargoes tested within a degree of 96°, the supposed trade standard. Samples below 95° peaked in May, and samples below 94° peaked in June. By October, over two-thirds of the samples were testing above 97°, as sugar arrived from the most advanced factories. It is worth noting that these data reflect the number of samples tested—one sample per mark, with a single cargo containing multiple marks—so they may not exactly reflect the distribution of polarizations across tons of sugar delivered. Data from C. A. Browne, "The Use of Temperature Corrections in the Polarization of Raw Sugars and Other Products upon Quartz Wedge Saccharimeters," *Journal of Industrial & Engineering Chemistry* 1, no. 8 (1909): 567–80.

Those facts together turned a convenient market fiction—that all sugar was the same and all 96—into a highly inconvenient one. The gap between physical reality and imagined standardization needed to be bridged for sugar to become a commodity fit for speculation and hedging on a modern exchange.

Outguess the Seller

The New York Coffee Exchange was created in 1882. For three decades it sold only beans, though occasionally it considered offering sugar contracts. But the Trust had conquered the raw sugar market.[11] As long as a monopoly or oligopoly controlled demand and set sugar prices, there was "an almost invariably steady tone of the market," according to its superintendent in 1931. A steady market meant no volatility, and without volatility nobody needed hedging, and hedging, not sales of physical goods, is the main business of an exchange.[12] There was no appetite for sugar contracts until the federal government began the process of breaking up the Trust in 1911.[13] The effort to add sugar futures was accelerated in late summer 1914, when war shut down the previous hubs for sugar contracts in Hamburg, London, Paris, and Amsterdam. By December, sugar futures were on the move, and within two years "and Sugar" had been added to the organization's name. Trading was suspended while the price of sugar was under government control, but by the Armistice the New York exchange was unassailably dominant, five times the size of any rival, and Hamburg, previously the largest, never even reopened.[14] Before the war, the speculative market in London had controlled the price of raw sugar in the United States; afterward, New Yorkers did, and "the prices of raw sugar that prevail in the Exchange are used as a basis for the prices of sugar in the markets of the world."[15] New exchanges were launching elsewhere at the same time, trying to profit from volatile postwar markets in agricultural and perishable goods.[16]

Most of the Sugar Exchange's contracts were used for hedging and speculation rather than as agreements for actual delivery. On every side of the sugar market, investors had reason to hedge. Planters plowed money into the ground now for uncertain sales later. Meanwhile, refiners wanted to lock in prices and supplies early to keep their margins tolerable, and the middlemen, the "operators" who both bought and sold sugar, wanted to cushion themselves against prices moving in either direction. Trading in commodities that

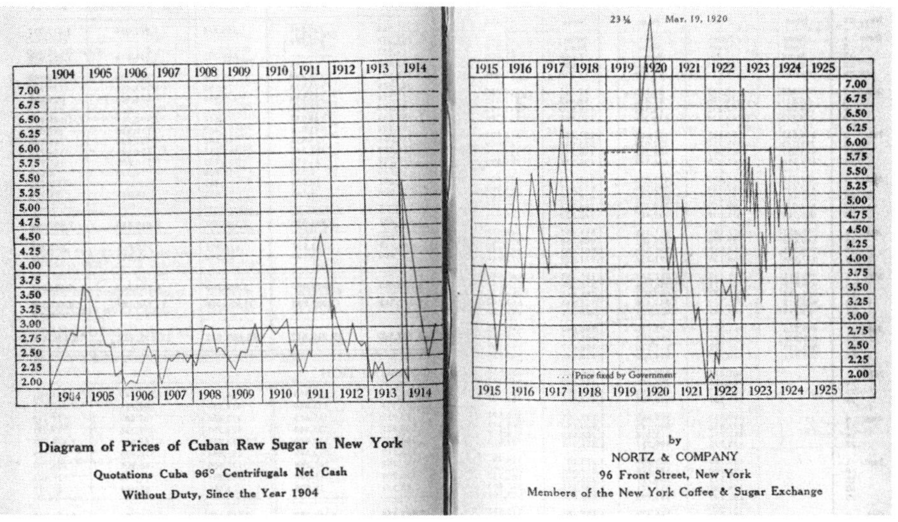

FIGURE 10.4. The price of 96° Cuban raw sugar in New York, 1904–1925. Nortz & Co., *Coffee and Sugar Facts: The New York Coffee and Sugar Exchange* (New York, 1925).

did not exist and were never meant to exist seemed to many critics an alien and repellent enterprise.[17] In part that was because these financial instruments had real-world effects. Contemporaries credited the ability to hedge with enabling greater production in Europe and Java, where managers now felt comfortable contracting for more cane and beet delivery, while American refineries and Cuban factories, even American-owned, had enough credit already. The quantity of sugar bought and sold on the exchange each year was still many multiples of the quantity of sugar Americans would actually eat. In the late 1920s, there was four times as much paper sugar as real.[18]

Meanwhile, less than 1 percent of the contracts traded on the exchange were settled by delivery, an even smaller proportion than some other contemporary commodity exchanges.[19] In 1922, for instance, about 6,000,000 tons of financialized sugar changed hands on the exchange, but only 55,000 tons were actually delivered through the exchange's mechanisms.[20]

The vast majority of actual physical sugar was sold what was known as "nearby" the exchange, on a cost-and-freight basis. In 1924, for instance, the New York Sugar Trade Laboratory made 38,680 tests, of which only 3,100 were for the exchange. The year before, they had made 33,803 tests, of which just 650 were for the exchange.[21] When the time arrived to possess or unload real sugar, a buyer or seller would settle their position on the exchange for cash, and then transact on the spot market, whose price closely, though not quite exactly, mirrored the exchange's.[22]

Yet there were reasons to take advantage of the exchange's delivery tools, as the large brokerage firm of Lamborn & Company assured current and potential Cuban clients. The small surcharge for exchange delivery was worth it for peace of mind, because "the Exchange provides certain guarantees to effectually safeguard the delivery and receipt of sugar. It provides certain machinery of delivery which is somewhat cumbersome in order that it be safe."[23] One source of that safety was the use of warehouses licensed by the exchange. Deliveries through the exchange came with warehouse receipts, while nearby deliveries frequently circumvented the warehouse.[24]

Futures contracts, however, derive their value from the fact that they are not quite pure bets but are entangled with the real world of goods. This entanglement, for instance, is what makes it possible to corner a market. By taking control of all the actual material due on a specific date, those running the corner acquire the power to name the price buyers must pay to avoid defaulting on their obligations.[25] Thus Sugar Exchange contracts still stipulated that physical stuff must be on a real ship, or on a real dock, on a specific date. In fact, the very inconvenience of the physical delivery of sugar was a tool used by buyers and sellers to force better prices for the rest of their speculative positions, even if it meant paying the exchange's delivery and handling surcharges.[26] Once a seller declared that he was fulfilling a contract through actual delivery, the buyer had just twenty minutes to figure out whether to sell on or accept it. As one broker explained to clients:

> For example, say a seller has sold 5,000 tons on the Exchange. The month for delivery arrives. If he cannot re-purchase at about the cost and freight market, he may issue delivery notice for say 500 tons. He will risk spending 11c to 16c per 100 pounds extra on 500 tons to improve his chances of buying back the other 4,500 tons at, or below, the cost and freight market. It may work. On the other hand, if the buyer thinks he can outguess the seller, he will accept delivery of the 500 tons. The buyer will risk spending 14c per 100 pounds extra on 500 tons, with the object of discouraging the seller and making him pay above the cost and freight market for the other 4,500 tons.[27]

In other words, each trade was a standoff, and the potential of actual delivery was a bullet. Whether buyer or seller blinked first depended on how the price had moved. That potential for real delivery was what distinguished the exchange from a casino. "Every future contract on the Exchange contemplates and provides for actual delivery, and actual delivery occurs unless the contract is off-set against another contract," its lawyers wrote.[28] The potential for real delivery also meant that what was formalized on the exchange was not idealized sugar but real sugar, with all its messy inconveniences and its "errors as yet incapable of exact control."[29] And one thing for which you could not buy a contract on the exchange was, it turns out, a simple substance called 96° sugar.

Cuban Basis

The No. 1 contract, the initial sugar contract traded on the New York Coffee and Sugar Exchange, was not a contract for 96° raw cane sugar. It was a contract for 96° raw centrifugal cane sugar *from Cuba*, "with additions or deductions for other grades according to the rates of New York Coffee and Sugar Exchange, Inc."

The contract defined sugar in terms of the Cuban crop because the island was the largest producer and supplier to the United States, and because it was by far the largest single producer of cane sugar in the world. On the eve of the First World War, when the exchange began trading sugar, Cuba accounted for over half the five million tons of sugar produced in the Americas, and more than a quarter of the world's crop. By the 1920s, it was a third of the world's crop, more than Java, Hawaii, Puerto Rico, the Philippines, Australia, Mauritius, and South Africa put together.[30] The scale of Cuban production meant that for sugar producers who wanted to guard against falling prices, or refiners who wanted to guard against rising ones, Cuba was the country to watch and fear. A bumper Cuban crop could crash the market, and a weak one could send it racing. The best way to hedge sugar was to hedge Cuban sugar, so it made sense to build the contract around it.[31] This followed the model that the exchange adopted for coffee, where it was Brazil's disproportionate supply that could wreck or make a market.[32]

A producer no longer had to consult Willett and Gray's *Weekly Statistical Sugar Trade Journal* but could simply look up the exchange's forward prices. The New York exchange incorporated so much of the Cuban crop that it broadcast a stronger informational signal. But the signal could sometimes be wrong, and speculative bubbles and collapses, like the notorious Dance of the Millions in 1920 and 1921, were facilitated by the fact that knowledge about the market passed through the exchange's pit. In 1923, the exchange blamed a 100 percent spike in sugar prices largely on a few adverse data points about the Cuban sugar crop.[33]

At the same time, Cuban sugar's idiosyncratic tariff position made this convenient contractual feature a little less convenient in reality. Sugar from Cuba paid a tariff, which domestic sugar from Louisiana did not, nor did sugar from Hawaii, Puerto Rico, or the Philippines, the areas the Supreme Court had notoriously concluded were "foreign in a domestic sense." But sugar from Cuba paid less of a tariff than sugar from countries such as Haiti or British Guiana, "foreign" in a foreign sense.

The contract in sugar was thus constructed to minimize tariff arbitrage. If someone bought fifty long tons of Cuban sugar at 4¢ per pound for delivery in one month's time, they did not want to discover that they had actually paid 3 1/2 ¢ per pound of Puerto Rican sugar, and half a cent to pad the seller's pocket. But the buyer didn't want to be on the hook for the extra duty that the government demanded on sugar from Brazil or Demerara. So sellers were obliged to pay a tariff allowance equal to the difference between what the sugar actually owed in tax, based on its polarization, and what the sugar would have owed, if that same polarization of sugar had come from Cuba.[34] The result was that not all sugar was bought and sold equally. Raw sugar from Puerto Rico arrived in New York first, so it was bought first and enjoyed "preference of price" from the refiners, yet that price was "always interpreted in terms of Cuban." Sugar from Haiti, the Dominican Republic, or elsewhere was usually sold in Europe or Canada, not the United States, even if it were owned by American capital. These sugars were only worth buying, and the burden of the tariff only worth assuming, if the Cuban crop had run out that year.[35]

The exchange also set other rules that the real goods had to follow. For coffee, these rules elaborately specified how beans could be delivered, how the institution would certify coffee samplers and graders, how they should sample and grade the coffee, and how members of the exchange could appeal graders' decisions to a board of arbitrators.[36] Coffee was graded on a scale of 1 through 8, where 8 was the worst coffee allowed to enter the United States, the average of Brazilian was never greater than 3, and no coffee was ever graded 1, as no coffee was ever perfect.[37]

Coffee was graded by a panel of coffee men. But, as the federal appeals court had noted, most sugar men did not know one end of a polariscope from another. So sugar's delivery and handling were specified by the rules as well, but as far as determining the polarization of this new commodity, the exchange deferred to outside and special expertise: the New York Sugar Trade Laboratory. The laboratory became, essentially, the exchange's house laboratory. According to the twenty-seventh sugar rule of the exchange's bylaws: "Raw Sugar samples shall be tested by two chemists, one selected by Seller and one by Buyer, and by the chemist of the New York Sugar Trade Laboratory as *the official representative* of the New York Coffee & Sugar Exchange, Inc" [emphasis added].[38] Even before the Sugar Exchange, these few rooms in lower Manhattan had exerted an outsized influence. And their authority and their credibility as issuers of scientific values depended on their legal and financial entanglement with the largest companies in the American trade. Now, they made the exchange possible.

The Allowances

For the Sugar Exchange to exist at all, the producers, importers, merchants, and refiners needed to have coalesced around a single number as the basis of trade. But it was hardly sufficient to have that number as a rough target. A futures contract was a binding agreement to deliver and to receive, even if it was unlikely to be consummated, and the rules of the exchange contract effectively governed the spot market in actual sugar. There could be no liquid market in sugar as a commodity, whether real or speculative, without buyers' "full confidence that, in the absence of occasional bankruptcy, fraud, or unintentional error, they would receive the specific grades contracted for."[39] But these grades could also not be too rigid. In the first decade of the Sugar Exchange, not once did the laboratory report that a year's sugar average as much as 96°. By 1923, the average had reached 95.98°, but the distribution still fluctuated month to month and cargo to cargo.[40]

Adjusting for this reality is what the "allowances" were for. The allowances were a list of standardized modifications to the contract price that dictated how much the buyer owed the seller, or the seller owed the buyer, if the sugar was more or less than 96° upon polariscopic testing. And, as we have seen, the sugar was never exactly 96°, so the allowances were not extraordinary measures but called upon in the ordinary course of business. The allowances were also sometimes called the "differentials," or even the "ups and downs."

By at least 1899, there were generally accepted allowances for 96°–basis sugar bought and sold in New York City. For every degree above 96° that the sugar tested, the buyer generally paid one thirty-second of a cent per pound, while the two degrees below 96° each cost the seller one-sixteenth of a cent per pound, and below that, three thirty-seconds.[41] These allowances lasted until the suspension of sugar trading by the US Sugar Equalization Board in 1918.[42]

Each degree of purity, in other words, did not cost the same. That was because refiners, counterintuitively, found it easier to clean impurities out of sugar below 96° than they did for sugar slightly above it. But this fact itself was not a natural property of sugar or of sugar-processing equipment. The basis had been made material; refinery machines were built to process sugar of 96° at maximum efficiency and minimum cost, and they were not as efficient if the sugar was much better or much worse.[43]

Although real sugar was rarely delivered through the exchange, it still needed all the rules of real sugar, including allowances. Hence the No. 1 contract's call for 96° Cuban centrifugal sugar "with additions or deductions for other grades according to the rates of New York Coffee and Sugar Exchange, Inc." Section 88a of the exchange's rules laid out those ups and downs. Unlike the previous allowances, these new rules accommodated a wide range of sugar prices. They were defined in fractions of a cent per pound, tabulated to the purchase price. Thus, after 1920, a pound of sugar priced from 3¢ to 3.99¢ at the 96° basis would cost .08¢ less for each degree it tested below 96° down to 92° (which was the minimum grade deliverable), .08¢ more for the degree from 96° up to 97°, and .04¢ more per degree above 97°. These increased as the price of sugar went up, though not as steeply as the core price. That is, for raw sugar that cost 6¢ per pound, each degree around 96° only cost .11¢ more or less, and if raw sugar spiked to 10¢ per pound, each degree was .15¢.[44] Although this accounted for variation in prices, the allowances had uncomfortable stepwise features, so that if the price of a raw pound moved from 3.99¢ to 4.01¢, the initial allowance jumped from .08¢ to .09¢.

The "nearby" sales were effectively controlled by the same rules as the exchange's allowances.[45] Sometimes this was unofficial, but it was often official.

Form 667

Raw Sugar Contract

Cubas—Cost and Freight

New York ______________ 192__

The American Sugar Refining Company

We have this day sold to you for account of

____

____

tons of 2,240 lbs. each of Cuba Centrifugal Sugar.
Delivery of five per cent. more or less than this amount is to be settled for at the market price of like sugars on day of arrival.

SHIPMENT to be made______________________________

DESTINATION by Steamer (or Steamers) to be named as soon as possible for______________

____

at buyers option to be declared before sailing.

AT A PRICE OF______________________________cents per pound, cost and freight, basis ninety-six degrees (96°.) average outturn polarization, net landed weights.

PAYMENT to ______________________________by ten (10) days sight draft for 95 per cent. of the invoice amount with shipping documents attached. Any balance to be paid after final settlement of weights and tests, and interest on same at rate of five per cent. to begin to run ten days from entry of steamer. Documents required to effect a prompt entry and discharge of cargo in the United States to be furnished by the Seller. Party in default in producing necessary papers for entry of sugar shall be liable for demurrage of the vessel and for actual expense incurred.

DELIVERY of the sugar to be made at a customary safe wharf or refinery, as directed by the buyer. If buyer orders the Steamer (or Steamers) to Boston to discharge, extra freight of ______________per 100 pounds over the New York rate to be for their account.

SAMPLES to be drawn mutually by buyer and seller. Three tests to be made of each sample of sugar, one by Seller's public chemist, one by Buyer's public chemist and one by the New York Sugar Trade Laboratory. The average of the two nearest polarizations to be taken as the final test. Settlement on each shipment to be made on the final tests, with the allowance per pound for each degree above the selling basis up to 98°, and per pound for each degree below the selling basis down to 91°, fractions in proportion, as per table on the reverse side hereof. But no sugar to be delivered below 91°, unless on discount terms mutually satisfactory to Consignee and Seller.

MARINE INSURANCE to be covered by______________________________from shore to shore including risk of lighters at ports of loading and discharge.

Accepted

____
Broker

____

____

Table of allowances at varying prices for raw sugar polarizing above and below 96° per degree. Fractions in proportion.

PRICE		96° to 91°		96° to 97°		Above 97°	
C & F	Duty Paid	C & F	Duty Paid	C & F	Duty Paid	C & F	Duty Paid
Cents	Cents						
1.00— 1.99	2.00— 2.99	6 pts.	8 pts.	6 pts.	8 pts.	3 pts.	5 pts.
2.00— 2.99	3.00— 3.99	7 "	9 "	7 "	9 "	3½ "	5½ "
3.00— 3.99	4.00— 4.99	8 "	10 "	8 "	10 "	4 "	6 "
4.00— 4.99	5.00— 5.99	9 "	11 "	9 "	11 "	4½ "	6½ "
5.00— 5.99	6.00— 6.99	10 "	12 "	10 "	12 "	5 "	7 "
6.00— 6.99	7.00— 7.99	11 "	13 "	11 "	13 "	5½ "	7½ "
7.00— 7.99	8.00— 8.99	12 "	14 "	12 "	14 "	6 "	8 "
8.00— 8.99	9.00— 9.99	13 "	15 "	13 "	15 "	6½ "	8½ "
9.00— 9.99	10.00—10.99	14 "	16 "	14 "	16 "	7 "	9 "
10.00—10.99	11.00—11.99	15 "	17 "	15 "	17 "	7½ "	9½ "

Allowances at prices higher than above to be on same proportionate basis.

FIGURE 10.5. A contract for delivery of sugar to the American Sugar Refining Company in the 1920s, virtually a copy of the No. 1 contract on the Exchange. The back listed the same table of allowances as established in the Exchange's rules. From Sugar (Polarization)—Refiners' Committee on Sugar Differentials, Sugar Institute—Correspondence, Braga Brothers Collection, George A. Smathers Libraries, University of Florida. Available online at Digital Library of the Caribbean, dloc.com.

The blank form of the American Sugar Refining Company's purchase contract from the 1920s shows the same instructions as the No. 1 contract on the exchange, and the table of allowances is printed on the back.

By the late 1920s refining technology had improved, and refiners were convinced they were paying too much for raw sugar that was, in essence, too pure. "Every month in 1927 showed an average polarization above 96, for the first time in the history of the Laboratory," marveled F. W. Zerban in his annual report for that year, and in the prior decade the average had gone up a full degree.[46] A research chemist rooted for purer sugar like a forecaster roots for a tornado.[47] But refiners were unhappy. "With the differentials now in effect," complained F. E. Sullivan of the Western Sugar Refinery in San Francisco, "sufficient raw sugar of 97° polarization to produce 100 pounds of refined costs us a little over three cents more per bag of refined than an equivalent amount of raws polarizing 96°"[48] For at least forty years, refiners had been attempting to escape the competitive trap of their business, as the economic historian Alfred Eichner wrote, "changing the rules of the game when they thought that the rules worked to their own personal destruction."[49] Now they did so again.

In 1927 fourteen of the fifteen cane sugar refiners formed a trade association called the Sugar Institute, the latest attempt to escape the competitive trap of refining by forming a cartel. This time, rather than conspiring in private as a trust, the institute form allowed the refiners to collude in public, through sharing information and "promot[ing] uniformity and certainty in business customs and practices," until it too unraveled before the courts.[50]

One of the first acts of the new institute was to form a committee to develop new differentials, soliciting input from some Cuban and Puerto Rican producers.[51] The importers proposed a flat rate, where every degree from 93° to 97° cost the same amount, and the minimum deliverable was lowered to 91°. But instead refiners adopted steeply sloping rules. They raised the minimum deliverable polarization to 93°, shifted the allowances to direct percentages of the price, and reduced the allowances as the sugar got better. That is, starting with a 96° basis, the first degree of loss cost 1.6 percent, the second 2 percent, and the third degree from 94° to 93° a full 2.5 percent. But sugar greater than 96° did not gain the seller in the same proportion: 1.5 percent up to 97°, and just 1.25 percent to 98°. Above that, there was no compensation at all.

The producers of raw sugar were not as coordinated as the refiners, and their lack of leverage translated to even these minute technical deliberations. "Some sellers will accept any kind of a contract that any refiner puts out, and all sellers will accept the worst kind of a contract at certain times," sighed C. J. Welch of Welch, Fairchild & Co., an American financier-owner. "The desire to sell is greater than the desire to buy, as a general thing, and consequently sellers are at the mercy of refiners."[52] Still, producers yelped that they needed months to accustom themselves to this new reality, since they had to puzzle out which sugar was no longer profitable. A factory's internal reporting, for instance, included not only how much sugar it had produced and the average polarization, but also how much sugar it would have produced, if it had been able to produce all of its sugar at exactly 96 percent sucrose content.[53] Calculations of profit and loss often hinged on impossibly detailed questions. Welch had once calculated that drying raw sugar in an aerating device called a granulator would increase its polarization and thus its price, but the change would actually cost more money than it earned, "because when sugars are dried in a granulator you can no longer pack 325 pounds in a bag, and the extra expense for bags entirely offsets the gain that would have been occasioned by the increased polarization."[54]

From New York, meanwhile, the president of the Sugar Exchange also cautioned the refiners that the trade in paper sugar now vastly exceeded their own consumption. "In the old days refiners were the only buyers," he reminded them, whereas "contracts on the Exchange for a period of thirteen months and any sudden change in allowances would affect unfavorably many thousands of tons now under contract."[55] The new rules were delayed, by universal consent of buyers and sellers, until May 1929. In their pamphlet addressed "To the trade," the committee claimed "that the above differentials are as nearly equitable as can be arrived at adjusting for polarization alone." In fact, despite their overwhelming leverage, the refiners professed benevolence. "Under certain conditions these differentials may be slightly more favorable to the producer than to the refiner in that they may discourage the production of sugars testing over 95°," said the committee, and it unsubtly warned those producers that in such a case the refiners would go back to the drafting table, this time probably without them.[56] "Refiners want a new contract," conceded Welch, "and they are going to get one."[57]

In order to write such standardized allowances, and in order then to have the data to modify and adjust them, the sugar refiners needed data that only the New York Sugar Trade Laboratory could supply. It was the sole institution with a panoptic view of the entire New York market (and even into the smaller markets like Boston or Philadelphia, which frequently sent samples for arbitration). The Trade Laboratory kept statistics about what amounts of which polarizations of sugar were entering when, and conducted research, especially during the slow seasons, into how sugar behaved under the actual conditions of its storage and transport. This research combined with those statistics to provide refiners, manufacturers, importers, and others with the basis to argue over the size of the allowances from 96°, and what would happen to the polarization of sugar over time. In the summer of 1925, for instance, enough high-grade Philippine sugars began to arrive in New York that its normal monthly polarization averages no longer effectively represented the Cuban and Puerto Rican crop. The directors thus ordered the laboratory to distinguish among cargoes from these three territories and "miscellaneous (Santo Domingo, British West Indies, etc.)."[58]

"The provisions of the by-laws and rules governing deliveries are designed to ensure proper grading, classification and uniformity," the exchange attested to the Supreme Court.[59] Without such rules, there was no commodity exchange. And there could be no such rules without an institution like the Sugar Trade Laboratory. The laboratory made the forward purchase of sugar, and thus futures contracts, possible, by playing two crucial roles. First, and most obviously, by its continued existence as a referee. The exchange—which joined the subscriber list of the Trade Laboratory as soon as sugar trading began—now had a single arbiter to which its rules could point.[60] As the Trade Laboratory had discovered in its initial assessment, chemists might be as far apart as 0.6° when simply testing sugars blind, not even with a patron's interest in mind. That margin would have overwhelmed any table of differentials.

Second, the refereeing function as imagined by the laboratory's organizers had already exerted an influence on the city's private chemists—and through the temperature room, on sugar chemists elsewhere. As we have seen, the laboratory's methods calibrated the polariscope's users more than they calibrated polariscopes themselves. The work between 1907 and 1914 had already policed the variability out of polariscopic testing to the point where it was within the individual personal equations and allowable margins of error. It was so tight that the Trade Laboratory's polarization came outside the other two fully one-sixth of the time.[61] The laboratory's policing of the trade had helped mitigate the fact that one cargo of sugar was not quite the same as another yet still had to be bought and sold in a comparable way.

A Question for the Refiners to Answer

Using purer raw sugar as an input was, within the limits of a degree or two, a shortcut to building a larger and more capacious refinery. "Our experience has been that as long as a refinery has a fair grain of raw sugar, the only particular advantage accruing to it from having sugars above 96° *is whatever increased capacity it may give*," wrote the Western Refinery's Sullivan.[62] In periods of relatively high demand, higher-polarization raws would benefit a refining company, allowing it to cube and box more refined sugar and sell more of it to consumers.

But the late 1920s was not such a period. The retail price of sugar was low and sliding slowly lower.[63] The reason that the refiners formed the Sugar Institute in 1927 was the same reason that refiners had joined the Sugar Trust in the 1880s and incorporated the American Sugar Refining Company a decade later. There was too much refining capacity in the United States, not too little—50 percent too much, according to refiners.[64] "What is needed is increased consumption and a halt in the growth of production," was the line that *Scientific American* passed on from the publicity department of the American Sugar Refining Company.[65] And the surest way to raise profits at a time of overcapacity was to coordinate the refineries' outputs. "Practical" sugar refiners knew this without even having to discuss it.[66] If they dialed back production at the same time, the industry as a whole would earn more, and then they could distribute the greater profits of oligopoly. The trouble was that each individual refinery had every incentive to cheat by ramping back up its own production to sell at the higher price. This incentive made cartels inherently unstable unless they could find a way to police themselves. So the tricky problem for the sugar industry and its clever lawyers was always finding a means of enforcing oligopoly while remaining on the sunny side of antitrust law.[67] The new Sugar Institute, even its own counsel admitted, practiced a degree of coordination that was legally questionable, and its members were terrified of even discussing prices at its meetings.[68]

But the Sugar Exchange was assuredly legal. The 1923 spike based on false information about Cuba had been "violent" enough that the federal government had prosecuted the officers of the exchange itself, charging them with an illegal combination to manipulate the price of sugar, and asserted that the rules were really designed to inhibit physical settlement rather than facilitate it. The swaps of fictitious sugar were simply speculative, the government said, with "artificial and unwarranted prices, not governed by the laws of supply and demand." It sought to enjoin the exchange from even "publishing the prices of raw or refined sugar in Exchange transactions as purporting to be its market price," so long as paper sugar was not directly correlated with real sugar. No sale was real "unless the person purporting to make such sale has in his possession or under his control a supply of sugar adequate to meet the requirements of such transaction, and the person purporting to purchase shall in good faith intend to buy and pay for such sugar and accept delivery as soon as the same can be made."[69] But the government had lost, and the Supreme Court had endorsed the exchange as a practically and philosophically legitimate enterprise.

In rewriting the ups and downs that would govern the paper sugar of the exchange—and thus the nearby spot sugar market in sugar cargoes—the refiners of the Sugar Institute had found a way to adjust, in a single stroke, the financial calculus of the entire industry. Lowering the allowance on sugar above 96° might appear to have been an attempt to make better raws cheaper for refiners to buy. But in fact it was an attempt to make better raws unavailable for refiners to buy. If it was no longer worthwhile to Cuban and Puerto Rican centrales to make sugar much above 96°, they would not make sugar much above 96°. Refiners could not buy something that was not for sale. And so, most importantly, refiners would not have to worry that other refiners were buying it nearby. By adjusting the polarization allowances to disincentivize the production of sugar much above 96°, the refiners, via their institute, had found a self-policing mechanism for limiting their industry's capacity across the board—both each refiner's own capacity and that of his competitors. No one could undercut the cartel by buying 97° raws if such raws did not exist. So they set out to "force the polarization down."[70]

Such a subtle conspiracy to shape raw sugar output was unlikely to suffer a legal challenge if it was filtered through the polarization allowances. The allowances, the Supreme Court had ruled, were part of a commodity exchange's mission of ensuring uniform quality and grades, the means by which this exchange fulfilled its economically important function to create a free market in sugar. The plot was unlikely even to be discovered. In decades of trying, the federal government had never been able to decipher the economics of refining. (Some critics even asked whether most refiners understood their own business given their persistent inability to turn a profit without a cartel.) Congress and various commissions had heard endless testimony on the matter, and asked countless questions of the country's refiners, and the consistent result was that the questioners were more confused at the end than at the beginning. When the Treasury had been charged with calculating the refining drawback, it had struggled to understand the business too. "What is sugar of different polarizations worth to a refiner?" Welch asked. "That is a question for the refiners to answer—and they will answer it in their own way—and it is pretty hard to reply to that."[71]

And indeed, the federal government does not appear to have figured it out. When the Justice Department sued the Sugar Institute for restraint of trade, in 1931, its complaint included the development of the polarization allowances. But the government only charged that the refiners illegally settled on "uniform and arbitrary differentials for raw cane sugars purchased for their respective refineries . . . and agreed to restrict their allowances on high quality cane sugar."[72] The charge was instead simply that the allowances had been developed unfairly, without due consultation with the raw producers. The district court judge did not concur even with that limited accusation. He noted that there had been plenty of discussion before the institute was formed. And substantively, the judge ruled "that the new arrangement was scientific and entirely fair, that any lack of representation was due to inadvertence and in any event was unimportant, [and] that scales of allowances similarly negotiated and agreed to were customary and practically necessary."[73]

The lower courts left the institute intact but prohibited the refiners from engaging in dozens of anticompetitive activities. The refiners appealed, and the government's subsequent brief to the Supreme Court did not even mention allowances at all.[74] It never explained why the refiners would have chosen to restrict the high-quality raws, nor why they would have wanted to alter the raw sugar produced in Cuba and Puerto Rico, and certainly not that they had done so in order to create mutual assurance that they would reduce the collective capacity of their own refining.

In 1936, the Supreme Court affirmed the lower court's injunctions and scotched for good the institute's experiment at evading the Sherman Antitrust

Act. But until then, the allowances seem to have worked. The refiners' committee had placed the first differential, on 97° sugar, at 1.5 percent. That was exactly the level of differential at which the producers had said they would not make sugar much above 96°.[75] And once the new allowances were in place, the average polarization of American raw sugar rose much more slowly than before, and stopped well short of 97°, just as the refiners intended.[76] Refiners' profits stabilized. In the early 1930s raw sugar tested by the New York Sugar Trade Laboratory leveled out around 0.2° shy of the mark, which in the fine margins of the business might as well have been a mile.

As of 2023, the No. 16 contract on the Intercontinental Exchange, the successor to the Sugar Exchange, still places testing in the hands of the New York Sugar Trade Laboratory.[77] Histories of commodities pivot on the point at which a powerful central organization sets the standards for distant suppliers and buyers. First nature is squeezed through these standards and extruded as the interchangeable objects of second nature, and that process makes a global trade possible. Yet the market in sugar and ultimately in sugar futures became liquid, and limpid, not because a laboratory set the terms for the similarity of things, but because it regulated how their dissimilarities would be measured. And it did so not by setting technical or material criteria for the qualities of commercial objects themselves. Rather, it socially policed and induced changes in behavior on the part of the fallible, and even consciously manipulative, human beings who assessed those qualities. As ever, there was an apparent simplicity to purity that hid the ambiguities of real sugar, and those who harnessed the ambiguities of measurement could take advantage of that appearance. The refiners understood sugar, they understood how to shape the market, and they understood how to hide their own knowledge behind the polariscope.

11

Final Receipt

Just Appreciation

But some of the refiners knew how to "pick their raws" better than others. Refiners, collectively, were always conspiring to suppress total capacity and share their increased profits. Yet cartels fear betrayal for good reason. Refiners were also always squirming against the very restrictions they imposed on themselves and manipulating the science and craft of sugar against each other. If they happened to defraud the government, that was collateral damage and an accidental bonus to their accounts. The suppression of raws above 96° was meant to reduce the capacity of the whole trade, and that simultaneously created opportunities for any refiner who could undermine the deal without others knowing.

"To grade a sugar on the polarization alone is, of course, all wrong," protested a superintendent at the Cuban-American Sugar Company during the writing of the new allowances, "because it is known that different factories turn out sugars which differ very much in refining value."[1] When Browne had warned that there was no single formula for temperature correction, he had argued that the various impurities in raw sugar meant that one cargo of 96° sugar was not in fact just like any other. This point was more than a chemist's pedantry. During the same differential discussions, an unnamed expert working for an unnamed East Coast refinery submitted a remarkable seven-page memorandum, titled "The Valuation of Raw Sugars," that explained exactly why.

This expert was troubled by the seductive simplicity of allowance tables because these adjusted for polarization, but for none of the other factors that determined its "refining value." "The refiner has therefore either to pay the market price for both good and poor raw sugars or else, when he can, refuse to buy such sugars as he knows from experience to be inferior."[2] So what were

FIGURE 11.1. Unloading bags of sugar in Brooklyn. American Sugar Refining Company receiving wharf, 1950 or 1951; *Brooklyn Daily Eagle* photographs, Brooklyn Public Library, Center for Brooklyn History.

those other factors? He listed seven aspects of raw sugar for refiners to pay attention to. First was the water content, because the wetter the sugar, the worse its "keeping quality." Next was color. Since refining was "in the main a separating of the coloring matter . . . from the chief part of the true sugar," and since there were no price differentials for color, "it becomes the refiners [*sic*] most important consideration in picking his raws." And when one lot of raw sugar contained bags of different color, washing the darker colors off meant overwashing the lighter bags. Cuban sugars were much harder to decolor than West Indian ones and exhausted the filters more quickly.

Then there was ash, the mineral remnant of the cane. Refining took care of it, but it tasted bad, and the more ash there was, the more costly it was to remove—and it was a headache to descale a refinery plant. (In fact, Claus Spreckels, of the Californian-Hawaiian sugar dynasty, had proposed that the differentials committee add allowances for ash, but the other members refused on the grounds that it would make testing impracticably complicated.) Insoluble cane fiber and "hardening in storage" were themselves harmless but served as an index of whether the manufacturer had been careful with the other properties. There was a property called "filtrabilty," covering the gummy and waxy substances that could clog filters. And finally, the "grist and quality of grain." Was the sugar hard and lumpy? Soft and mushy? Anything but "free running" sugar, flowing smoothly through the refinery from bags to boxes, represented a cost and a hassle. A wise refiner knew to look out for all of these in his raw sugar purchases. But none of them were captured in the standard contract.

Here was his summary of the situation:

> The term "96° Centrifugals" under which Cuban Raw Sugars are sold, in common with those of other origin generally met with, was in the original instance not only descriptive of the sugar but definitive of its characteristics. The comparative simplicity of the method of raw sugar production of those times when this term "96° Centrifugals" was chosen, allowed of a quantitative measure, that is, 96° polarization, being also an expression of the raw sugars [*sic*] qualitative characteristics.

In other words, back in the late nineteenth century, the polarization of a sugar had been more than just an impersonal numerical representation of its sucrose content. The number was an index of the sugar's character, a shorthand for a full description of its many valuable properties. But by now "96°" had lost these complex meanings. The farther and more widely it traveled, the more it shed its nuances and connotations that depended on the original context. In 1940, recall from the first chapter, Fernando Ortiz wrote that "for both sugar-grower and refiner the aim is the most" of everything, including "the most indifference as to quality."[3] Here, in 1927, this unnamed official made the same point in jargon, when he lamented that his industry was "largely concerned with the yields."

> Within the last twenty to twenty-five years there has been an enormous development largely concerned with the yields which the raw sugar factories obtain and through which there has come about a retention of the 96° polarization as *a quantative* [*sic*] *measure without just appreciation of the qualitative characteristics which, in the first instance, were defined by this same term. . . .* [The] methods of raw sugar production now permit of wide ranges in quality within raw sugars of 96° polarization, which variations in quality are, and have been without just measure of comparative value. [Emphasis added]

Without "proper measures of quality" by refiners, he foresaw, their suppliers would all be "reduced to the dead level of mediocrity if not actual inferiority."[4] No raw sugar maker would have any reason to improve their product along any axis but polarization. Refiners would get what they measured, and bad sugar would drive out good.[5] The owners of many refineries were also, through "interlocking directorates," the owners of Caribbean sugar producers, and had reasons to support the price of raw sugar too.[6] The fact that even so they chose to incentivize worse sugar is further evidence of their goal of reducing overall capacity.

"Without just appreciation" was simultaneously passive and personal. If a person could read a sugar's character, it told them about the sugar's refining value and it told them about where and by whom it had been made. But if a person could read a sugar's character, it also told others something about them. The fact that this unnamed expert felt the need to lay out how to value raw sugars suggests that not all refiners understood value as well as others. A refiner named Claus Doscher had run the Brooklyn Sugar Refining Company in the 1880s and had jumped back into the business in 1897 with the New York Sugar Refinery, lured by the attractive premium earned by refined sugar. Testifying before the US Industrial Commission in 1899, all that the experienced inquisitor was able to get out of Doscher on the matter of refining value was this: "Sugars vary, you know . . . when you take the common sugars they have more or less impurities."[7] He claimed not to know the running costs of his own refinery. It was possibly not a coincidence that the Brooklyn had hemorrhaged cash, and Doscher mostly seems to have lost money in the business, except by selling out to Henry Havemeyer twice.[8] In the seventeenth century, the "greatest and most reputable merchant" in northwest England had failed as a sugar refiner because "he did not him selfe understand the art or mistery of it."[9] Sugars vary, you know.

Even if the twentieth-century refiners' cartel had conspired to force the polarization down, some among them knew how to buy better raw sugar than others. As always, refining knowledge was usually tacit, and it was always tactile, sensory, and somewhat mysterious, just as it had been for centuries. Not for the first time, however, a technical word like "producer" or "manufacturer" hid the human reality of sugar.

It is important to remember that the qualities that sharp refinery buyers looked for were indebted to other knowledges. These knowledges were equally tacit, equally tactile, equally sensory, and even more skillful and embodied and difficult to execute, and they were the property of the workers who actually made the sugar, although those refinery buyers would never have acknowledged such a debt. They would have credited plantation owners, probably engineers and superintendents, possibly chemists, and perhaps even a boiler, if only because they employed "pan men" on their own production lines.[10] They would not have given thought to their dependence on the workers who took and tested the samples that permitted chemical control, who kept trash out of crushers, or who kept machines clean and free of invisible contaminants that caused sugar to spoil. What if anything a worker in Cuba or the West Indies thought of refinery buyers in New York is harder to imagine.

Yet it took all of their skills to make a character of sugar that a refiner could pride himself on justly appreciating. "There was some pleasure and satisfaction in the old days in exercising the knowledge that comes from practical experience and selecting with care and skill the best article at the lowest possible price," sighed a British sugar economist and anti-beet bounty activist in 1918. "That part of expert knowledge in the sugar industry has gone."[11] Such eulogies for craft were premature, in the fields, in the factories, and in the refineries. Not all sugar was the same, which some sugar people still knew and others had forgotten.

The House Becoming Sour

But was sugar even the same as itself? All raw sugar could be "changed by keeping." Before centrifuges, sugar often lost enough moisture in transit that prices might be agreed in advance conditional on a certain desiccation.[12] English shoppers in the seventeenth century asked grocers whether their sugars were "new" or had been whiling away in a storeroom.[13]

Compared with what came before, however, centrifugals really were remarkably stable. This relative permanence helped transform buying and selling. Practically speaking, refiners could now pile up stocks of raws when sugar was cheap and deplete those stocks when it was dear.[14] Material stability underlay the futures market, since contracts provided for dates of delivery, not dates of production. The market worked because participants could now store sugar for months before delivery. As two economists put it in a 1949 treatise on speculative exchanges, "The commodity should lend itself to the warehouse's economic function of providing time utility. . . . The surplus supply of the present must be capable of becoming the essential supply of the future."[15] More plainly, if an exchange is to smooth demand and supply over time, then the goods put into storage today must be the same goods when they come out tomorrow.[16]

Centrifugal sugar was rarely stored for long. There were enough ships to load it quickly, and avoiding unnecessary storage before shipping was a principle of cane factory management. On the US end, meanwhile, refiners rarely kept more raw sugar on hand than they would need in a month, or perhaps two at peak summer demand.[17] As a research chemist at the Louisiana Sugar Experiment Station in 1904, Charles Browne canned samples of a dozen 96° sugars to test deterioration. Microbes ate three degrees of sugar, but it took them nine months.

FIG. 14
Pile of fermenting slime on floor of a sugar warehouse.

FIGURE 11.2. Especially after World War I, it was difficult not to notice that centrifugal sugar deteriorated. C. A. Browne, "The Deterioration of Raw Cane Sugar: A Problem in Food Conservation," *Journal of Industrial & Engineering Chemistry* 10, no. 3 (1918): 178–90.

The deterioration of centrifugals in the Atlantic suddenly became more important when steamers were drafted into war service in 1914.[18] In the first year of fighting, 80 percent of France's beet industry was destroyed, and the Central Powers' sugar was off limits.[19] The western allies needed oceangoing vessels to refill their sugar bowls. Sugar began to accumulate in warehouses "under heavy expense," and there it began to decay. By the end of the war, rot was on every mill owner's mind.[20] Browne estimated in 1918 that "deliquescence" cost Cuba alone more than a million dollars a year.[21] "The shipping situation has been acute for the past three years, but sufficient storage has not yet been provided, consequently the sugar was piled up all about the mills," a consultant engineer reported to the Cuba Cane Corporation in 1919. Before the war, Cuba Cane, which controlled more Cuban output than any other firm, had considered using its size to hold sugar in storage until it could get the best price. Instead, sugar sitting still cost the company money. "Deterioration and rehandling charges, due to lack of storage, are items that will have a bearing on this year's profits."[22] The average polarization of New York raws was actually lower in 1920 than it had been in 1914.[23]

Unsurprisingly, with their centralized technical surveillance, Hawaiian plantations had gotten there first. In 1895, the very first year of the Hawaiian Sugar Planters' Association experiment station, its director Walter Maxwell issued a report to all the association's members about the fermentation of their sugars.[24] Some planters believed that the problem originated in the fields, or possibly in the properties of the cane itself. But Maxwell concluded the fault lay in the mills. Up through the vacuum pans he found everything generally clean, but he was appalled by the state of the wooden troughs that carried sugar afterward, which he found soaked with fermenting molasses. One manager called his own plantation a "stink hole."[25] As portions of the No. 1 sugars, the first through the pipes, were recirculated to seed No. 2, No. 3, and No. 4, fermenting agents went with them. Comparing Hawaiian mill books with tests in San Francisco refineries, and as far away as New York, Maxwell found that good sugars lost only fractions of a percent, while bad ones could lose more than 6 percent.

Maxwell expressed sentiments similar to those the anonymous refinery engineer would later put into his memorandum about the meaninglessness of 96°. He berated planters who looked at the polarization allowances, projected profit and loss, and then "urged and practiced the keeping down of all sugars close to the 96 per cent. test." They were "demoralizing" their workforce, "for it became foolish to urge cleanliness in methods when the object required was a dirty product." He included a letter from a Cuban contact condemning some factories there that made "dirty" sugars intentionally polarizing no higher than 94°, and who were then punished when those dirty sugars lost three degrees. Nor was this a new problem. Maxwell inspected records from a decade earlier that showed the same losses. "The truth is, last year was the beginning of a systematic chemical control of sugars shipped from the islands, and the discovery of this fermentation was a first important result."[26] He recommended disinfecting equipment more often, keeping better sugars apart, and shipping poorer sugars faster before they lost value.

Thirteen years later, his frustrated successors found themselves issuing many of the same recommendations.[27] By then, though, the tighter corporate link between the archipelago's raw producers and the refineries in California had become an advantage. It allowed close observation and analysis of the same lots of sugars, it facilitated the sharing of otherwise proprietary chemical-control data, and it gave producers reason to be on the same page as refiners about the uncleanliness of their sugars, as the East Coast refiners noted with some envy.[28] Deterioration anywhere reduced profits for everyone, especially if anything fermented inside the liquor pipes of the refinery, for "in refinery work," wrote the Californian-Hawaiian executive George Rolph, "what is to be feared more than anything is the house becoming 'sour.'"[29]

Hawaiian sugar took so much longer to arrive on the mainland that chemists and refiners noticed deterioration earlier. In the Caribbean, it remained an occasionally lethal curiosity. The same year as Maxwell conducted his factory study, a shipful of Demerara centrifugals fermented right in the Georgetown harbor. After three days, there was enough carbon dioxide in the hold to asphyxiate three poor dockworkers. Harrison puzzled that the polarization appeared to have gone up, not down, as the sugar became lethal.[30]

In the original 1912 edition of Browne's *Handbook of Sugar Analysis*, deterioration received one sentence and a summary table in 977 pages.[31] Deterioration became an economic concern to capital only after 1914, which was about the point at which deterioration really became visible for the first time. As with temperature, the systematic surveillance and control of polarizations, by the New York Sugar Trade Laboratory and by conglomerates like the Cuban-American and Cuba Cane, made it possible to label deterioration of sugars in shorter Atlantic transit as a phenomenon distinct from all the other ways sugar might change. Confusingly, sometimes decaying sugar lifted its polarization but dropped in refining value, partly because it lost moisture as well as sucrose. As we have seen, a different polarization could be blamed on many things. It was hard to disentangle whether a polarization changed because the temperature varied, because the sugar had decayed, because there was a physical problem with the polariscopes, or because, from the perspective of those working around the 20° isotherm, the tropical polariscopist himself had become sour.

In the late spring of 1919 the management of the Hormiguero Central Company in Cuba, one of the subscribers to the New York Sugar Trade Laboratory, complained to the trustees that the lab was always testing its sugars too low. "The results have absolutely no common basis of comparison," Browne responded icily. Neither Hormiguero nor the private chemists had constant-temperature rooms and were still resorting to arithmetic corrections. Office temperature when the Hormiguero sugars arrived was 32°C. "The chemists for buyers and sellers have evidently added too much of a temperature correction, or else the samples of sugar are very uneven in quality." The Hormiguero company claimed it always made its sugars to test greater than 96°, but "if this is the case," Browne advised them, "recent shipments of the Hormiguero mark have undergone a serious loss in deterioration."[32] Calculating the deterioration of sugar depended not only on measurements taken at the beginning of its journey and at the end, but on the power to accept or reject the commensurability of those measurements.

A million deteriorated dollars a year had gotten everyone's attention. Once the war ended, there was a flurry of research into deterioration, which interested executives as well as chemists and botanists.[33] Some hypothesized that the problem had something to do with how moisture attached to the molasses coating on each crystal, providing a happy base for microbes to invert the sucrose. The smaller the crystals, the larger the surface area to moisten. The unknown inverting agents also seemed to like warmer temperatures, so deterioration sped up in the summer and slowed in the winter. It even mattered how and when the bag was packed: Bag the sugar right out of the centrifuge, and "favoring warmth" would push moisture to the outside of the bag, creating zones where microbes thrived. As bags sweated inside warehouses, condensation on the ceiling rained back onto the tops of the piles.[34]

The Colonial Sugar Company of Australia settled on an empirical formula: Divide the moisture content by 100 minus the polarization, and get a number called "the factor of safety." Crude or not, this measure was adopted across the Pacific and Atlantic industries.[35] A factor of safety less than 0.33 or so, and the sugar usually seemed safe. But this delicate fraction was no guarantee, and sugars above it might mysteriously keep while those below it fell apart. The most patient and elaborate studies were conducted by a married partnership, Lillian and Nicholas Kopeloff, working together at Louisiana State University. Between 1918 and 1920, they inoculated sugars with various molds and bacteria and discovered that the most destructive were mold strains present in virtually every raw sugar already. The fungi contained an enzyme that inverted sucrose. Given enough spores, raw sugar of any moisture would lose polarization and refining value. A low factor of safety was not an impregnable defense, the Kopeloffs reported. There was no such thing as stable raw sugar. All one could do was roughly predict the keeping quality, try to maintain a clean house, and get raw sugar out of warehouses as fast as possible.

In Cuba and Puerto Rico at least, the efforts seemed to work within a few years. "Little deteriorated sugar" arrived in New York in 1923, according to Zerban's New York Sugar Trade Laboratory, partly because the timing of the arrivals meant they spent less time in storage, and likewise in 1924 "deterioration was not excessive." By 1927, the average polarization for every month was above 96°, and the overall average had jumped 0.3° in a year. Zerban felt sure that "the problem of deterioration is gradually being overcome," and polarizations were going up, which may be why the Sugar Institute forced them down again.[36]

My Readings Being Naturally Wild

Measurements underlay the empires that turned nature into commodities. The people with a lot of money at stake, whether that money was plowed into Caribbean ground or invested in futures, had an interest in making those measures seem stable and robust. One argument of this book is that, instead of being robust, they were always fragile and open to question, especially by people who knew that organic matter and commodities were never quite the same thing, and that there was advantage in the cracks between second nature and first. The sugar trade had also depended on the differences between tropical environments and temperate ones, and in some ways it even depended on maintaining those differences as they applied to knowledge and to technologies. At the same time, those differences also made it hard to standardize the measurements on which the business of a commodity was built. And disputes over measurements are always disputes over money and labor, no matter how picayune and technical those measurements may seem.

Perhaps no story shows this fragility more than the collapse of one chemist's reputation. It was a German polariscope that did him in, making the most authoritative sugar chemists in America think "that I am so incompetent as to use a saccharimeter and not know whether it is accurate or not."[37] Chemists took credit for a well-run sugar house, but they could find ways to blame others when things went awry.

The culprit instrument came from the esteemed workshops of Schmidt & Haensch, "the best and standard" for sugar.[38] This one was a double-wedge saccharimeter of the same model that the Customs Service had instructed its employees to use. As an instructor at MIT, George Rolfe had bought this specimen, serial number 3078, on behalf of the American Sugar Refining Company in 1899. But the refinery didn't need it, so he purchased it for himself at cost. Rolfe calibrated it against the Institute's own Schmidt & Haensch double-wedge, serial number 2880, and he flipped it right away to the management of Aguirre, the giant sugar central on the southern coast of Puerto Rico. A few years later, Rolfe followed the instrument to become the factory's superintendent. It may have been the central's first polariscope, and it served the company until 1919.[39]

If Rolfe followed his own published advice for managers and chemists, he probably tended toward micromanagement at Aguirre. His father had been a Cambridge schoolmaster, a Shakespearean who had the foresight to install one of America's first high school chemistry laboratories.[40] Rolfe younger had earned a solid chemical reputation through two decades in the Caribbean and at MIT. He had literally written the book about the polariscope, the book that Harvey Wiley and Ferdinand Wiechmann had reviewed and then used their reviews to blast each other over temperature corrections. And in 1910, he had published his short paper in the *Louisiana Planter* comparing polarizations at tropical and "standard" temperatures, the paper Browne had praised for its care and consistency. For that research, which we discussed in chapter 8, he had made a thousand polarizations on no. 3078, and he spent much of the paper describing the meticulous conditions at Aguirre that made his comparisons credible.

Shortly afterward Rolfe parted ways with Aguirre management, and an assistant and former student named McCarthy replaced him as superintendent. Before the 1912 crop, McCarthy brought no. 3078 back to Boston from Aguirre to have Rolfe regraduate the marks on its scales. On this instrument the scales were made of ivory, which already by 1909 had been worn so smooth as to be nearly unreadable. By 1912 they were "nearly obliterated," and McCarthy wanted "the divisions made plainer if possible."[41] Rolfe had the instrument repaired in MIT's workshops, calibrated it himself, and sent it back.

The South Porto Rico Sugar Company made a habit of sending its polariscopes north when in need of repairs. In keeping with the history of sugar companies, they tried to evade tariffs on instruments as they did so.[42] By the end of the 1919 season, McCarthy decided that no. 3078 needed another checkup, and this time he chose to send the saccharimeter to the Bureau of Standards in Washington. Rolfe, by now working for Boston financiers with Cuban interests, called it to the attention of the bureau's prickly polarimetry expert, Frederick Bates. Rolfe was confident, he boasted, that "this instrument will read 100 if the polarization is carried out according to [Bates's] standard, providing the scale is undamaged." Still, it was a risk. An ivory scale "is held to be inaccurate I understand," he admitted, but "while these scales buckle somewhat in humid atmosphere . . . I have never found appreciable linear error."[43]

By 1919 there were two competing normal weights for saccharimeters, one of which Bates championed and had optimistically called the International scale.[44] He framed it as a correction—by one-tenth of one percent—of the other, which had been developed and promulgated by German chemists. Even after the armistice, colleagues from Britain, France, the United States and the Low Countries were disinclined to favor anything German. Even the International Commission on Uniform Methods of Sugar Analysis could not politically stay in Berlin.[45] Rolfe was not alone, however, in dismissing Bates's correction simply as an artifact of "the tinkering of German instrument makers." Half the S. & H. saccharimeters in Boston worked one way, he said, half the other way, both "consistently within the limit of error." Rolfe reckoned his own practiced eye could pick out which was which. Rolfe's eye was good enough that Browne suggested him as an independent auditor of the Trade Laboratory's methods.[46] MIT's saccharimeters, against which he had calibrated his own, were as sound as any. "For twenty years," he judged, "to the limit of the precision of the instrument, the instruments at the Institute have given constant values."

In the end, he only learned of the Bureau of Standards evaluation because another chemist passed through Aguirre, saw its assessments, and brought copies to Browne, who forwarded them on. Poor no. 3078 had been in awful shape.[47] All the glass and crystal elements needed removing and cleaning. The scale was barely visible and the ivory was so mushy the scales could not even be re-engraved. After the grinding season, the Bureau staff offered to take the time to make new scales out of a nickel alloy. For now, the $20 fee they charged was "by no means comparable with the actual expense of putting the instrument in working condition." In the meantime, "owing to the inaccuracy and unreliability of this instrument, we feel it would not be proper for us to render a certificate." In the range of raw sugar, the instrument was more than a third of a degree off.

When a chemist arrived at a new sugar house, Rolfe had written, he should take stock of the equipment, resources, and people at his disposal, devoting particular scrutiny to the men who would take samples in his name, "for negligence and incompetence here cannot be made good by the most careful laboratory work."[48] Sloppy work in the present would discredit everyone in the future. Now it was his own negligence and incompetence that made his results turn bad. So much for his paper's finding of remarkable concordance between his factory polarizations and the Trade Laboratory's results. "I did not pretend to know why this was so," he admitted. It might have been chance that "local conditions" had led them to line up, and in any case, "giving the details I invited intelligent criticism of my tests."[49] Had all of Aguirre's production under his superintendence been off as well?

Over twenty years of letters to colleagues, long into retirement on Martha's Vineyard, Rolfe repeatedly revisited the day in 1912 he had declared no. 3078 to be calibrated and let it go back to Puerto Rico with his approval. Initially he blamed "our instrument maker Mr. Selig" at MIT, "a very good and careful workman" ordinarily, but one whose "dividing engine" had perhaps engraved the markings 0.3 percent too distant.[50] Or perhaps it was that McCarthy had to leave unexpectedly for the start of the grinding season, and so Rolfe had barely half an hour to check no. 3078, and it went from a cold room into a hot one, "my readings being naturally wild" as a result. Or perhaps it was the fact that European instrument makers, like Selig, preferred to engrave divisions with a V-shaped tool rather than the American U-shaped one, so that as the scales wore down the lines became fainter. Rolfe lamented that "a German mechanic always works just as he was originally taught, just as a Chinaman."[51] For decades, Chinese workers had filled many of the most technically sensitive positions in the Caribbean sugar industry, but sugar-factory owners spoke of their diligence while deriding their ingenuity. Now others had taken Rolfe's disciplinarian approach to the invisible technicians toiling beneath him—whether instrument repairmen or samplers or Chinese laborers on brutally exploitative contracts—and turned them on his own work. He had failed to adapt his instrumental practice to local conditions.[52]

Yet what is noteworthy, reading these letters, is that none of Rolfe's fellow sugar chemists seemed to hold this miscalibration against him nearly as much as he held it against himself. That is because the failure of Schmidt & Haensch no. 3078 is not exceptional but exemplary. The case should remind us of a different lesson: that the polariscope, the pivotal scientific instrument on which the whole purity machine turned, actually fell out of whack all the time. This should be no surprise, as the natural state of instruments is in fact to be in disrepair. "Faults are defaults, yet instruments perform," writes Simon Schaffer.[53] Like the larger and more visible machines that industrialized sugar, during the grinding season, if a polariscope came out of alignment, it might be patched together by the local instrument technician with whatever skills and tools were available.[54]

This Schmidt & Haensch model was notably fickle. To set the zero point, which the polariscopist did several times a day, he slipped a small key onto a "nipple" on its left side and turned the key until he regained a neutral-looking field. But, as the Customs Service warned its examiners, "This nipple must not be confounded with a similar nipple on the *right*-hand side"—about an inch and a half away—"which it fits as well, but which *must never be touched*, as the adjustment of the instrument would be seriously disturbed."[55] Hundreds of polarizations a day, but one easy slip from disaster. The sugar world was littered with seriously disturbed polariscopes, assuming that, amid the smoke of controversies over normal weights and temperatures, one could even figure out which were in order and which in disorder. MIT's own sugar laboratory had turned a Schmidt & Haensch model down, just for teaching purposes and even at a bargain price, because it was askew by 0.1 percent. Two others in Boston were known to be "wrongly regraduated." After all, not just the observer's personal equation mattered, but also it was "a great fault in saccharimeter construction is that there is a personal error on the part of some workman in setting the 100° point."[56] Rolfe did not challenge Bates's assessment of no. 3078, though he wanted to. "Of course the figures of the

Bureau of Standards are beyond criticism," he had deferentially clarified, "the only point is, on what saccharimeter!"[57]

In his comprehensive 1942 Bureau of Standards circular on "Polarimetry, Saccharimetry, and the Sugars," Bates did not cite any of Rolfe's papers on the polariscope at all. But he did offhandedly note of the turn of the century, "It was not known at the time, as it is now, that the Schmidt & Haensch instruments of those days were in error by about 0.2 percent."[58] All of them were off by a margin almost as big as the correction that had propelled the refiners to the Supreme Court.

Arbitrary Test

Once upon a time, the American Sugar Refining Company and its subsidiaries had shipped seven million barrels of sugar a year: actual barrels, not just units of volume. The company had built 130 miles of railroads to carry eighty million board feet a year from proprietary forests that would have covered more than half of Rhode Island, and on the eve of the First World War, its crews had planted a hundred thousand spruce and white pine a year.[59] Most of those trees never became heads and staves, because after the next war, on November 1, 1945, all the eastern refiners gave their customers two weeks' notice that they would no longer be selling sugar in barrel form.

The trade journal *Sugar* devoted half a column to the barrel's eulogy. Wartime rationing of wood had killed it, although it had already been weakened by its relative inconvenience, inefficiency, and uncleanliness compared to the "conquering progress" of simple paper packaging. Nostalgists recalled how homespun democracy had been built around that "pillar of American social life" in every country store.[60] Not everyone had such fond memories, though. In 1888, *Puck* had imagined a "small retail grocer" interviewing the barrel in his own shop on the tariff question. The barrel spoke for the Trust and its ruthless power. "I am sure I wish I had never seen you," the grocer retorted.[61]

Such symbolism was probably not what *Sugar* meant when it lamented that "in the days when Americans took their politics seriously great issues were fought out and the fate of parties sealed from the vantage point of the sugar barrel." The demise of the barrel was one more small step away from the messiness of first nature and one further attenuation of the connection between sugar's materiality and power. The month after the last sugar was barreled, another elastic snapped, when the members of the Coffee and Sugar Exchange revised their bylaws as they prepared for postwar reopening of trading.

The contracts in the ring by this point were the No.3, which covered deliveries to the United States and fulfilled on an American wharf, and the No. 4,

FIGURE 11.3. A barrel of refined sugar waiting on a scale. Reginald Sexton Hall et al., "Production Control in the Revere Sugar Refinery" (Massachusetts Institute of Technology, Department of Engineering Administration, 1922).

which covered everywhere else and was fulfilled by delivery in a Cuban warehouse. In December 1946, exchange members made the No. 4 conform to US polarization allowances, and effectively shut down the No. 3 and replaced it with a new No. 5. The new No. 5 contract was meant to accommodate the fact that the typical cargo of sugar was now changing hands much more often, on average, before it reached a refiner, even if it was not actually staying in storage much longer than it once had done.[62] Each of these transactions meant that bags had to be weighed again. Each meant paying samplers to open a percentage of the bags, plunge their triers, and assemble a representative sample. And each meant paying commercial chemists to report how much of the sugar's value had been eaten away, then paying the Trade Laboratory to arbitrate between the first two, at a dollar per polarization.[63] All this measuring of weight and quality was getting frustrating and expensive. "The objective in framing the new No. 5 contract," the exchange's president wrote in another article for *Sugar*, "was to reduce to a minimum the cost of delivering and receiving by eliminating all unnecessary handling, weighing, sampling, and polarizing of the actual warehoused sugar."[64] How many of those procedures were unnecessary? It turned out the answer was: all of them.

Previously, when a member of the exchange delivered a cargo of raw sugar to a bonded warehouse, he had needed to submit a certificate to the exchange itself, indicating the exact weight and polarization of the lot. The No. 5 still required these certificates. But now they would contain numbers that were, essentially, made up. The "Original Deliverer," as the president explained, "puts a lot into circulation by tendering documents that establish that a 'lot' of sugar is in a licensed warehouse and which gives [*sic*] this lot an arbitrary weight of 112,000 lbs. (50 long tons) and an arbitrary test of 96° polarization." It was a legal use of the word "establish," rather than a natural one. The Original Deliverer delivered a cargo of sugar to the warehouse with a piece of paper stating that the sugar weighed an amount it almost certainly did not weigh, and that it had been measured at a level of quality at which it had definitely not been measured. Nothing was "established" except by fiat, but it was very convenient.

As a particular lot changed hands, the buyers and sellers based their transactions on more fictions. "This arbitrary statement of weight and polarization," the president continued, "eliminates all weighing, sampling, and testing until the lot finally is taken up into consumptive channels by the Final Receiver." So long as the sugar sat in the warehouse, it could be sold and resold without any real testing. Each new owner simply carried forth the assumption that the cargo weighed exactly 50 long tons and used an "assumed" standard figure for deterioration, that it lost six-tenths of a degree polarization

FIGURE 11.4. Sacks of raw sugar piled high at the Revere Sugar Refinery's warehouse in Boston. Reginald Sexton Hall et al., "Production Control in the Revere Sugar Refinery" (Massachusetts Institute of Technology, Department of Engineering Administration, 1922).

per month in storage. (By 1941, the exchange had adopted 0.25 as its own factor of safety for "safe keeping.")[65] The verb chosen for this change in matter was "depreciate," a tidy conflation of market value and chemical composition. Multiply the "assumed monthly depreciation," 0.06°, by the number of months since Original Delivery. Subtract the result from the fictitious initial polarization of 96°. Suddenly speculators had a polarization upon which to trade, without any of the fuss of actually polarizing anything. Meanwhile, the actual sugar continued to deteriorate in its own idiosyncratic way.

Eliminating the costs of sampling and weighing revealed deeper inefficiencies that could now be economized. Previously, each lot of fifty long tons needed to be segregated from others in the warehouse, so that bags could be dragged out for sampling and weighing. But now that they only needed to be hauled in once and hauled out once, warehouses could optimize their space. Bags began to pile much higher, and fewer workers were paid to handle raw sugar fewer times, as refiners had long sought to do.[66] The sugar warehouse became another site in the history of capitalism where metrological choices obscured conflicts over human labor.

The "arbitrary" features of the No. 5 contract made it possible for a member of the exchange to speculate in sugar while knowing less than ever before about the sugar that he owned. The refiner Lawson Fuller's ancient prediction had come to pass. "The seller will have his goods sampled and tested," he had told Congress in 1878, "and then the buyer will have them sampled and tested . . . and then both parties knowing how much they are cheating each other, will come to a square understanding and strike a balance just [as] though they had not sampled or tested the goods at all."[67]

Only at the end of the cargo's time in storage, when a refiner actually meant to take possession of it, did its arbitrary qualities need to be marked against observable ones. "It is now time to determine if the original certificate arbitrarily declaring the weight at 112,000 lbs. and the test as 96 degrees polarization, less the assumed monthly depreciation of .06 degree per month, actually reflects the physical condition of the sugar he is about to receive." The cargo was weighed and the samples sent off to three chemists, whereupon the allowances kicked in. If the two closest tests averaged below 96°, or if the cargo weighed less than 50 long tons, then the Original Deliverer had to pay up through the exchange's clearinghouse. If they averaged above, or the sugar weighed more, the Final Receiver owed the difference. The exchange's president pitched this moment as high drama. But just as no 96° sugar was ever 96°, no cargo would ever come out of storage with a physical condition that matched its arbitrary condition.

The certificates filed with the initial delivery "established" a cargo's weight and polarization by establishing nothing. Everyone who bought and sold that cargo acknowledged that they were buying and selling 50 long tons of 96° sugar that did not weigh 50 long tons nor measure 96°. Through this complete surrender to the fictions of second nature, even a product as variable as sugar could stay in the warehouse as it was bought and sold, hedged and shorted. As long as you were not the Final Receiver, you were no longer betting on the qualities of any particular lot of sugar but rather on the fiction of universal sugar. Each physical sample of sugar, each bag, each barrel, and each cargo was uncooperative and unknowable in its own way. Only when it sat unmeasured in a warehouse did sugar finally behave like an ideal commodity.

Acknowledgments

First my teachers. As an undergraduate at Columbia, I was introduced by Elizabeth Blackmar to the nineteenth-century America of Herman Melville and Henry Adams, and Matt Jones showed me the power and humor of the history of science. I'm grateful for Ashli White's patience and friendship, and Barbara Fields's laser rigor.

At Cambridge University's Department of the History and Philosophy of Science, Simon Schaffer is an exemplary thinker, reader, adviser, teacher, mentor, and friend. He has an uncanny ability to persuade his students that they have taught him something brilliant and original, even when we suspect he has gotten there long before us. I hope this book is more than just an impoverished simulacrum of what he would write.

It is impossible to do justice to Dave Kaiser's unfailing kindness and generosity as my dissertation adviser at MIT. Despite doing the work of at least three professors, he was never too busy, answering emails at 2 a.m. and 6 a.m., reading whatever I sent him, door always open, always responding with thoughtful comments and an insistence on seeing the next draft. I have only ever seen him tired once. Harriet Ritvo and Chris Capozzola, the other members of my committee, have provided more than a decade of tough criticism, unflagging support, and humor about the past and present. And in that wobbly first postdoctoral year, I was lucky to have James Delbourgo and Toby Jones at Rutgers University for guidance. I'm forever indebted to Humberto Garcia Muñiz for his hospitality and mentorship in Río Piedras.

At the University of Virginia, I've been fortunate to work under supportive and collaborative chairs in two departments: Karen Parshall, Claudrena Harold, and Tom Klubock in History, and Anna Brickhouse, Sylvia Chong, Jennifer Greeson, and Penny von Eschen in American Studies. I'm grateful for the intellectual and personal welcome of colleagues including Roberto Armengol, Brian Balogh, Allison Bigelow, Kevin Driscoll, Elizabeth Fowler, Fiona Greenland, Grace Hale, Paul Halliday, John Handel, Ben Hays, Matt Hedstrom, Andrew Kahrl, Kyrill Kunakhovich, Kathryn Laughon, Ted Lendon, Erik Linstrum, Victor Luftig, Anne Garland Mahler, Chuck Mathewes, Elena McGrath, Christian McMillen, Elizabeth Meyer, Brian Owensby, Claire Payton, Joseph Seeley, Jen Sessions, Janet Spittler, Samhita Sunya, Lana Swartz, Sylvia Tidey, and Caitlin Wylie. My co-organizers at MADCAP, Ira Bashkow, Fahad Bishara, and Sarah Milov, deserve special mention here, as do Dan Gingerich and Sandip Sukhtankar for inviting me to join their anti-corruption CLEAR lab as a new faculty member. And I've learned from feedback from all the participants in MADCAP, the CLEAR lab, and the faculty writing group. Keisha John, Christian McMillen, and Jenn Bair made a point of watching out for joint-appointed faculty on the tenure track. Every junior faculty deserves that level of support. I don't know where you thank an office fish, but thanks and RIP to Gulliver Mobius Travels.

All this research and writing is pretty pointless if it doesn't wind up in the classroom. Teaching is the part of being a professor that really matters. I would like to thank the staff of the Center for Teaching Excellence, especially Adriana Streifer, Dorothe Bach, and Michael Palmer, for turning me into a far better and more reflective teacher for my students. Better teaching has led to better thinking and that's in this book too.

Professors would be helplessly beached without the staff who make departments run. At MIT, Karen Gardner and Bianca Singletary pulled all the strings. At the University of Virginia, thanks to Kathleen Miller, Karly Kovalcik, Carol Anne Brown, Latricia Washington, Nicole Jones, Ashley Herring, Pamela Pack, Kent Merritt, Tim Dunne, Brandon Block, Kelly Robeson, and Caterina Eubanks. The staff at UVA Libraries, especially Keith Weimer, has helped me and my students through many problems. Over the years, student research assistants have helped piece together key parts of this story: Thanks to Lauren McGlothlin, Nick Scott, and Sophie Wagner. Aaron Leon Freedman and Bethany McGlyn found most of the images and permissions on tight deadlines.

Speaking of students: There is no intellectual community quite like MIT's Program in History, Anthropology, and STS. I sometimes think of myself as the squarest scholar to come out of an edgy and innovative program. My comrades in E51—and ever since—have included Etienne Benson, Nick Buchanan, Marie Burks, Alex Chan, Peter Doshi, Xaq Frolich, Chihyung Jeon, Amy Johnson, Shreeharsh Kelkar, Yoshi Kikuchi, Shekhar Krishnan, Nicole Labruto, Lisa Messeri, Lucas Mueller, Teasel Muir-Harmony, Peter Oviatt, Canay Ozden-Schilling, Rebecca Perry, Hilary Robinson, Mike Rossi, Ryan Shapiro, Shira Shmuely, Ellen

Spero, Alma Steingart, Michaela Thompson, Nate Deshmukh Towery, Ben Wilson, Rebecca Woods, and Ben Wurgaft. I have the best writing and baking companions in Mary Brazelton, Joy Rankin, and Emily Wanderer. Up Mass Ave were Anouska Bhattacharyya, Eli Cook, Philip Lehmann, Jamie Martin, Yael Merkin, Caitlin Rosenthal, Josh Specht, and Jeremy Zallen, plus all the people who passed through the Center for History and Economics teas. During my year in Philadelphia, I got to know Evan Hepler-Smith, Noam Maggor, and Ariel Ron. From the other Cambridge, I'm glad to know Jenny Bangham, Richard Carr, Nicky Reeves, and Dan Wilson. In San Juan, Teófilo Espada-Brignoni showed me the best coffee, food, drink, bookstores, and music. Stephen Mullen knew how to take serious history very seriously but not as seriously as Glasgow Celtic. Many of the people I've just mentioned knew Cecilia Cardenas-Navia and miss her too.

As a student at MIT, I also depended on the advice and company of many faculty, including Ann McCants, Will Deringer, Vincent Lepinay, Stefan Helmreich, Leo Marx, Heather Paxson, Robin Scheffler, Hannah Rose Shell, Craig Wilder, and Rosalind Williams. Emma Rothschild supported my work even before we met and gathered an amazing group of students around her weekly teas. Among colleagues elsewhere, thanks to Carin Berkowitz, Ezra Feldman, Paul Israel, Peter Mandler, Seth Rockman, Dan Rood, and Anya Zilberstein. Deborah Warner provided crucial early backing for my work. Rob McGreevey and Katy Dawley kindly allowed me to glimpse into Alan Dawley's unfinished work. In Havana, what good fortune to make a friend in José Solá, and to have the chance to learn from Óscar Zanetti. It has been a delight to get to know Nicolas Barreyre in Charlottesville and in Paris, and to talk sugar and history with Reinaldo Funes Monzote in Princeton and beyond.

The members of the fall 2020 Agraphia writing accountability group, especially Love Ronnelid, made sure this project (and my employment prospects) cleared the halfway mark.

It was a revelation, at the Acquired Tastes workshop in summer 2018, to discover other historians who thought about writing, how much we could learn from each other, and how much we could learn from the pros. Thanks to Ben Cohen, Mookie Kideckel, and Anna Zeide for organizing, Helen Betya Rubinstein for coaching, and everyone who participated.

At the University of Chicago Press, my editor Tim Mennel has enthusiastically supported this project almost as long as I have, and been gracious, unflappable, perceptive, and above all patient. Andrea Blatz's care and diligence got it over the line. I am grateful to the two anonymous reviewers of the proposal; to the editorial board of the Synthesis series for welcoming the project; to Charles Dibble for his meticulous copyediting, Dave Luljak for the index, and Tamara Ghattas for supervising the editing and proofreading; to the Synthesis editor who reviewed the full manuscript; and, especially, to Dan Rood, fellow polariscopist, whose comments on the manuscript as a reviewer were so helpful, perceptive, and encouraging.

I spent time at the following archives over the course of this project, and the staff everywhere were patient, kind, knowledgeable, and unflappable: the National Institute for Standards and Technology, especially Keith Martin and Mary-Deirdre Coraggio; the Archivo General de Puerto Rico, and especially to Juan Roman; the New York Public Library; the Library of Congress Manuscript Division; the American Philosophical Society; the University of California Berkeley Bancroft Library; the Columbia University Rare Book and Manuscript Library, the National Archives and Records Administration in College Park; the University of Glasgow Archives, especially Alma Topen and Rachael Egan; the Glasgow School of Art; the Mitchell Library of Glasgow; the University of Strathclyde Archives and Special Collections; the Pennsylvania Historical Society; the Chemical Heritage Foundation; the Archivo de Arquitectura y Construcción de la Universidad de Puerto Rico; the University of Hawai'i at Manoa. To any graduate students reading this, here is some advice: Take librarians and archivists out to lunch.

This project depended on support, over many years, from the National Science Foundation, the Social Science Research Council, the Chemical Heritage Foundation (now Science History Institute), the Joint Center for History and Economics at Harvard and Cambridge, MIT's Luis Francisco Verges fellowship and Center for International Studies, the American Philosophical Society, the Centre for Business History in Scotland, the Instituto de Estudios del Caribe in Rio Piedras, New York Public Library, and UVA's Institute of the Humanities and Global Cultures. I was and am honored that the Business History Conference awarded my dissertation its Krooss Prize and the Association of Business Historians in the UK its Coleman Prize

The book was immeasurably shaped by feedback from audiences at the Cambridge University History and Philosophy of Science Department Seminar, the Princeton Davis Center, the École des hautes études en sciences sociales, the Treasury History Association, the Workshop on the History of Environment, Agriculture, Technology, and Science (WHEATS), the US Political Economy Lab, the Business History Conference, the Organization of American Historians, the American Society for Environmental History, the Association of Caribbean Historians, the University of California at Santa Barbara, and the John Carter Brown Library. My students in HIST 3501, "The Global History of Sugar," in spring 2023 were among the most perceptive readers I've had for any part of this project.

Special thanks to Rafe Bonvillian, Harry Havemeyer, Jonathan Kingsman, and Charley Richard for sharing their expertise about the way the sugar commodity business works now.

Final thanks. Everyone should have a friend who knows the tricks you play on yourself as well as Hilary Robinson and Amy Johnson know mine. Ben Cohen, Augustine Sedgewick, Tom Ozden-Schilling, and Laura Martin have tolerated an endless stream of John McPhee chat and career and life questions.

Thanks to John McPhee, for writing back to my email with the perfect John McPhee email.

Had I not stopped on a canal path to ask two strangers about a recycling center, I would never have discovered the joyous friendship of Peter and Chris Jimack, whom I instantly felt I'd known my whole life.

Speaking of which, it helps to have people who have known you for more than twenty years: thinking especially of Matt Christiansen, Kathy Gilsinan, Sasha Katsnelson, Ani Ravi, Kate Reber, Alex Rolfe, Vanessa Schneider, Sean Sheffler-Collins, Lia Simon, and JJ Stranko.

Some years in Cambridge, I may have spent more time riding bikes to Concord and Dover with the MIT Cycling Team, particularly Seth Behrends, Alex Chaleff, Nate Dixon, and Zach LaBry, than I spent working on my PhD. In retrospect, I wish I'd spent even more time riding and racing with them.

In Charlottesville, thanks to Christina Black, Evan Bruno, Ben Cullop, Sally Hudson, Jonathan Katz, Josiah Luftig, and Andy Mahler, along with others mentioned elsewhere here. Merci à Prof. Manon Soulet, y gracias a Dra. Paola Monteros-Freeman. My family is grateful to the community at CBI Forest School, and to Lorie, Mimi, and everyone else at Atlas Coffee. A lot of this book was thought up while running on shoes provided by the Lorenzoni family at the Ragged Mountain Running Shop.

Thanks to the steering committee of Indivisible Charlottesville with whom I spent many evenings and weekends for several years: Joe Calhoun, Catherine Cayce, Corinne Cayce, Leanne Fox, Ken Horne, Patrick Jackson, Nancy Kliewer, Kat Maybury, Caroline Melton, Michael Payne, and Guillermo Ubilla.

The traveling bilberries, Boris Jardine and Josh Nall, have helped me survive Neolithic flint mines, abandoned radio telescopes, nettly ditches, dodgy bicycles, stiperstones, and the Iron Bridge, not to mention HPS. If not for the quantity and quality of material circulated over the Please Stop Now email list, however, this book would have been done years ago. PSN is hereby un-acknowledged.

Many important decisions have depended on our friendships with Ben Siegel and Caterina Scaramelli, even though Cat still owes me a citation to the *Women's Petition Against Coffee*.

I'm blessed to be part of a loving, feisty, erudite, argumentative extended family who are also expert cooks. Helen and Jim and Anya, Lilly and Nate and Skylar, Martha and Bryon, Joe, and Jeff. My grandparents, Mike and Anne and Sid and Molly, steered me with their fundamental ethics, goodness, and kindness. At almost exactly the moment that I finished my dissertation, I was welcomed into another family. Jeanie and John Kuhn have been deep sources of warmth, friendship, and advice ever since, and they are also remarkably game to be woken up by preschoolers. It's a joy to count as family Noah, Kate, Anna, Brooks, Mindy, Paul, Megan, Brian, and Kevin too.

My brother Adam understands me like no one else. It's impossible to properly express my gratitude to my parents, Katalin Roth and Phillip Singerman, or to express my delight at seeing them become Nana and Gup to my own wild and wonderful and kind daughters, Marguerite and Piera. Of course the book is for Mary, who sent me an email from Houghton Library at what she couldn't have known was exactly the right time.

Some material in this book was previously published in the following publications:

"Science, Capitalism, and Corruption in the Gilded Age," *Journal of the Gilded Age and Progressive Era* 15, no. 3 (2016): 278–293.

"The Limits of Chemical Control in the Caribbean Sugar Factory," *Radical History Review* 2017, no 127 (2017): 39–61.

"Modern Food as Ranked Food: Who's Afraid of the Dark Sugar?" in *Acquired Tastes*, edited by Benjamin R. Cohen, Michael S. Kideckel, and Anna Zeide. The MIT Press, 2021. © 2021 by Massachusetts Institute of Technology.

"Sugar Machines and the Fragile Infrastructure of Commodities in the Nineteenth Century," *Osiris* 33, no. 1 (2018): 63–84. © 2018 by The History of Science Society.

Notes

Prologue

1. Louis A. Pérez, *Cuba: Between Reform and Revolution*, 3rd ed., Latin American Histories (Oxford University Press, 2006), 113–19.

2. Wilfrid Skaife to Edwin F. Atkins, April 30, 1894, box II.1, Edwin F. Atkins Papers, Massachusetts Historical Society, Boston, Massachusetts.

3. Skaife to Atkins, May 7, 1894, box II.1 in Atkins Papers.

4. Skaife to Atkins, June 8, 1894, box II.1 in Atkins Papers.

5. Skaife to Atkins, June 24, 1894, box II.1 in Atkins Papers.

Chapter 1

1. Pennsylvania Sugar Company, *The Story of Cane Sugar* (Philadelphia, 1925).

2. Benjamin Allen, *A Story of the Growth of E. Atkins & Co. and the Sugar Industry in Cuba* (Boston, 1925), 35–36.

3. H. C. Prinsen Geerligs, *The World's Cane Sugar Industry, Past and Present* (Norman Rodger, 1912), 383; C. A. Browne, "Sugar Industry in Foreign Lands," *Industrial and Engineering Chemistry* 25, no. 1 (1933): 61–68.

4. Paul H. Moore et al., "Sugarcane: The Crop, the Plant, and Domestication," in *Sugarcane: Physiology, Biochemistry, and Functional Biology*, ed. Paul H. Moore and Frederik C. Botha (Wiley-Blackwell, 2013), 1–17.

5. John Daniels and Christian Daniels, "Sugarcane in Prehistory," *Archaeology in Oceania* 28, no. 1 (1993): 1–7.

6. Rai Bahadur Joges Chandra Ray, "Sugar Industry in Ancient India," *Journal of the Bihar and Orissa Research Society* 4, no. 4 (1918): 441.

7. Christian Daniels and Nicholas K. Menzies, *Science and Civilisation in China*, vol. 6, *Biology and Biological Technology*, part 3, *Agro-Industries and Forestry, Agro-Industries: Sugarcane Technology, Forestry*, 3 (Cambridge University Press, 2010), 88; Ulbe Bosma, *The World of Sugar: How the Sweet Stuff Transformed Our Politics, Health, and Environment Over 2,000 Years* (Belknap Press of Harvard University Press, 2023), 8.

8. Sucheta Mazumdar, *Sugar and Society in China*, Peasants, Technology, and the World Market, Harvard-Yenching Institute Monograph Series 45. (Harvard University Asia Center, 1998), 22.

9. Jason W. Moore, "Madeira, Sugar, and the Conquest of Nature in the 'First' Sixteenth Century, part 2: From Regional Crisis to Commodity Frontier, 1506–1530," *Review* (Fernand Braudel Center) 33, no. 1 (2010): 1–24; Moore et al., "Sugarcane"; Jason W. Moore, "Sugar and the Expansion of the Early Modern World-Economy: Commodity Frontiers, Ecological Transformation, and Industrialization," *Review* (Fernand Braudel Center) 23, no. 3 (2000): 409–33 the Plant, and Domestication.

10. Bosma, *The World of Sugar*, 32.

11. Sidney W. Mintz, *Sweetness and Power: The Place of Sugar in Modern History* (Viking, 1985); Sophia Nguyen, "Kara Walker 'Sweet Talks' Harvard," *Harvard Magazine*, December 11, 2014; Kara Walker, *A Subtlety; or, the* Marvelous Sugar Baby: *An Homage to the unpaid and overworked Artisans who have refined our Sweet tastes from the cane fields to the Kitchens of the New World on the Occasion of the demolition of the Domino Sugar Refining Plant*, 2014, http://creativetime.org/projects/karawalker/.

12. Carole Shammas, *The Pre-Industrial Consumer in England and America* (Clarendon Press, 1990), 82.

13. This is the latest estimate from slavevoyages.org (accessed March 10, 2022, https://www.slavevoyages.org/assessment/estimates), In 1950 Noël Deerr opened the second volume of his *History of Sugar* with a chapter on the scale of slavery. Explaining what was knowledge and what was speculation, he arrived at a bit less than 12.5 million people enslaved for sugar manufacture. But he guesstimated "the tale and toll of the slave trade" at twenty million. Of that total, he said, sugar could be discredited with two-thirds. Noël Deerr, *The History of Sugar*, 2 vols. (Chapman and Hall, 1950), 2:284.

14. Sucheta Mazumdar, "China and the Global Atlantic: Sugar from the Age of Columbus to Pepsi-Coke and Ethanol," *Food and Foodways* 16, no. 2 (2008): 137–40.

15. Mervyn Ratekin, "The Early Sugar Industry in Española," *Hispanic American Historical Review* 34, no. 1 (1954): 1–19; Sidney W. Mintz, "The Plantation as Socio-Cultural Type," in *Plantation Systems of the New World: Papers and Discussion Summaries*, Social Science Monographs 7 (Pan American Union, 1959); Philip D. Curtin, *The Rise and Fall of the Plantation Complex: Essays in Atlantic History*, 2nd ed., Studies in Comparative World History (Cambridge University Press, 1998); B. W. Higman, "The Sugar Revolution," *Economic History Review* 53, no. 2 (2000): 213–36.

16. Manuel Moreno Fraginals, *The Sugarmill: The Socioeconomic Complex of Sugar in Cuba, 1760–1860*, trans. Cedric Belfrage (Monthly Review Press, 1976), 116.

17. John E. Crowley, "Sugar Machines: Picturing Industrialized Slavery," *American Historical Review* 121, no. 2 (2016): 403–36.

18. In 1944 Eric Williams argued that the profits from slave plantations financed the industrial revolution, a claim that historians and economists have been investigating ever since. Eric Eustace Williams, *Capitalism & Slavery* (University of North Carolina Press, 1944).

19. J. H. Galloway, *The Sugar Cane Industry: An Historical Geography from Its Origins to 1914* (Cambridge University Press, 1989).

20. See, e.g., Rebecca J. Scott, *Degrees of Freedom: Louisiana and Cuba After Slavery* (Belknap Press of Harvard University Press, 2005); Rebecca J. Scott, *Slave Emancipation in Cuba: The Transition to Free Labor, 1860–1899* (1985; repr., University of Pittsburgh Press, 2000); Moreno Fraginals, *The Sugarmill*; Daniel B. Rood, *The Reinvention of Atlantic Slavery: Technology, Labor, Race, and Capitalism in the Greater Caribbean* (Oxford University Press, 2017); Dale Tomich, "The Second Slavery and World Capitalism: A Perspective for Historical Inquiry," *International*

Review of Social History 63, no. 3 (2018): 477–501; Lawrence Helfgot Kessler, "Planter's Paradise: Nature, Culture, and Hawai'i's Sugarcane Plantations" (PhD diss., Temple University, 2016).

21. Deerr, *The History of Sugar*, 2:460. When Deerr, a cane factory chemist, dedicated a dry technical handbook to his "admiration of the stubborn fight made by sugar cane planters in all parts of the world against a State-aided system of suppression," he did not mean the legacy of slavery and actual state-aided oppression all around him, but European governments' subsidies for beet agriculture and refining. Noël Deerr, *Sugar House Notes and Tables: A Reference Book for Planters, Factory Managers, Chemists, Engineers, and Others Employed in the Manufacture of Cane Sugar* (E. & F. N. Spon, 1900), vi.

22. Manuel Moreno Fraginals, "Plantations in the Caribbean: Cuba, Puerto Rico, and the Dominican Republic in the Late Nineteenth Century," in *Between Slavery and Free Labor: The Spanish-Speaking Caribbean in the Nineteenth Century*, ed. Manuel Moreno Fraginals et al. (Johns Hopkins University Press, 1985), 3–8; Manuel Moreno Fraginals, "Plantation Economies and Societies in the Spanish Caribbean, 1860–1930," in *The Cambridge History of Latin America*, vol. 4, *c. 1870 to 1930*, ed. Leslie Bethell (Cambridge University Press, 1986), 192.

23. Rebecca Solnit, *River of Shadows: Eadweard Muybridge and the Technological Wild West* (Viking, 2003); Mary Kuhn, *The Garden Politic: Global Plants and Botanical Nationalism in Nineteenth-Century America*, America and the Long Nineteenth Century (New York University Press, 2023).

24. Charles Grier Sellers, *The Market Revolution: Jacksonian America, 1815–1846* (Oxford University Press, 1991); Benjamin R. Cohen, *Pure Adulteration: Cheating on Nature in the Age of Manufactured Food* (University of Chicago Press, 2022), 16.

25. Henry Adams, *The Education of Henry Adams*, 3rd ed. (Modern Library, 1999), 4.

26. Ken Alder, *The Measure of All Things: The Seven-Year Odyssey and Hidden Error That Transformed the World* (Free Press, 2002).

27. Simon Schaffer, "Late Victorian Metrology and Its Instrumentation: A Manufactory of Ohms," in *Invisible Connections: Instruments, Institutions, and Science*, Proceedings of SPIE–The International Society for Optical Engineering , April 1–2, 1991, ed. Robert Bud and Susan E. Cozzens, (SPIE, 1991); Peter Louis Galison, *Einstein's Clocks, Poincaré's Maps: Empires of Time* (W. W. Norton, 2003).

28. Michael Kershaw, "The International Electrical Units: A Failure in Standardisation?," *Studies in History and Philosophy of Science* 38, no. 1 (2007): 108–31; Ken Alder, "Scientific Conventions: International Assemblies and Technical Standards from the Republic of Letters to Global Science," in *Nature Engaged: Science in Practice from the Renaissance to the Present*, ed. Mario Biagioli and Jessica Riskin, Palgrave Studies in Cultural and Intellectual History (Palgrave Macmillan, 2012), 19–39; Raymond Cochrane, *Measures for Progress: A History of the National Bureau of Standards* (National Bureau of Standards, US Dept. of Commerce, 1966).

29. Nicholas Dew, "Vers la ligne: Circulating Measurements Around the French Atlantic," in *Science and Empire in the Atlantic World*, ed. James Delbourgo and Nicholas Dew (Routledge, 2008), 53–54.

30. See, for instance, Kapil Raj, *Relocating Modern Science: Circulation and the Construction of Knowledge in South Asia and Europe, 1650–1900* (Palgrave Macmillan, 2008); James Delbourgo, *Collecting the World: The Life and Curiosity of Hans Sloane* (Allen Lane, 2017); Simon Schaffer, "Newton on the Beach: The Information Order of Principia Mathematica," *History of Science* 47, no. 3 (2009): 243–76; E. J. Browne, *Charles Darwin*, vol. 1, *Voyaging* (Princeton University Press, 1996); James Poskett, *Horizons: The Global Origins of Modern Science* (Mariner Books, 2022).

31. Sven Beckert et al., "Commodity Frontiers and the Transformation of the Global Countryside: A Research Agenda," *Journal of Global History* 16, no. 3 (2021): 435–50; Moore, "Sugar and the Expansion of the Early Modern World-Economy."

32. Prakash Kumar, *Indigo Plantations and Science in Colonial India* (Cambridge University Press, 2012); Rebecca J. H. Woods, *The Herds Shot Round the World: Native Breeds and the British Empire, 1800–1900*, Flows, Migrations, and Exchanges (University of North Carolina Press, 2017).

33. Jeremy Zallen, *American Lucifers: The Dark History of Artificial Light, 1750–1865* (University of North Carolina Press, 2019).

34. Augustine Sedgewick, *Coffeeland: One Man's Dark Empire and the Making of Our Favorite Drug* (Penguin Press, 2020).

35. Paul Lucier, *Scientists & Swindlers: Consulting on Coal and Oil in America, 1820–1890*, Johns Hopkins Studies in the History of Technology (Johns Hopkins University Press, 2008).

36. Jeffrey C. Williams, "The Origin of Futures Markets," *Agricultural History* 56, no. 1 (1982): 315.

37. William Cronon, *Nature's Metropolis: Chicago and the Great West* (W. W. Norton, 1991), 116.

38. Cronon, *Nature's Metropolis*, 123.

39. Cronon, *Nature's Metropolis*, 118.

40. James C. Scott, *Seeing Like a State: How Certain Schemes to Improve the Human Condition Have Failed* (Yale University Press, 1999), 8.

41. Sven Beckert, *Empire of Cotton: A Global History* (Alfred A. Knopf, 2014), 210–11; Walter Johnson, *River of Dark Dreams: Slavery and Empire in the Cotton Kingdom* (Belknap Press of Harvard University Press, 2013), 248–59.

42. Sedgewick, *Coffeeland*, 122.

43. Cronon, *Nature's Metropolis*, 133.

44. Cronon, *Nature's Metropolis*, 141.

45. Beckert, *Empire of Cotton*, 210–11.

46. Johnson, *River of Dark Dreams*, 248–51.

47. Donald A. MacKenzie, *Inventing Accuracy: A Historical Sociology of Nuclear Missile Guidance* (MIT Press, 1990); Simon Schaffer, "Accurate Measurement Is an English Science," in *The Values of Precision*, ed. M. Norton Wise (Princeton University Press, 1995), 135–72; Simon Schaffer, "Late Victorian Metrology and Its Instrumentation: A Manufactory of Ohms," in *The Science Studies Reader*, ed. Mario Biagioli (Routledge, 1999), 457–78.

48. The anthropologist of science Bruno Latour argued that laboratories are special because they have techniques for making natural phenomena more easily visible. More counterintuitively, he argued that what makes scientific laboratories powerful is that scientists, so to speak, try to make "society conform to the inside of laboratories so that the latter's activity can be made relevant to the society." And there is no better example for Latour of this process than *métrologie*, the network of constants that makes all claims to universality possible in the first place. Bruno Latour, "Give Me a Laboratory and I Will Raise the World," in *Science Observed: Perspectives on the Social Study of Science*, ed. Karen D. Knorr-Cetina and Michael Mulkay (Sage, 1983), 166–67.

49. One of the great puzzles of early thermodynamics was measuring the ratio at which mechanical work converted to heat energy. The man who solved the puzzle was another brewer, James Joule. His 1845 paper to the Royal Society modestly described his remarkable experimenter's technique and carefully crafted measuring device. When the historian Otto Sibum tried to

reproduce the experiment in the 1990s, he found that Joule had disguised and even lied about his work, making the experiment and his skills seem far less idiosyncratic than they really were. He had to obscure his singularity, Sibum concluded, because an experiment that only Joule could execute was no basis for an absolute universal measurement of heat .H. O. Sibum, "Reworking the Mechanical Value of Heat: Instruments of Precision and Gestures of Accuracy in Early Victorian England," *Studies in History and Philosophy of Science* 26, no. 1 (1995): 73–106.

50. With some values, the ship-in-bottle-ness is obvious, such as with standards used in a particular corner of industry. The eighteenth-century brewing technologist John Richardson invented both a quantity called the "pounds-per-barrel extract" and an instrument called the "saccharometer" which, he said, was the only way of measuring it. Richardson claimed that this package of unit and device was the true way to measure the strength of beer, and did so persuasively enough for the industry to adopt it. Even though such a quantity seems obviously arbitrary, the fragility of standard values is still visible in the most fundamental values of modern science. Values that seem derived from nature itself are still made persuasive by the same tricks. James Sumner, "John Richardson, Saccharometry and the Pounds-Per-Barrel Extract: The Construction of a Quantity," *British Journal for the History of Science* 34, no. 3 (2001): 255–73.

51. Joseph O'Connell, "Metrology: The Creation of Universality by the Circulation of Particulars," *Social Studies of Science* 23, no. 1 (1993): 129–73.

52. As Galison showed, what gave these objects power was not their unprecedentedly precise metallurgy but the manner of their burials. Men in suits, international delegates to the relevant treaty, watched other men in suits entomb M and K. They locked two doors with three keys, then sealed those keys inside envelopes entrusted to various chairpeople and directors. Technicians had spent tens of thousands of painstaking hours engineering these objects. They had also created official copies of M and K for distribution to the signatory nations, using a meticulous device called a universal comparator. Without the suits and keys and treaties, they would just have been painstakingly engineered objects in a French basement. Only the somber movements made them international reference objects. Galison, *Einstein's Clocks, Poincaré's Maps*, 91.

53. Nate Shaw and Theodore Rosengarten, *All God's Dangers: The Life of Nate Shaw* (Knopf, 1974), 188.

54. Moore, "Sugar and the Expansion of the Early Modern World-Economy."

55. The historian Daniel Roche called sugar "the totemic food of the Enlightenment civilization" because it signified both pleasure and violence: Daniel Roche, *A History of Everyday Things: The Birth of Consumption in France, 1600–1800*, trans. Brian Pearce (Cambridge University Press, 2000), 245. Other substances had different powers, of which Europeans were less certain than sugar's sweetness. Some of Locke's contemporaries jested that whatever coffee (for instance) was, it had the effect of disappointing wives by sapping their husbands' libido. *The Women's Petition Against Coffee: Representing to Publick Consideration the Grand Inconveniencies Accruing to Their Sex from the Excessive Use of That Drying, Enfeebling Liquor* [. . .] (London, 1674).

56. John Locke, *An Essay Concerning Humane Understanding* (London, 1694), 207; Eric Otremba, "Inventing Ingenios: Experimental Philosophy and the Secret Sugar-Makers of the Seventeenth-Century Atlantic," *History and Technology* 28, no. 2 (2012): 124–35.

57. Jean-Baptiste-Pierre Le Romain, "Sucre," in *Encyclopédie ou Dictionnaire raisonné des sciences, des arts et des métiers* (Paris, 1765).

58. Mintz, *Sweetness and Power*, 5. "Sucrose—what we call 'sugar'—is an organic chemical of the carbohydrate family" (19).

59. Bosma, *The World of Sugar*, 1. Some other examples: "Sucrose is indistinguishable in taste and chemical composition no matter its source, and it yields a qualitatively uniform product. The great variety of types and grades of sugar are all the same substance and differ only in chemical purity and the degree of crystallization." Dale W. Tomich et al., *Reconstructing the Landscapes of Slavery: A Visual History of the Plantation in the Nineteenth-Century Atlantic World* (University of North Carolina Press, 2021), 87–88. Or "In turning the pith of the sugar cane plant into white crystalline sucrose, the ingenios of the Atlantic basin relied upon a series of mechanical and chemical processes which were highly sophisticated for their day" Otremba, "Inventing Ingenios." Or "The German chemist Andreas Marggraf (1709–1782) discovered in 1747 that the sweetness was sucrose, which when extracted from pulverized beetroots was identical in all properties to the sugar derived from sugarcane." Noël Kingsbury, *Hybrid: The History and Science of Plant Breeding* (University of Chicago Press, 2009). Or see Galloway, *The Sugar Cane Industry*, 1.

60. Wilfrid Skaife, "Sugar Producing Plants," *Canadian Record of Science* 3, no. 8 (1889): 474–75.

61. Mintz, *Sweetness and Power*, 21.

62. Neil L. Pennington and Charles W. Baker, *Sugar: A User's Guide to Sucrose* (Van Nostrand Reinhold, 1990), 1.

63. William Allen Miller, *Elements of Chemistry: Theoretical and Practical*, part 3, *Organic Chemistry* (J. W. Parker and Son, 1857); *Oxford English Dictionary*, s.n. "Sucrose" (modified July 2023), https://doi.org/10.1093/OED/6670170844.

64. *Oxford English Dictionary*, s.n. "Carbohydrate" (modified September 2023), https://doi.org/10.1093/OED/2207088674.

65. Theresa Levitt, *Elixir: A Parisian Perfume House and the Quest for the Secret of Life* (Harvard University Press, 2023).

66. As Allison Bigelow writes of mining in the Iberian empires, "technical languages indexed broader cultural patterns of life, thought, meaning, and value." Allison Margaret Bigelow, *Mining Language: Racial Thinking, Indigenous Knowledge, and Colonial Metallurgy in the Early Modern Iberian World* (University of North Carolina Press, 2020), 3.

67. The shift to a molecular understanding of the world was a revolution in epistemic regimes, as profound as the emergence of probabilistic thinking or objectivity itself. For more on epistemic regimes in history, see Thomas S. Kuhn, *The Structure of Scientific Revolutions* (University of Chicago Press, 1962); Lorraine Daston and Peter Galison, *Objectivity* (Zone Books, 2007); Ian Hacking, *Historical Ontology* (Harvard University Press, 2002); Ian Hacking, "Making Up People," *London Review of Books* 28, no. 16 (2006); Ursula Klein and Wolfgang Lefèvre, *Materials in Eighteenth-Century Science: A Historical Ontology* (MIT Press, 2007), 9. These epistemic changes are so large that, as Martin Kusch observed, the ordinary mechanisms of historical cause and effect no longer seem satisfying. Martin Kusch, "Objectivity and Historiography," *Isis* 100, no. 1 (2009): 127–31.

68. Antoni Van Leeuwenhoek, "Other Microscopical Observations, Made by the Same, about the Texture of the Blood, the Sap of Some Plants, the Figure of Sugar and Salt, and the Probable Cause of the Difference of Their Tasts," *Philosophical Transactions of the Royal Society of London* 10, no. 117 (1675): 380–85; for virtual witnessing, see Steven Shapin and Simon Schaffer, *Leviathan and the Air-Pump: Hobbes, Boyle, and the Experimental Life* (Princeton University Press, 1985).

69. Van Leeuwenhoek, "Other Microscopical Observations," 383–84.

70. Witold Kula, *Measures and Men*, trans. R. Sretzer (Princeton University Press, 1986), 99.

Because measures "are replete with important, concrete social meaning," Kula wrote, the job of the historian is to uncover those meanings, not just "translate" previous units into today's.

71. D. Quélus, *Histoire naturelle du cacao et du sucre* (d'Houry, 1719), 196.

72. Miguel de Baeza, *Los quatro libros del arte de la confitería*, ed. Antonio Pareja (1592; repr., Toledo, 2014), 25; Juan de la Mata, *Arte de reposteria* [. . .] (En la Imprenta de Josef Herrera, 1786), 4; Eddy Stols, "The Expansion of the Sugar Market in Western Europe," in *Tropical Babylons: Sugar and the Making of the Atlantic World, 1450–1680*, ed. Stuart B. Schwartz (University of North Carolina Press, 2004), 245.

73. Jon Stobart, *Sugar and Spice: Grocers and Groceries in Provincial England, 1650–1830* (Oxford University Press, 2016), 56.

74. Edwin Farnsworth Atkins, *Sixty Years in Cuba: Reminiscences of Edwin F. Atkins* (Privately printed at the Riverside Press, 1926). Of course, the sugar on Americans' and Europeans' tables in 1925 really was noticeably different from the sugar on the tables of their great-grandparents. Yet it is also true that when historians talk about the extraordinary changes to the sugar industry, they are echoing the language of many contemporary observers. The original historians to write about the first sugar revolution did so at the same time Atkins wrote his memoirs of the second. See Higman, "The Sugar Revolution," 215.

75. Fernando Ortiz, *Cuban Counterpoint: Tobacco and Sugar*, trans. Harriet De Onís (Duke University Press, 1995), 24.

76. Elizabeth Massa Hoiem, "The Progress of Sugar: Consumption as Complicity in Children's Books about Slavery and Manufacturing, 1790–2015," *Children's Literature in Education* 52, no. 2 (2021): 162–82.

77. "The Valuation of Sugars," in Sugar (Polarization)—Refiners' Committee on Sugar Differentials, Sugar Institute—Correspondence, University of Florida (hereafter UFL).

78. Michel Callon, *Markets in the Making: Rethinking Competition, Goods, and Innovation*, ed. Martha Poon, trans. Olivia Custer (Zone Books, 2021), 29–36.

79. Ortiz, *Cuban Counterpoint*, 7.

80. Sugar "comes into the world without a last name, like a slave," he wrote, and is "refined until it can pass for white, travel all over the world, reach all mouths, and bring a better price." Ortiz, *Cuban Counterpoint*, 42.

81. On the choice of sugar as a subject, the historical philosopher Hasok Chang explains why he chose the subject of his book *Is Water H_2O?*: "I have deliberately chosen as the subject of my study one of the most familiar substances in human life and one of the most basic facts about that substance to make us all aware of the challenges involved in building scientific knowledge, no matter how simple or taken for granted." Hasok Chang, *Is Water H_2O?* (Springer Netherlands, 2012), xv.

82. As James Scott wrote in *Seeing Like a State*, "formal schemes of order are untenable without some elements of the practical knowledge that they tend to dismiss" (@scottSeeingStateHow1999, 7). Not all forms of standardization work in the same way, and resistance to standardization, appropriately enough, takes local and idiosyncratic forms. Scott reminds us that what look like anomalies are actually what make the system work, and as Miruna Achim notes, what is important historically about commodities is not just how close they are to an idealized perfect commodity, abstracted from the place and means and people of its production, but also how they are not identical Scott, *Seeing Like a State*, 335; Miruna Achim, "The Art of the Deal, 1828: How Isidro Icaza Traded Pre-Columbian Antiquities to Henri Baradère for Mounted Birds and Built a National Museum in Mexico City in the Process," *West 86th: A Journal of Decorative Arts, Design History, and Material Culture*, July 19, 2015.

Chapter 2

1. Patricia Mainardi, "Grandville, Visions, and Dreams," *Public Domain Review*, September 26, 2018, https://publicdomainreview.org/essay/grandville-visions-and-dreams/; John Tresch, *The Romantic Machine: Utopian Science and Technology After Napoleon* (University of Chicago Press, 2012), 176–84.

2. J. J. Grandville and Taxile Delord, *Un autre monde: Transformations, visions, incarnations, ascensions, locomotions, explorations, pérégrinations, excursions, stations, cosmogenies, fantasmagories, réveries, lubies, facéties, folatreries, métamorphoses, boomorphoses, lithomorphoses, métempsycoses, apothéoses, et autres choses* (H. Fournier, 1844), 64.

3. Holly Case, *The Age of Questions; or, a First Attempt at an Aggregate History of the Eastern, Social, Woman, American, Jewish, Polish, Bullion, Tuberculosis, and Many Other Questions Over the Nineteenth Century, and Beyond* (Princeton University Press, 2018), 6.

4. Louis Napoleon Bonaparte, "Analysis of the Sugar Question," in *The Political and Historical Works of Louis Napoleon Bonaparte*, vol. 2 (Published at the Office of the Illustrated London Library, 1852), 36.

5. Jonna M. Yarrington, "Droits and Frontières: Sugar and the Edge of France, 1800–1860" (master's thesis, University of Arizona, 2014).

6. For an exciting counterexample, see Bernadette Pérez's dissertation and forthcoming book, "Before the Sun Rises: Contesting Power and Cultivating Nations in the Colorado Beet Fields" (PhD diss., University of Minnesota, 2017).

7. John Perkins, "The Political Economy of the Sugar Beet in Imperial Germany," in *Crisis and Change in the International Sugar Economy, 1860–1914*, ed. Bill Albert and Adrian Graves, (ISC Press,1984); George S. Vascik, "Sugar Barons and Bureaucrats: Unravelling the Relationship Between Economic Interest and Government in Modern Germany, 1799–1945," *Business and Economic History* 21 (1992): 336–42.

8. Justus von Liebig, *Familiar Letters on Chemistry, in Its Relations to Physiology, Dietetics, Agriculture, Commerce, and Political Economy*, ed. John Blyth, 4th ed. (Walton and Maberly, 1859), 159.

9. Bosma, *The World of Sugar*, 94.

10. Roland Treillon and Jean Guérin, "La guerre des sucres," *Culture Technique* 16 (1986): 224–35, comes closest to telling a story in which the identity of the substances is not given.

11. Sidney W. Mintz, *Tasting Food, Tasting Freedom: Excursions into Eating, Culture, and the Past* (Beacon Press, 1996), 85.

12. Klein and Lefèvre, *Materials in Eighteenth-Century Science*, 1–2.

13. Klein and Lefèvre, *Materials in Eighteenth-Century Science*, 199.

14. Martin S. Staum, "Marggraf, Andreas Sigismund," in *Dictionary of Scientific Biography*, ed. Charles Gillispie (Scribner, 1980).

15. Ursula Klein, "Apothecary Shops, Laboratories and Chemical Manufacture in Eighteenth-Century Germany," in *The Mindful Hand: Inquiry and Invention from the Late Renaissance to Early Industrialisation*, ed. Lissa Roberts et al. (Koninkliijke Nederlandse Akademie van Wetenschappen, 2007), 247–79; Valentina Pugliano, "Natural History in the Apothecary's Shop," in *Worlds of Natural History*, ed. Emma C. Spary et al. (Cambridge University Press, 2018), 44–60.

16. Delbourgo, *Collecting the World*; Lorraine Daston and Katharine Park, *Wonders and the Order of Nature, 1150–1750*, rev. ed. (Zone Books, 2001).

17. Lissa Roberts, "The Death of the Sensuous Chemist: The 'New' Chemistry and the Transformation of Sensuous Technology," *Studies in History and Philosophy of Science, Part A*, 26, no. 4 (1995): 503–29.

18. Klein and Lefèvre, *Materials in Eighteenth-Century Science*, 202–3.

19. George H. Coons, "The Sugar Beet: Product of Science," *Scientific Monthly* (Business Press, 1949), 151.

20. François de Commerell, *An Account of the Culture and Use of the Mangel Wurzel, or Root of Scarcity* (C. Dilly and J. Phillips, 1787).

21. Benjamin Rush to George Washington, April 26, 1788, https://founders.archives.gov/documents/Washington/04-06-02-0204; Thomas Jefferson to George Logan, November 24, 1816, https://founders.archives.gov/documents/Jefferson/03-10-02-0411.

22. Klein, "Apothecary Shops, Laboratories and Chemical Manufacture in Eighteenth-Century Germany," 256.

23. Staum, "Marggraf, Andreas Sigismund."

24. Andreas Sigismund Marggraf, "Expériences chymiques faites dans le dessein de tirer un véritable sucre de diverses plantes, qui croissent dans nos contrées," *Mémoires de l'Académie royale des Sciences et Belles Lettres*, 1747, 79–90.

25. Olivier de Serres, *Le théâtre d'agriculture et mesnage des champs*, 2 vols. (Paris, 1805), 2:247; Deerr, *The History of Sugar*, 2:473.

26. Miles Ogborn, "Vegetable Empire," in *Worlds of Natural History*, ed. Emma C. Spary et al. (Cambridge University Press, 2018), 271–86.

27. Yarrington, "Droits and Frontières," 42.

28. Ursula Klein, "Contexts and Limits of Lavoisier's Analytical Plant Chemistry: Plant Materials and Their Classification," *Ambix* 52, no. 2 (2005): 107–57.

29. Louis Hasbrouck, "Notes from Lectures on Chemistry Delivered by Doctor John Maclean, Princeton College, in 1796 & 1797," American Philosophical Society, Philadelphia, PA, Ms. 540.H27, 67–85.

30. J. B. Gough, "Achard, Franz Karl," in *Dictionary of Scientific Biography*, ed. Charles Coulston Gillispie (Scribner, 1980).

31. See Richard Pares, "The London Sugar Market, 1740–1769," *Economic History Review* 9, no. 2 (1956): 254–70, for sugar prices in the mid-eighteenth century.

32. Franz Carl Achard, *Traité complet sur le sucre européen de betteraves: Culture de cette plante considérée sous le rapport agronomique et manufacturier*, ed. Ch. Derosne, trans. D. Angar (Paris, 1812); see the "introduction préliminaire" by the chemist and sugar-machinery entrepreneur Charles Derosne.

33. Klein, "Apothecary Shops, Laboratories and Chemical Manufacture in Eighteenth-Century Germany," 256.

34. Vascik, "Sugar Barons and Bureaucrats," 336; Deerr, *The History of Sugar*, 2:474–75.

35. John Quincy Adams to Thomas Boylston Adams, July 20, 1800, https://founders.archives.gov/documents/Adams/04-14-02-0163-0002.

36. M. A. N. Scherer and Jean-Baptiste Van Mons, "Note sur l'extraction du sucre de la betterave: Extraite d 'une lettre de M. A. N. Scherer, au citoyen Van Mons," *Annales de chimie, ou, Recueil de mémoires concernant la chimie et les arts qui en dépendent* 30 (1799): 299–302.

37. John Quincy Adams to Thomas Boylston Adams, August 9, 1800. https://founders.archives.gov/documents/Adams/04-14-02-0163-0006.

38. Alexander von Humboldt and Aimé Bonpland, *Relation historique du voyage aux régions équinoxiales du nouveau continent*, 3 vols. (F. A. Brockhaus, 1825), 3:417–18,

39. Thomas Jefferson to Alexis Marie Rochon, December 14, 1813. https://founders.archives.gov/documents/Jefferson/03-07-02-0025.

40. Thomas Jefferson to André Thoüin, December 14, 1813. https://founders.archives.gov/documents/Jefferson/03-07-02-0026.

41. See April Merleaux, *Sugar and Civilization: American Empire and the Cultural Politics of Sweetness* (University of North Carolina Press, 2015).

42. Treillon and Guérin, "La guerre des sucres," 225–26.

43. Ruth Scurr, *Napoleon: A Life Told in Gardens and Shadows* (Liveright, 2021), 142.

44. Maurice P. Crosland, *The Society of Arcueil: A View of French Science at the Time of Napoleon I* (Harvard University Press, 1967), 61.

45. Louis Napoleon Bonaparte, "Analysis of the Sugar Question," 3; Deerr, *The History of Sugar*, 2:487.

46. Crosland, *The Society of Arcueil*, 63.

47. Pierre Vigreux, "Aux origines du savoir agro-alimentaire: La création de l'École Nationale des Industries Agricoles (Douai, 1893)," *Revue du Nord* 72, no. 285 (1990): 258; Crosland, *The Society of Arcueil*, 36.

48. Deerr, *The History of Sugar*, 2:494; Yarrington, "Droits and Frontières," 41.

49. Deerr, *The History of Sugar*, 2:476.

50. Jules Helot, *Le sucre de betterave en France de 1800 à 1900* (Fernand et Paul Deligne, 1900), 198.

51. Berlin Institute for Sugar Technology pamphlet, Papers of Dan Gutleben, Banc MSS 70/39, carton 2, Sugar Research Extracts, Bancroft Library, University of California Berkeley.

52. Crosland, *The Society of Arcueil*, 36.

53. Helen Nearing and Scott Nearing, *The Maple Sugar Book: Together with Remarks on Pioneering as a Way of Living in the Twentieth Century* (Schocken Books, 1970). To radical-minded Americans of the late twentieth century, tapping a sugar maple meant striking a blow against capitalism. The tree and its flowing syrup represented a different way of organizing society and the economy, one that had remained more or less unaltered and inalterable for centuries if not millennia. From the Maine woods, simple-living advocates Helen and Scott Nearing wrote *The Maple Sugar Book* in 1950, and subtitled it "*Being a plain practical account of the Art of Sugaring designed to promote an acquaintance with the Ancient as well as the Modern practise, together with remarks on Pioneering as a way of living in the twentieth century.*" In 1979, Pete Seeger turned "maple syrup time" into a jaunty anti-industrial tune, exhorting his listeners to "seize the minute" without letting the syrup seize on the fire. "As in life or revolution /rarely is there a quick solution / anything worthwhile takes a little time," he sang, ending with a shout: "split wood, not atoms!" Pete Seeger, "Maple Syrup Time," track A7 on *Circles & Seasons*, Warner Brothers Records, BSK 3329, 1979.

54. Thomas Jefferson to Benjamin Vaughan, June 27, 1790, https://founders.archives.gov/documents/Jefferson/01-16-02-0342.

55. Thomas Jefferson to John Holmes, April 22, 1820, https://founders.archives.gov/documents/Jefferson/03-15-02-0518.

56. Robert Boyle, *Some Considerations Touching the Usefulnesse of Experimental Naturall Philosophy: The Second Part* (Oxford, 1663), 112.

57. Pierre-François-Xavier de Charlevoix, *Histoire et description générale de la nouvelle France, avec le journal historique d'un voyage fait par ordre du roi dans l'Amérique septentrionnale*, 6 vols. (Paris, 1744), 5:181.

58. Esther Louise Larsen and Peter Kalm, "Peter Kalm's Description of How Sugar Is Made from Various Types of Trees in North America," *Agricultural History* 13, no. 3 (1939): 153–54.

59. Benjamin Rush, *An Account of the Sugar Maple-Tree, of the United States, and of the Methods of Obtaining Sugar from It* [. . .] (R. Aitken & Son, 1791), 4–9.

60. Rush, *An Account of the Sugar Maple-Tree*, 8.

61. Alan Taylor, *William Cooper's Town: Power and Persuasion on the Frontier of the Early American Republic* (Alfred A. Knopf, 1995), 120–25.

62. Carroll Davidson Wright, *Comparative Wages, Prices, and Cost of Living: From the Sixteenth Annual Report of the Massachusetts Bureau of Statistics of Labor, for 1885* (Wright & Potter, 1889), 117.

63. Taylor, *William Cooper's Town*, 134.

64. William Short to Thomas Jefferson, January 16, 1791, https://founders.archives.gov/documents/Jefferson/01-18-02-0174.

65. William Short to Thomas Jefferson, July 17, 1791, https://founders.archives.gov/documents/Jefferson/01-20-02-0296; and January 24, 1792, https://founders.archives.gov/documents/Jefferson/01-23-02-0064.

66. Marie Nicolas Bouillet, *Atlas universel d'histoire et de géographie* (Librairie de L. Hachette et Cie, 1865), 68.

67. Thomas Jefferson to James Madison, July 6, 1791, https://founders.archives.gov/documents/Jefferson/01-20-02-0253.

68. James Madison to Thomas Jefferson, July 10, 1791, https://founders.archives.gov/documents/Madison/01-14-02-0035.

69. James Madison to Thomas Jefferson, July 24, 1791 https://founders.archives.gov/documents/Jefferson/01-20-02-0314; Thomas Jefferson to James Madison, July 28, 1791, https://founders.archives.gov/documents/Madison/01-14-02-0050.

70. Thomas Jefferson to William Short, November 25, 1791, https://founders.archives.gov/documents/Jefferson/01-22-02-0305.

71. William Cooper, "A Letter on the Manufacture of Maple Sugar," *Transactions of the Society for the Promotion of Agriculture, Arts and Manufactures, Instituted in the State of New York.*, 2 vols. (1792), 1:84; Thomas Jefferson to William Short, November 25, 1791, https://founders.archives.gov/documents/Jefferson/01-22-02-0305.

72. Le Romain, "Sucre."

73. Achard, *Traité complet sur le sucre européen de betteraves*, iv.

74. Napoleon Bonaparte, "Analysis of the Sugar Question," 2–3.

75. Scurr, *Napoleon*, 154.

76. Scherer and Van Mons, "Note sur l'extraction du Sucre de la betterave."

77. Von Liebig, *Familiar Letters on Chemistry*, 18.

78. Joseph Proust, "Lettre du Professeur Proust a Jean-Christophe Delamétherie sur le sucre du raisin," *Journal de physique, de chimie et d'histoire naturelle*, 1802, 113; Crosland, *The Society of Arcueil*, 34.

79. Jill H. Casid, *Sowing Empire: Landscape and Colonization* (University of Minnesota Press, 2004), 108.

80. Scherer and Van Mons, “Note sur l’extraction du sucre de la betterave.”

81. Crosland, *The Society of Arcueil*, 36.

82. Coons, “The Sugar Beet.”

83. Achard, *Traité complet sur le sucre européen de betteraves.*

84. Jean-Antoine Chaptal, *Mémoire sur le sucre de betteraves* (Huzard, 1818), 50.

85. Chaptal, *Mémoire sur le sucre de betteraves*, 51.

86. Chaptal, *Mémoire sur le sucre de betteraves*, 52.

87. Berlin Institute for Sugar Technology pamphlet, Papers of Dan Gutleben, Banc MSS 70/39, Carton 2, Sugar Research Extracts, Bancroft Library, University of California Berkeley.

88. Charles Kostelnick, “Visualizing Technology and Practical Knowledge in the Encyclopédie’s Plates: Rhetoric, Drawing Conventions, and Enlightenment Values,” *History & Technology* 28, no. 4 (2012): 443–54; Celina Fox, *The Arts of Industry in the Age of Enlightenment* (Yale University Press for the Paul Mellon Centre for Studies in British Art, 2009).

89. I’m grateful to Kyrill Kunakhovich for help with this translation.

90. Roche, *A History of Everyday Things*, 245.

91. Jane Haldimand Marcet, *Conversations on Chemistry, in Which the Elements of That Science Are Familiarly Explained and Illustrated by Experiments and Plates* (Sidney’s Press, 1813), 235.

92. Treillon and Guérin, “La guerre des sucres,” 228–32.

93. Humboldt and Bonpland, *Relation historique du voyage aux régions équinoxiales du nouveau continent*, 417–18. Muchas gracias a Reinaldo Funes Monzote por informarme a esta discusion en el libro de Humboldt.

94. Lyon Playfair and Wemyss Reid, *Memoirs and Correspondence of Lyon Playfair* (Cassell and Company, 1899), 33.

95. Pierre Macherey, *The Object of Literature* (Cambridge University Press, 1995), 206.

96. William Brock, *Justus von Liebig: The Chemical Gatekeeper* (Cambridge University Press, 1997), 118–19.

97. Deerr, *The History of Sugar*, 2: 476; Karl Christoph Vogt, *Aus meinem Leben: Erinnerungen und Rückblicke* (E. Nägele, 1896), 130. (“Ein jeder esel kann schließlich eine Zuckerfabrik dirigieren.”)

98. Justus Liebig, *Chemistry, and Its Application to Physiology, Agriculture, and Commerce.* (New York, 1847), v.

99. Von Liebig, *Familiar Letters on Chemistry*, 159.

100. Vascik, “Sugar Barons and Bureaucrats,” 336.

101. David Lee Child, *The Culture of the Beet, and Manufacture of Beet Sugar* (Weeks, Jordan & Co.), 18.

102. Perkins, “The Political Economy of the Sugar Beet in Imperial Germany.”

103. Rood, *The Reinvention of Atlantic Slavery*, 19–41.

104. Theresa Levitt, *The Shadow of Enlightenment: Optical and Political Transparency in France, 1789–1848* (Oxford University Press, 2009), 5.

105. Levitt, *Elixir.*

106. Levitt, *The Shadow of Enlightenment*, 12; Gerard L’Estrange Turner, *Nineteenth-Century Scientific Instruments* (University of California Press, 1983), 222.

107. Jean-Baptiste Biot, “Comparaison du sucre et de la gomme ara bigue dans leur action sur la lumière polarisée,” *Bulletin des sciences par la société philomatique de Paris*, 1816, 125.

108. Jean-Baptiste Biot, “Examen comparatif du sucre de maïs et du sucre de betterave, soumis aux épreuves de la polarisation circulaire,” *Comptes rendus hebdomadaires des séances de l’Académie des Sciences* 2 (1836): 464–67.

109. Jean-Baptiste Biot, "Sur l'utilité que pourraient offrir les caractères optiques dans l'exploitation des sucreries et des raffineries," *Comptes rendus hebdomadaires des séances de l'Académie des Sciences* 10 (1840): 265.

110. Michael Fakhri, *Sugar and the Making of International Trade Law*, Cambridge Studies in International and Comparative Law (Cambridge University Press, 2014).

111. Grandville and Delord, *Un autre monde*, 64.

112. Deerr, *The History of Sugar*, 2:474.

Chapter 3

1. Samuel Martin, *An Essay on Plantership: Inscribed to Sir George Thomas, Bart. As a Monument to Ancient Friendship* [. . .] (Robert Mearns, 1785), 18.

2. *Ingenio* could also refer to other advanced extractive industrial processes, such as mining. See Allison Bigelow and Pablo Cruz, "*Ingenios* and Ingenuity: Rethinking Indigenous Histories of Silver in the Colonial Andean Mining Industry," *Colonial Latin American Review* 30, no. 4 (2021): 520–44.

3. Rebekah Mitsein, "Humanism and the Ingenious Machine: Richard Ligon's True and Exact History of the Island of Barbados," *Journal for Early Modern Cultural Studies* 16, no. 1 (2016): 104.

4. John Daniels and Christian Daniels, "The Origin of the Sugarcane Roller Mill," *Technology and Culture* 29, no. 3 (1988): 504.

5. Richard Pares, *A West-India Fortune* (Longmans, Green and Co., 1950), 115.

6. Quélus, *Histoire naturelle du cacao et du sucre*, 158.

7. Edward Littleton, *The Groans of the Plantations, or, a True Account of Their Grievous and Extreme Sufferings by the Heavy Impositions Upon Sugar and Other Hardships Relating More Particularly to the Island of Barbados* (London, 1689), 19–20.

8. Elizabeth Maddock Dillon, "The Cost of Sugar: Narratives of Loss of Life and Limb," plenary talk at *Beyond Sweetness: New Histories of Sugar in the Early Atlantic World*, John Carter Brown Library, Brown University, October 26, 2013.

9. Littleton, *The Groans of the Plantations*, 17.

10. Caitlin Rosenthal, *Accounting for Slavery: Masters and Management* (Harvard University Press, 2018).

11. Moreno Fraginals, *The Sugarmill*, 39.

12. Rood, *The Reinvention of Atlantic Slavery*, 18–19.

13. Quélus, *histoire naturelle du cacao et du sucre*, 173.

14. Richard Ligon, *A True and Exact History of the Island of Barbados* (Humphrey Moseley, 1657), 3, 92–95.

15. Henry Bessemer, *On a New System of Manufacturing Sugar from the Cane* (London: Privately printed, 1852), 24.

16. Deerr, *The History of Sugar*, 2:537–50.

17. A. J. Wallis-Tayler, *Sugar Machinery: A Descriptive Treatise Devoted to the Machinery and Processes Used in the Manufacture of Cane and Beet Sugars* (W. Rider and Son, 1895), 16.

18. Deerr, *The History of Sugar*, 2:546.

19. José Antonio Piqueras, "The Discovery of Progress in Cuba: Machines, Slaves, Businesses," in *New Frontiers of Slavery*, ed. Dale W. Tomich, trans. Paul Edgar (State University of New York Press, 2016), 64.

20. Lizette Cabrera Salcedo, *De los bueyes al vapor: Caminos de la tecnología del azúcar en Puerto Rico y el Caribe* (La Editorial Universidad de Puerto Rico, 2010), 303–13; Daniel B. Rood, review of *De los bueyes al vapor: Caminos de la tecnología en Puerto Rico y el Caribe*, *Caribbean Studies* 39, no. 1/2 (2011): 243–48.

21. Rood, *The Reinvention of Atlantic Slavery*, 36–37; Deerr, *The History of Sugar*, 2:565–70. Rood notes that the Rilieux vacuum pan then "had to be reverse-creolized to function in the peculiarities of the European climate."

22. Deerr, *The History of Sugar*, 2:573–76.

23. Teresita Martínez Vergne, *Capitalism in Colonial Puerto Rico: Central San Vicente in the Late Nineteenth Century* (University Press of Florida, 1992); Andrés Ramos Mattei, *La sociedad del azúcar en Puerto Rico, 1870–1910* (Universidad de Puerto Rico, Recinto de Río Piedras, 1988).

24. Carlos Peñaranda, *Cartas Puertorriqueñas: Dirigidas al célebre poeta Don Ventura Ruiz Aguilera, 1878–1880* (Sucesores de Rivadeneyra, 1885), 87. "El azúcar, depurada de toda miel, que filtra á otro depósito exterior, lavada por el vapor, y seca y compacta por la rapidez del aire que la despide con fuerza á las paredes laterales del tambor, forma apretadas masas, y es conducida en andas de blanca madera al almacén general, *dando envidia con su pureza y transparencia al cristal y á la nieve*." Translation my own.

25. Nadia Fernández-de-Pinedo and David Pretel, "Circuits of Knowledge: Foreign Technology and Transnational Expertise in Nineteenth-century Cuba," in *The Caribbean and the Atlantic World Economy: Circuits of Trade, Money and Knowledge, 1650–1914*, ed. Adrian Leonard and David Pretel, Cambridge Imperial and Post-Colonial Studies Series (Palgrave Macmillan, 2015), 269.

26. *Glasgow*, vol. 1, *Beginnings to 1830*, ed. T. M. Devine and Gordon Jackson (Manchester University Press, 1995); Richard Pares, *The Historian's Business, and Other Essays* (Clarendon Press, 1961), 210.

27. Stephen Mullen, *The Glasgow Sugar Aristocracy: Scotland and Caribbean Slavery, 1775–1838*, New Historical Perspectives (University of London Press, 2023).

28. Michael S. Moss and John R. Hume, *Workshop of the British Empire: Engineering and Shipbuilding in the West of Scotland* (Heinemann, 1977), 3–5. Fernández-de-Pinedo and Pretel show that the large majority of sugar equipment overall came from Britain between 1844 and 1857, although during that period slightly more (264 against 233) grinding mills came from the United States.

29. John Mayer, *Notices of Some of the Principal Manufacturers of the West of Scotland* (Blackie & Son, 1876), 66.

30. *Puerto Rico en la mirada extranjera: La correspondencia de los cónsules Norteamericanos, Franceses e Ingleses, 1869–1900*, ed. Gervasio L. García Rodríguez and Emma Aurora Dávila Cox (Centro de Investigaciones Históricas, Decanato de Estudios Graduados e Investigación, Universidad de Puerto Rico, Recinto de Río Piedras, 2005), 165; Andrés Ramos Mattei, "Technical Innovations and Social Change in the Sugar Industry of Puerto Rico," in *Between Slavery and Free Labor: The Spanish-Speaking Caribbean in the Nineteenth Century*, ed. Manuel Moreno Fraginals et al. (Johns Hopkins University Press, 1985), 158–78.

31. Galloway, *The Sugar Cane Industry*, 261; Deerr, *The History of Sugar*, 2:610. The index for J. H. Galloway's *Sugar Cane Industry* does not even include an entry for "Glasgow," and while that for Noël Deerr's standard and encyclopedic *History of Sugar* does mention the names of specific machinery producers, it includes only an entry for Glasgow as a center of refining.

32. Harold John Cook, *Matters of Exchange: Commerce, Medicine, and Science in the Dutch Golden Age* (Yale University Press, 2007).

33. James A. Secord, "Knowledge in Transit," *Isis* 95, no. 4 (2004): 654–72; Raj, *Relocating*

Modern Science. Jim Secord and Kapil Raj, among others, have called for histories of science in transit, lest an emphasis on the specific local cultures of scientific knowledge distract from the ultimate objective of understanding how knowledge moves beyond that culture.

34. Introduction, in *Instruments, Travel and Science: Itineraries of Precision from the Seventeenth to the Twentieth Century*, ed. Marie-Noëlle Bourguet, et al. (Routledge, 2002), 3–9.

35. Moreno Fraginals, "Plantations in the Caribbean," 8; Rood, *The Reinvention of Atlantic Slavery*, 34.

36. Ortiz, *Cuban Counterpoint*, 49.

37. Gabrielle Hecht, *Being Nuclear: Africans and the Global Uranium Trade* (MIT Press, 2012).

38. Robert Harvey, *Early Days of Engineering in Glasgow* (Aird & Coghill, 1919), 8.

39. Michael Pacione, *Glasgow: The Socio-Spatial Development of the City* (J. Wiley, 1995), 75–77; Harvey, *Early Days of Engineering in Glasgow*, 9.

40. S. G. Checkland and Anthony Slaven, eds., *Dictionary of Scottish Business Biography, 1860–1960* (Aberdeen University Press, 1986), 188–89.

41. "The Implement and Machinery Review," September 1, 1886, in "Newspaper cuttings," TD185/5, Records of Messrs. Duncan Stewart & Co., Mitchell Library, Glasgow.

42. The Story of Duncan Stewarts, and Their Association with the British Beet Sugar Industry," Address Given by Mr. K. S. Arnold at the General Managers' Annual Conference of the British Sugar Corporation at Peterborough, June 1, 1955. UGD052/1/4/3, Records of Duncan Stewart & Co Ltd, University of Glasgow Archive Center, Glasgow (hereafter UGA).

43. Wallis-Tayler, *Sugar Machinery*, 36.

44. R. A. Buchanan, "The Diaspora of British Engineering," *Technology and Culture* 27, no. 3 (1986): 501–24.

45. Mayer, *Notices of Some of the Principal Manufacturers of the West of Scotland*, 116.

46. *Beardmore News*, November 1923, Records of Duncan Stewart & Co, Ltd., UGD052/1/4/2, UGA.

47. Jonathan Curry-Machado, *Cuban Sugar Industry: Transnational Networks and Engineering Migrants in Mid-Nineteenth Century Cuba* (Palgrave Macmillan, 2011), 73–100.

48. Fernández-de-Pinedo and Pretel, "Circuits of Knowledge," 273; Job No. 386, Mirrlees Watson Order Book, UGD118/2/4/10, UGA.

49. Edwin Atkins to W. H. Ross Esq., August 22, 1885, II.6.287, Atkins Family Papers, Massachusetts Historical Society (hereafter MHS).

50. Atkins to Ross, April 17, 1888, II.7.394, Atkins Family Papers, MHS.

51. Atkins to Ross, n.d., II.8.260, and August 14, 1889, II.8.287–88, MHS.

52. Curry-Machado, *Cuban Sugar Industry*, 87–90.

53. Duncan Stewart Cost Book 1881–1901, p. 18, UGD052/1/1/1, UGA.

54. It is not clear when firms began using progress photographs; Volume 1 of A. & W. Smith's photograph albums is dated 1907. UGD118/1/5/1, UGA.

55. The Maryland planter Charles Carroll complained in 1776, "Descriptions of machines are seldom accurately made, and when done with exactness are seldom understood." Charles Carroll, *Journal of Charles Carroll of Carrollton During His Visit to Canada in 1776, as One of the Commissioners From Congress: With a Memoir and Notes (Printed by J. Murphy for the Maryland Historical Society, 1876*, 56–57. Thanks to Craig Wilder for pointing this out to me.

56. "Copy minute of agreement between N. P. Nathan's Sons and Robert Gilbert (Nov. 1883)—Forbes and Bryson, writers, Glasgow," UGD052/1/7/2, UGA.

57. Letter, July 7, 1884, from David H. Andrew, UGD052/1/7/4, UGA.

58. Letter, September 15, 1884, to Arucas, Canary Islands, UGD052 1/7/5, UGA.

59. Letter May 21, 1885, to Arucas, Canary Islands, UGD052/1/7/6, UGA.

60. Sugar Department Letterbooks, Mirrlees Watson & Co., UGD062 1/3/1, UGA.

61. Mirrlees Watson Mill and Krajewskis Order Book, UGD118/2/4/37, UGA. While in St. Lucia in 1973, then-Prince Charles stumbled across a rusting Mirrlees mill that was over a century old. "Mirrlees, Tait & Watson, Eglinton Works, Cook St., Engineers," AGN 479, Mitchell Library, Glasgow, UK.

62. Bookers Sugar Estates, Ltd., *Bookers Sugar: Supplement to the Accounts of Booker Brothers, McConnell & Co., Limited* (Georgetown, British Guiana, 1954), 123.

63. Moss and Hume, *Workshop of the British Empire*, 162.

64. David Edgerton, *The Shock of the Old: Technology and Global History Since 1900* (Oxford University Press, 2006), 77; Richard S. Dunn, *Sugar and Slaves: The Rise of the Planter Class in the English West Indies, 1624–1713* (University of North Carolina Press, 1972), 200–201.

65. Simon Schaffer, "Easily Cracked: Scientific Instruments in States of Disrepair," *Isis* 102, no. 4 (2011): 707.

66. Curry-Machado, *Cuban Sugar Industry*, 80–95.

67. Andrew Russell and Lee Vinsel, "Hail the Maintainers," *Aeon*, April 7, 2016.

68. Mirrlees Watson Mills and Krajewskis, book no. 1 (1841–1912) and no. 2 (1883–1964), UGD118/2/4/37 and /2/4/38, UGA.

69. Atkins, *Sixty Years in Cuba*, 93.

70. Mirrlees Watson Mill & Krajewskis, book no. 1 (1841–1912), UGD118/2/4/37, UGA.

71. Mirrlees Watson Mill & Krajewskis, book no. 2 (1883–1964), UGD118/2/4/38, UGA.

72. Moss and Hume, *Workshop of the British Empire*, 31.

73. Robert Harvey, "The History of the Sugar Machinery Industry in Glasgow," *International Sugar Journal*, 1917, 112.

74. Harvey, "The History of the Sugar Machinery Industry," 113.

75. Drawing from C. A. Hasche in Mayaguez, dated July 8, 1864, UGD202/2430, UGA.

76. UGD118/2/4/37. See also Descriptive Drawings Index, "Horizontal Sugar Mills," p. 502, UGD118/2/7/12, UGA.

77. Chain link drawing from Thomas Dodd, September 16, 1859, UGD202/1555, UGA (emphasis added).

78. Mirrlees Order Book, 1891, Job no. 304, UGD118/2/4/10, UGA.

79. Harvey, "The History of the Sugar Machinery Industry in Glasgow," 59.

80. Harvey, *Early Days of Engineering in Glasgow*, 11.

81. Peter Jeffrey Booker, *A History of Engineering Drawing* (Chatto & Windus, 1963), 134.

82. Steven Shapin, "The Invisible Technician," *American Scientist* 77, no. 6 (1989): 554–63; Stewart, 2008; Larry Stewart, "Assistants to Enlightenment: William Lewis, Alexander Chisholm and Invisible Technicians in the Industrial Revolution," *Notes and Records of the Royal Society of London* 62, no. 1 (2008): 17–29; Steven Shapin and Barry Barnes, "Science, Nature and Control: Interpreting Mechanics' Institutes," *Social Studies of Science* 7, no. 1 (1977): 31–74; Maxine Berg, *The Machinery Question and the Making of Political Economy, 1815–1848* (Cambridge University Press, 1980), 147–55.

83. See John Laidlaw letters, T-HB 72, Mitchell Library, Glasgow. For apprentices at Duncan Stewart and their salaries see TD 158/11, Apprenticeship Book, Mitchell Library.

84. Frances Robertson, "Manufacturing the Visual Economy in Nineteenth-Century Britain," conference presentation, International Society for Cultural History, Lunéville, July 2–5, 2012.

85. Moss and Hume, *Workshop of the British Empire*, 160; Booker, *A History of Engineering Drawing*.

86. Frances Robertson, "Delineating a Rational Profession: Engineers and Draughtsmen as 'Visual Technicians' in Early Nineteenth-Century Britain," conference presentation, 3 Societies Meeting, Philadelphia, July 11–14, 2012.

87. "Inventory and Valuation of Machinery Plant, and Tools at Eglinton Engine Works, Cook Street, Glasgow, made by Messrs John Turnbull Jr. & Sons, Consulting Engineers, 18 Blythswood Sq. Glasgow, 20th February 1897," UGD118,1/7/1, UGA.

88. Sarah Fayen Scarlett, "The Craft of Industrial Patternmaking," *Journal of Modern Craft* 4, no. 1 (2011): 27–48.

89. Harvey, *Early Days of Engineering in Glasgow*, 60–61.

90. Drawing dated June 24, 1857, UGD202/0044, UGA.

91. Drawing dated December 24, 1861, UGD202/2148, UGA.

92. Drawing dated May 25, 1862, UGD202/1890, UGA.

93. David McGee, "From Craftsmanship to Draftsmanship: Naval Architecture and the Three Traditions of Early Modern Design," *Technology and Culture* 40, no. 2 (1999): 214; Jonathan Zeitlin, "Between Flexibility and Mass Production: Strategic Ambiguity and Selective Adaptation in the British Engineering Industry, 1830–1914," in *World of Possibilities: Flexibility and Mass Production in Western Industrialization*, ed. Charles F. Sabel and Jonathan Zeitlin (Cambridge University Press, 1997), 247.

94. "Tenemos diseños y plantillas de todas la centrífugas fabricadas por nosotros desde la fundación de nuestra casa en 1870 hasta la fecha, así que podemos atender á cualquier pedido de repuestos." Watson, Laidlaw, & Co. catalog, "Centrifugas," in Colección Central Mercedita, caja "Católogos Comerciales 1920s–1940s," Archivo General de Puerto Rico, San Juan.

95. Bruno Latour, *Science in Action: How to Follow Scientists and Engineers Through Society* (Harvard University Press, 1987), 232–37.

96. Matthew S. Hull, *Government of Paper: The Materiality of Bureaucracy in Urban Pakistan* (University of California Press, 2012), 23–24.

97. Harvey Engineering Company advertisement at the end of Llewellyn Jones and Frederic I. Scard, *The Manufacture of Cane Sugar* (Edward Stanford, 1909).

98. John Alfred Heitmann, *The Modernization of the Louisiana Sugar Industry, 1830–1910* (State University Press, 1987); Glasgow & West of Scotland Technical College annual report, 116th session (1912), p. 20. OE/4, University of Strathclyde Archives. For the Audubon Sugar School in Baton Rouge, see Heitmann, chapter 10.

99. Thomas H. P. Heriot, "Technical Training for the Sugar Industry," *International Sugar Journal* 28 (1916): 173–79; Glasgow & West of Scotland Technical College annual report, 116th session (1912), p. 64, University of Strathclyde Archives. The salary for 1911 is found in OE/6/1/1, GWSTC staff card index, University of Strathclyde Archives.

100. This information comes from the GWSTC annual reports for 1912 and 1913; I was unable to find entries for these donations in the relevant books of the various companies.

101. Glasgow & West of Scotland Technical College annual report, 117th session (1913), p. 24, University of Strathclyde Archives, OE/4.

102. Heriot, "Technical Training for the Sugar Industry," 175–76.

103. Heriot used a derogatory term that I have not reproduced.

104. Thomas Hawkins Percy Heriot, *Science in Sugar Production: An Introduction to Methods of Chemical Control* (N. Rodger, 1907), 7.

105. Heriot, "Technical Training for the Sugar Industry," 178.

106. Bessemer, *On a New System of Manufacturing Sugar from the Cane*, 57.

107. Mary S. Morgan and Marcel Boumans, "Secrets Hidden by Two-Dimensionality: The Economy as a Hydraulic Machine," in *Models: The Third Dimension of Science*, ed. Soraya de Chadarevian and Nick Hopwood (Stanford University Press, 2004), 369–401; for an example of the limits of physical models representing complex systems.

108. Heriot, "Technical Training for the Sugar Industry," 173 (emphasis added).

109. Thomas H. P. Heriot, "The Sugar Industry After the War," *Proceedings of the Royal Society of Glasgow* 49 (1918): 31.

110. . E10/2/2, Prospectus of Day Classes, 1912–13, 29, and E10/2/6, "Guide to Evening Classes in Science and Technology," 1912–13, University of Strathclyde Archives.

111. . Glasgow & West of Scotland Technical College annual report ,116th session (1912), p. 20; E10/2/2, Glasgow & West of Scotland Technical College Prospectus of Day Classes, 1911–12, University of Strathclyde Archives (emphasis added).

112. William Garvie Hall, "Future of Sugar Is in the Orient," *The Trans-Pacific: A Magazine of International Service Covering the Far East and Australasia*, September 1921, 59; Angus McLean, *Local Industries of Glasgow and the West of Scotland* (British Association for the Advancement of Science, 1901), 64.

113. Francis Amasa Walker, ed., *United States Centennial Commission. International Exhibition, 1876: Reports and Awards, Groups III–VII*, vol. 4 (Government Printing Office, 1880), 63.

114. Hall, "Future of Sugar Is in the Orient," 59.

115. "Honolulu Iron Works Scores in Formosa," *Planter's Monthly* 28 (1909): 346–47.

116. George F. Nellist, *The Story of Hawaii and Its Builders, with Which Is Incorporated Volume III Men of Hawaii* (Honolulu Star-Bulletin, 1925), 497–500.

117. Nellist, *The Story of Hawaii*, 327.

118. United States Geological Survey, "Ancient Hawaiians: Plenty of Oars but No Ores" (Volcano Watch, September 24, 2009), https://www.usgs.gov/center-news/volcano-watch-ancient-hawaiians-plenty-oars-no-ores; John Rickman, *Journal of Captain Cook's Last Voyage, to the Pacific Ocean: On Discovery: Performed in the Years 1776, 1777, 1778, 1779, and 1780* [. . .] 2008, 197, Eighteenth Century Collections Online; William L Worden, *Cargoes: Matson's First Century in the Pacific* (University Press of Hawai'i, 1981), 4. Worden quotes a twentieth-century shipping executive: "if a thing can't be made of lava, coral, air or semitropical plants, then the chances are Hawaii must import it."

119. Nellist, *The Story of Hawaii and Its Builders*, 497.

120. "To Revive Sugar in Mexico, Government and Honolulu Iron Works Plan Cooperatives," *New York Times*, January 17, 1925, https://timesmachine.nytimes.com/timesmachine/1925/01/17/101633759.pdf?pdf_redirect=true&ip=0.

121. "Remodelling Brooklyn Refinery," *New York Times*, March 17, 1926, https://timesmachine.nytimes.com/timesmachine/1926/03/17/119063682.pdf?pdf_redirect=true&ip=0.

122. Hall, "Future of Sugar Is in the Orient."

123. D. Bradley, "Fletcher & Stewart Ltd: A Business History" (MPhil thesis, University of Nottingham), pp. 118, 122, 211, UGD118/11/1/48, UGA.

124. Nellist, *The Story of Hawaii and Its Builders*.

125. Hall, "Future of Sugar Is in the Orient,"

126. Entries for October 24 1921, and May 3, 1922, A. & W. Smith Minute Book no. 2, UGD118 1/1/1, UGA.

127. "The Adeline Sugar Factory," *American Sugar Industry and Beet Sugar Gazette*, January 1912.

Chapter 4

1. Walter Rodney, *A History of the Guyanese Working People, 1881–1905*, Johns Hopkins Studies in Atlantic History and Culture (Johns Hopkins University Press, 1981), 161.

2. Atkins, *Sixty Years in Cuba* 109–10; for views of Chinese laborers among Cuban planters, see Scott, *Slave Emancipation in Cuba*, 34.

3. Christian Schnakenbourg, "From the Sugar Estate to Central Factory: The Industrial Revolution in the Caribbean (1840–1905)," in Albert and Graves, *Crisis and Change in the International Sugar Economy*.

4. Edwin F. Atkins to J. George Lumilius, November 11 1989, II.8.361, Atkins Papers, Massachusetts Historical Society.

5. Rood, *The Reinvention of Atlantic Slavery*, 34.

6. Humberto García-Muñiz, "Louisiana's 'Sugar Tramps' in the Caribbean Sugar Industry," *Revista/Review Interamericana* 29, no. 1–4 (1999).

7. Mervyn David Lincoln, "The Culture of the South African Sugarmill: The Impress of the Sugarocracy" (PhD diss., University of Cape Town, 1985), 250.

8. University of Hawaii (Honolulu), *The Story of Cane Sugar, as Told to an Imaginary Class in Agriculture*. (University of Hawaii, 1928), 16.

9. Heriot, *Science in Sugar Production*, 14.

10. Otremba, "Inventing Ingenios," 122–25; Hans Sloane, *A Voyage to the Islands Madera, Barbados, Nieves, S. Christophers and Jamaica, with the Natural History of the Herbs and Trees, Four-Footed Beasts, Fishes, Birds, Insects, Reptiles, &c.* [. . .], 2 vols. (Printed by B. M. for the author, 1707), 1:lxi; John Proculus Baker, *An Essay on the Art of Making Moscovado Sugar Wherein a New Process Is Proposed* (Joseph Weatherby, 1775), 34; Martin, *An Essay on Plantership*, 23. Other Europeans found convoluted ways to deny the ingenuity of the enslaved. Hans Sloane, initiator of the British Museum's collections, relayed a ridiculous story about the origins of clayed sugar in the early pages of his narrative of a voyage to Jamaica. "Claying Sugar, as they report here," he dutifully recorded, "was first found out in Brazil, a Hen having her Feet dirty, going over a Pot of Sugar by accident, it was found under her tread to be whiter than elsewhere." Perhaps, to Sloane's all-collecting mind, an experimentalist chicken was no curiouser than any of the other wondrous objects or stories he amassed for his cabinets in London. Or possibly Sloane is letting his readers in on the joke, the phrase "as they report here" quietly signaling his personal skepticism of the planters' story. For centuries, writers on sugar exhorted the owners of plantations to devise rules on how to boil it, the most basic problem of their commodity's manufacture. Even disagreement could be spun as reflecting the superiority of Europeans over Africans. One eighteenth-century writer lamented that "few Planters agree in any Rule for judging when the Subject is sufficiently evaporated to strike," but also that this reflected a healthy experimental philosophy, since by contrast "it is certain, the Negro-Boilers have no rule at all." Their "long Experience" led them to "guess by the Appearance of the Liquor," and the reality that this unruly anti-method might produce sugar was almost as remarkable as the wonder of the sugar itself. Martin suggested planters use their fingers to test by touch as well as by eye, just "as two witnesses are better than one."

11. Pares, *A West-India Fortune*, 117.

12. Sloane, *A Voyage to the Islands*, lxii.

13. Benjamin McMahon, *Jamaica Plantership* (E. Wilson, 1839), 38–46.

14. Moreno Fraginals, *The Sugarmill*, 116.

15. Martin, *An Essay on Plantership.*, 23.

16. Moreno Fraginals, *The Sugarmill*, 116.

17. María M. Portuondo, "Plantation Factories: Science and Technology in Late-Eighteenth-Century Cuba," *Technology and Culture* 44, no. 2 (2003): 251.

18. Martin, *An Essay on Plantership*, 24.

19. Pares, *A West-India Fortune*, 188.

20. Rosenthal, *Accounting for Slavery*, 30.

21. Boiling House Book, Newton Plantation (1798), MS523, Senate House Library, University of London, provided digitally. (x-devonthink-item://B57373BB-5023-489D-8050-45C221F9CBBA)

22. Heitmann, *The Modernization of the Louisiana Sugar Industry*, 25–32; Rood, *The Reinvention of Atlantic Slavery*.

23. Deerr, *The History of Sugar*, 2:588.

24. Biot, "Sur l'utilité que pourraient offrir les caractères optiques dans l'exploitation des sucreries et des raffineries," 265.

25. Jean-Baptiste Biot, "Sur l'application des propriétés optiques à l'analyse quantitatives des mélanges liquides ou solides [. . .]," *Comptes rendus hebdomadaires des séances de l'Académie des Sciences* 16 (January–June 1843): 639.

26. Ligon defended the "slothfulnesse or sluggishnesse" of planters as a combination of earned rest and understandable torpor: "a decay of their spirits, by long and tedious hard labour, sleight [*sic*] feeding, and ill lodging, which is able to wear out and quell the best spirit of the world."

27. Biot, "Sur l'utilité que pourraient offrir les caractères optiques dans l'exploitation des sucreries et des raffineries," 265.

28. Moreno Fraginals, *The Sugarmill*, 142–44.

29. R. W. Beachey, *The British West Indies Sugar Industry in the Late 19th Century* (Blackwell, 1957).

30. Moreno Fraginals, *The Sugarmill*, 145–48; Portuondo, "Plantation Factories," 253–54.

31. Pérez, *Cuba*, 87, 103.

32. Skaife to Atkins, April 30, 1894; Skaife to Atkins, May 24, 1894; Atkins to Skaife, May 7, 1894, II.12.59, MHS.

33. Atkins to Terry, II.8.471, December 31, 1889, MHS.

34. Noël Deerr, *Sugar and the Sugar Cane: An Elementary Treatise on the Agriculture of the Sugar Cane and on the Manufacture of Cane Sugar* (N. Rodger, 1905); Oscar Zanetti Lecuona et al., "Nöel Deerr en la Guayana Británica, Cuba y Puerto Rico (1897–1921). Memorándum para la Historia del Azúcar en el Caribe," *Revista Mexicana del Caribe* 6, no. 11 (2001): 57–154.

35. Deerr, *Sugar and the Sugar Cane*, 281.

36. Deerr, *Sugar and the Sugar Cane*, 296.

37. Deerr, *Sugar and the Sugar Cane*, 282–83.

38. Atkins to Skaife, May 4, 1894, II.12.50, MHS.

39. Skaife to Atkins, May 14, 1894, MHS.

40. Deerr, *Sugar and the Sugar Cane*, 282–83.

41. Heriot, *Science in Sugar Production*, 14.

42. Hubert Edson, *Sugar, from Scarcity to Surplus* (Chemical Publishing Co., 1958), 95.

43. Deerr, *Sugar and the Sugar Cane*, 295–96.

44. Deerr, *Sugar and the Sugar Cane*, 329.

45. Edson, *Sugar, from Scarcity to Surplus*, 52.

46. Deerr, *Sugar and the Sugar Cane*, 327.

47. Deerr, *Sugar and the Sugar Cane*, 281.

48. Edson, *Sugar, from Scarcity to Surplus*, 95.

49. Guilford Lawson Spencer, *A Hand-Book for Sugar Manufacturers and Their Chemists*, 2nd ed. (New York, 1893), 16.

50. Heriot, "Technical Training for the Sugar Industry."

51. Edwin Atkins to J. George Lumilius, November 11, 1889, II.8.361, MHS.

52. . Skaife to Atkins, June 8, 1894, MHS (emphasis in the original).

53. Wilfrid Skaife, "The Field for Chemical Improvement in the Manufacture of Sugar," *Journal of the Franklin Institute* 127, no. 3 (1889): 215–26; Wilfrid Skaife, "Future Industrial Opportunities in Cuba," *Engineering Magazine* 15, no. 3 (1898): 363–70; Skaife, "Sugar Producing Plants"; Skaife to Atkins, June 24, 1894, MHS. While extolling the investment possibilities in Cuba after its independence from Spain, Skaife was far less sanguine about its Creole peoples. "The whites control the island, and outnumber the colored races," he wrote to the readership of the *Engineering Magazine* in 1898. "There may be trouble yet between the races, but the whites will win in any struggle, and preserve the island from the fate which has overtaken Jamaica, San Domingo, and others. The domination of the blacks means industrial and political death . . . Any investor or settler in Cuba must, from the beginning, fight heart and soul against anything that will give power into the negroes' hands, for they will use it mercilessly against the whites" Skaife, "Future Industrial Opportunities in Cuba," 370.

54. George William Rolfe, *The Polariscope in the Chemical Laboratory: An Introduction to Polarimetry and Related Methods* (Macmillan, 1905), vi. Prior to taking up that position, he had served for almost twenty years as "technical chemist in the sugarhouses of the West Indies [and] in the glucose industry of the West."

55. "Yields & Losses for the Crop of 1908," in Central Aguirre chemical reports, Colección Mercedita, Archivo General de Puerto Rico (AGPR).

56. Although the report itself is not explicit, it is clear these three "mills" are three distinct tandems independently grinding cane rather than a triplet of three-roller mills linked into a single tandem. This is because the amount of cane each is listed as having ground—4,129,000 lb., 45,429,920 lb., and 181,435,490 lb., respectively—adds up to the 230,994,410 lb. ground by the whole central.

57. "Chemist's Report for the Crop of 1908," in Central Aguirre chemical reports, Colección Mercedita, AGPR.

58. But because Mill No. 2 had remained online for 119 of 132 days (though of less than ten hours apiece, as opposed to No. 3's eighteen and a half), the central as a whole had ground nearly 179,000 tons of cane. The yield report for the next year, the crop that ran from December 23, 1909 to June 30, 1910, did not even bother to provide spaces for the days, hours, cane, bagasse, and sucrose calculations for No. 1, but saved space by printing "NOT WORKING." Central Aguirre chemists' reports, 1910 and subsequent years, Colección Mercedita, AGPR.

59. Moreno Fraginals, *The Sugarmill*, 153. For a discussion of the surprising relationship between slavery and skill in Cuban plantations, see Scott, *Slave Emancipation in Cuba*, 27.

60. Pérez, *Cuba*, 86; Scott, *Slave Emancipation in Cuba*, 6–41.

61. Atkins, *Sixty Years in Cuba*, 109.

62. Scott, *Slave Emancipation in Cuba*, 82–87.

63. Rolfe, *The Polariscope in the Chemical Laboratory*, 136–44.

64. Atkins to Konig, October 26, 1887, II.7.329; Atkins to August Heyne, May 11, 1888, II.7.415 and May 18, 1888, II.7.430; Atkins to Arthur Siebert, May 23, 1888, II.7.435, MHS.

65. Sergei Alasheev, "On a Particular Kind of Love and the Specificity of Soviet Production," in *Management and Industry in Russia: Formal and Informal Relations in the Period of Transition*, ed. Simon Clarke (Edward Elgar, 1995), 69–98; David Montgomery, *The Fall of the House of Labor: The Workplace, the State, and American Labor Activism, 1865–1925* (Cambridge University Press, 1987), 116.

66. Gillian McGillivray, *Blazing Cane: Sugar Communities, Class, and State Formation in Cuba, 1868–1959*, American Encounters / Global Interactions (Duke University Press, 2009), 138; P. S. Smith, "Electricity Applied to the Manufacture of Sugar," *General Electric Review*, August 1913; A. I. M. Winetraub, "The Electrification of Cane Sugar Factories." *General Electric Review*, January 1914, 74–78. A factory was increasingly likely, after 1900, to be driven by an electric plant burning bagasse rather than directly by steam engines.

67. Fernández-de-Pinedo and Pretel, "Circuits of Knowledge," 271.

68. Seaforth M. Bellairs, "Twenty Years' Improvements in Demerara Sugar Production," *The Sugar Cane*, October 1892.

69. Reginald Sexton Hall et al., "Production Control in the Revere Sugar Refinery" (Massachusetts Institute of Technology, Department of Engineering Administration, 1922), 51.

70. Rolfe, *The Polariscope in the Chemical Laboratory*, 131.

71. Atkins to Jose Montalvo, May 27, 1885, II.6.113–15; Atkins to Fowler, June 10, 1885, II.6.147–54, MHS.

72. Rolfe, *The Polariscope in the Chemical Laboratory*, 132.

73. Deerr, *Sugar and the Sugar Cane*, 347.

74. Edith A. Browne, *Peeps at Industries: Sugar* (A. & C. Black, 1911), 85.

75. Browne, *Peeps at Industries*, 44.

76. Skaife to Atkins, April 30, 1894, MHS.

77. Skaife to Atkins, May 7, 1894, MHS.

78. Ramiro Guerra y Sánchez, *Sugar and Society in the Caribbean: An Economic History of Cuban Agriculture*, trans. Marjory M. Urquidi (Yale University Press, 1964).

79. César J. Ayala, *American Sugar Kingdom: The Plantation Economy of the Spanish Caribbean, 1898–1934*. University of North Carolina Press, 1999.

80. Alan Dye, "Avoiding Holdup: Asset Specificity and Technical Change in the Cuban Sugar Industry, 1899–1929," *Journal of Economic History* 54, no. 3 (1994): 628–53.

81. Alan Dye, *Cuban Sugar in the Age of Mass Production: Technology and the Economics of the Sugar Central, 1899–1929* (Stanford University Press, 1998), chap. 6.

82. Dye, *Cuban Sugar*, 253–54.

83. Guilford Lawson Spencer, *General Instructions and Methods of Analysis and Chemical Control: For Use in the Factories of the Cuban-American Sugar Company* (Cuban-American Sugar Company, 1916), 16.

84. Guerra y Sánchez, *Sugar and Society in the Caribbean*, 198–99. The "model contract" that Guerra y Sánchez reproduces at the end of *Sugar and Society in the Caribbean* stipulates that the colono's cane "will be paid for on the basis of its weight as indicated by the scale of the mill yard of the plantation where it is ground." The colono's representative could check this weight (though not explicitly the scale itself) "at his own expense." The terms of the payment are somewhat unclear, stating that "the Company [i.e., the central] will pay the colono the value in official currency of five and a quarter (5¼) arrobas of centrifugal sugar, with juice [?] of a polarization

of 96 degrees," based on the previous fortnight's average price as quoted by the Association of Havana Sugar Brokers.

85. Brookings Institution, *The Sugar Problem of Puerto Rico* (Association of Sugar Producers of Puerto Rico, 1936), 16.

86. Brookings Institution, *The Sugar Problem of Puerto Rico*, 18.

87. Scott, *Degrees of Freedom*, 117; Ayala, *American Sugar Kingdom*, 4, 133.

88. McGillivray, *Blazing Cane*, 168–71.

89. McGillivray, *Blazing Cane*, 141.

90. Brookings Institution, *The Sugar Problem of Puerto Rico*, 17–19.

91. José Solá, "'The Funnel System in Which His Is the Little End': The Technological Transformation of the Sugar Industry and American Protectionism in the Emergence of the Colonos in Caguas, Puerto Rico, 1898–1928" (PhD diss., University of Connecticut, 2004).

92. Brookings Institution, *The Sugar Problem of Puerto Rico*, 19.

93. Arthur D. Gayer et al., *The Sugar Economy of Puerto Rico* (Columbia University Press, 1938), 137 (emphasis in the original).

94. Solá, "'The Funnel System in Which His Is the Little End,'" 7.

95. See *Laws of Puerto Rico Annotated* (Butterworth Legal Publishers, Equity Pub. Division, 1954), 1925, Act no. 10; 1927, Act no. 31; 1930, Act no. 32; 1932, Act no. 31.

96. See *Laws of Puerto Rico Annotated*, 1934, Act. no. 71, sec. 1.

97. See *Laws of Puerto Rico Annotated*, 1934, Act no. 71, sec. 4.

98. Gayer et al., *The Sugar Economy of Puerto Rico*, 143. This was because unless the central were able to extract 10.5 percent or more sugar, the colono would not receive the 7 percent by weight that was still the expected norm. Gayer and his fellow researchers collected the records of payment covering 50 percent of the island's production for the preceding decades, and found that, measured by weight of cane, the average percentage paid to the colonos was 7.3 percent, and very few were above 7.5 percent or below 6.5 percent. "It would appear," they wrote, that seven pounds of sugar per 100 pounds of cane was still considered the norm, "and that the various systems are designed to average out at about that rate for cane of average good quality. *In other words, the more refined methods of liquidation appear to make the older rough-and-ready system their point of departure.*"

99. Humberto García-Muñiz, *Sugar and Power in the Caribbean: The South Porto Rico Sugar Company in Puerto Rico and the Dominican Republic, 1900–1921* (Editorial Universidad de Puerto Rico, 2010), 81.

100. Gayer et al., *The Sugar Economy of Puerto Rico*, 275–77.

101. See *Laws of Puerto Rico Annotated*, 1937, Act no. 112.

102. This was in order that the inspecting chemical engineer, whose office had been created in 1934, could calculate the factor in each factory for the coming season, thereby making sure that the estimated yield did not exceed the real yield.

103. Gayer et al., *The Sugar Economy of Puerto Rico*, 137.

104. See *Laws of Puerto Rico Annotated* 1937, Act no. 112, sec. 14.

Chapter 5

1. Arthur Hill Hassall, *Food and Its Adulterations; Comprising the Reports of the Analytical Sanitary Commission of "the Lancet" for the Years 1851 to 1854 Inclusive, Revised and Extended* (London, 1855), 17–19.

2. Staffan Müller-Wille, "Linnaean Paper Tools," in *Worlds of Natural History*, ed. Emma C. Spary et al. (Cambridge University Press, 2018), 216.

3. Deborah Blum, *The Poison Squad: One Chemist's Single-Minded Crusade for Food Safety at the Turn of the Twentieth Century* (Penguin Press, 2018), 66.

4. Robert Niccol, *The Sugar Insect: "Acarus Sacchari," Found in Raw Sugar* (De Armond & Goodrich, 1868), 329. The HathiTrust copy scanned by Harvard University Libraries indicates that it was given to the library in 1869 by Charles Sumner.

5. "Sugar," in *The New American Cyclopædia: A Popular Dictionary of General Knowledge*, ed. George Ripley and Charles A. Dana (D. Appleton & Co., 1869).

6. E. R. Leland, "Mites, Ticks, and Other Acari," *Popular Science Monthly*, February 1879.

7. Merleaux, *Sugar and Civilization.*

8. *Tenth Census of the United States, Report on the Manufactures of the United States*, 22 vols. (Government Printing Office, 1883), 2:77, https://catalog.hathitrust.org/Record/000441415.

9. Richard White, *The Republic for Which It Stands: The United States During Reconstruction and the Gilded Age, 1865–1896*, Oxford History of the United States (Oxford University Press, 2017), 6; Richard Franklin Bensel, *The Political Economy of American Industrialization, 1877–1900* (Cambridge University Press, 2000); Lewis L. Gould, "Tariffs and Markets in the Gilded Age," *Reviews in American History* 2, no. 2 (1974): 266–71; Andrew Wender Cohen, "Smuggling, Globalization, and America's Outward State, 1870–1909," *Journal of American History* 97, no. 2 (2010): 373, 376.

10. Wendy A. Woloson, *Refined Tastes: Sugar, Confectionery, and Consumers in Nineteenth-Century America* (Johns Hopkins University Press, 2002).

11. *Memorial Presented to the Committee of Ways and Means on the Sugar Tariff, Advocating a Uniform Rate of Duty up to No. 13 Dutch Standard, with Arguments in Favor of Same by John E. Searles, Jr., and Edward P. Eastwick* (T. McGill & Co., 1880), 22; Bensel, *The Political Economy of American Industrialization, 1877–1900*, chap. 7; *Annual Report of the Secretary of the Treasury on the State of the Finances for the Year 1879* (Government Printing Office, 1879), iii–iv.

12. *A Review of the Case of the United States vs. 712 Bags of Dark Demerara Centrifugal Sugars, A. W. Perot & Co., Claimants: Tried at the September Term of the District Court of the U.S. for the District of Maryland* (Lucas Brothers, 1878), 49.

13. W. H. Perot to Charles Chandler, January 2, 1878, Charles Chandler Papers, Rare Books and Manuscript Library, Columbia University.

14. Walter Rodney, *Guyanese Sugar Plantations in the Late Nineteenth Century: A Contemporary Description From the "Argosy"* (Release Publishers, 1979), 3–31.

15. Gideon E. Moore, *Statement Relative to the Artificial Coloring of Imported Sugars* (Government Printing Office, 1881), 20.

16. Tariff of 1861 (Morrill Tariff), chap. 48, 12 Stat. 178, 179 (March 2, 1861).

17. Revenue Act of 1861, chap. 45, 12 Stat. 292 (August 5, 1861) (emphasis added).

18. "Important to Sugar Dealers," *New York Times*, July 6, 1862.

19. Moore, *Statement Relative to the Artificial Coloring of Imported Sugars*, 16.

20. *The Sugar Cane*, November 1892, 575.

21. Robert White Stevens, *On the Stowage of Ships and Their Cargoes, Freights, Charter-Parties, &c.* (Longmans, 1859), 278.

22. Stevens, *On the Stowage of Ships and Their Cargoes*, 228–32.

23. Letter from the Secretary of the Treasury, Transmitting the Report of Certain Commissioners Appointed to Examine Custom-houses, and Recommending Appropriations for Their Pay (45th Congress, 1st Session, October 25, 1877), 7.

24. For character and food, see Cohen, *Pure Adulteration*, 34.

25. "The Custom-House Inquiry: Further Evidence by Examiners," *New York Times*, May 23, 1877.

26. US Treasury Department, *Annual Report of the Secretary of the Treasury on the State of the Finances for the Year 1878* (Government Printing Office, 1878), xxviii.

27. *Testimony in Relation to the Sugar Frauds, Taken by the Subcommittee of the Committee of Ways and Means of the House of Representatives, New York, September, 1878* (Brooklyn Daily Times Print, 1878), 36.

28. *Testimony in Relation to the Sugar Frauds*, 39.

29. *Testimony in Relation to the Sugar Frauds*, 8.

30. Charles Chandler, "Sugar," in *Johnson's New Universal Cyclopædia* (New York, 1877).

31. Chandler, "Sugar," 632.

32. Moore, *Statement Relative to the Artificial Coloring of Imported Sugars*, 19.

33. Chandler, "Sugar," 641.

34. *Testimony in Relation to the Sugar Frauds, Taken by the Subcommittee of the Committee of Ways and Means of the House of Representatives, New York, September, 1878*, 97; Chandler, "Sugar," 641.

35. *Reports from the Secretary of the Treasury, of Scientific Investigations in Relation to Sugar and Hydrometers, Made Under the Superintendence of Professor R. S. McCulloh* (Wendell and Van Benthuysen, 1848), 49.

36. *Papers Relative to Drawback Rates on Exported Sugar* (Evening Post, 1877); *Reports from the Secretary of the Treasury, of Scientific Investigations in Relation to Sugar and Hydrometers.*

37. Rodney, *Guyanese Sugar Plantations in the Late Nineteenth Century*, 18.

38. *Annual Report of the Secretary of the Treasury on the State of the Finances for the Year 1877* (Government Printing Office, 1877), xxvi; F. W. Taussig, *The Tariff History of the United States*, 5th ed. (G. P. Putnam's Sons, 1910; repr. Mises Institute, 2010),142.

39. *Annual Report of the Secretary of the Treasury on the State of the Finances for the Year 1877*, xxvi.

40. *Letter from the Secretary of the Treasury, Transmitting the Report of Certain Commissioners Appointed to Examine Custom-houses, and Recommending Appropriations for Their Pay, Their Pay*, 45th Congress, 1st Session, October 25, 1877, 1–10.

41. "The Coloured-Sugar Question in America," *The Sugar Cane*. February 1, 1878.

42. *A History of Enforcement in the United States Customs Service, 1789–1875*, Special Customs bicentennial reissue, Customs Publication no. 522 (US Customs Service, 1988), chap. 5; Nicholas R. Parrillo, *Against the Profit Motive: The Salary Revolution in American Government, 1780–1940* (Yale University Press, 2013), chapter 6.

43. *A Review of the Case of the United States vs. 712 Bags of Dark Demerara Centrifugal Sugars*, 4–14.

44. *A Review of the Case of the United States vs. 712 Bags of Dark Demerara Centrifugal Sugars*, 10, 66, 9.

45. [untitled article], *The Sugar Cane*, February 1, 1878.

46. Perot to Chandler, October 1, 1878, Charles F. Chandler papers, box 260, folder 7, Rare Book and Manuscript Library, Columbia University, New York.

47. *A Review of the Case of the United States vs. 712 Bags of Dark Demerara Centrifugal Sugars*, 30.

48. *A Review of the Case of the United States vs. 712 Bags of Dark Demerara Centrifugal Sugars*, 53.

49. Morton to Chandler, April 11, 1878, Charles F. Chandler Papers, box 260, folder 7, Rare Book and Manuscript Library, Columbia University, New York.

50. *A Review of the Case of the United States vs. 712 Bags of Dark Demerara Centrifugal Sugars*, 28.

51. *Annual Report of the Secretary of the Treasury on the State of the Finances for the Year 1878*, 29.

52. S. E. Chamberlin et al., *Report on the Methods of Manufacturing Sugars in West India Islands and British Guiana* (Government Printing Office, 1880), 3–4.

53. Chamberlin et al., *Report on the Methods of Manufacturing Sugars*, 12–16.

54. Chamberlin et al., *Report on the Methods of Manufacturing Sugars*, 38.

55. Chamberlin et al., *Report on the Methods of Manufacturing Sugars*, 16–21.

56. Chamberlin et al., *Report on the Methods of Manufacturing Sugars*, 23.

57. *Tenth Census of the United States: Report on the Manufactures of the United States*, 77; *Sugar Tariff: Speech of Hon. James A. Garfield, of Ohio, Delivered in the House of Representatives, Wednesday, February 26, 1879* (Washington, DC, 1879), 5–6.

58. Garfield, *Sugar Tariff*, 12.

59. Robert Howe, *The Labor Side of the Great Sugar Question* (New York, 1878).

60. Harry W. Havemeyer, *Merchants of Williamsburgh: Frederick C. Havemeyer, Jr., William Dick, John Mollenhauer, Henry O. Havemeyer* (H. W. Havemeyer, 1989), 39.

61. Howe, *The Labor Side of the Great Sugar Question*, 4.

62. Howe, *The Labor Side of the Great Sugar Question*, 20.

63. Pérez, *Cuba*, 94.

64. Howe, *The Labor Side of the Great Sugar Question*, 25.

65. *Review of the Efforts of the Forty-fifth Congress of the United States for the Revision of the Sugar Tariff, Together with Speeches Made by Members of the House of Representatives* (New York, 1879), 9.

66. Garfield, *Sugar Tariff*, 8.

67. Henry Alvin Brown, *Statements Made Before the Committee of Ways and Means on the Sugar Question in the Interests of American Consumers, Home Industries, and Revenue* (Judd & Detweiler, 1880), 22–24.

68. Garfield, *Sugar Tariff*, 12.

69. Moon-Ho Jung, *Coolies and Cane: Race, Labor, and Sugar in the Age of Emancipation* (Johns Hopkins University Press, 2006); "Suppression of the Coolie Trade," *New York Times*, February 26, 1877.

70. Donica Belisle, "Eating Clean: Anti-Chinese Sugar Advertising and the Making of White Racial Purity in the Canadian Pacific," *Global Food History* 6, no. 1 (2020): 41–59.

71. David Ames Wells, *The Sugar Industry of the United States, and the Tariff. Report on the Assessment and Collection of Duties on Imported Sugar: On the Results of an Economic and Financial Inquiry into the Relation of the Sugar Industry of the United States in Its Several Departments of Production, Importation, Refining and Distribution of Product, to the Existing Federal Tariff* (New York, 1878), 114.

72. *Testimony in Relation to the Sugar Frauds, Taken by the Subcommittee of the Committee of Ways and Means of the House of Representatives, New York, September, 1878*, 99.

73. Bill James, *Popular Crime: Reflections on the Celebration of Violence* (Scribner, 2011), 39; "Charley Ross—Another of Him Found," *New Orleans Daily Democrat*, January 20, 1878.

74. "Charley Ross," *Alexandria Gazette and Virginia Advertiser*, February 5, 1878; "Charley Ross," *New York Herald*, February 5, 1878.

75. "Is the Boy Charley Ross?" *New York Times*, February 3, 1878.

76. "Charley Ross," February 5, 1878.

77. "Not Charley Ross," *New York Herald*, February 6, 1878.

78. Charles A. Browne, "Laboratory Journal of Interviews," December 3, 1919, box 18, Browne Papers, New York Public Library.

Chapter 6

1. "Funeral of W. H. Grace," *Brooklyn Daily Eagle*, December 27, 1897; "Ruffianism in Broadway," *New York Times*, July 21, 1877.

2. "A Custom House Fracas," *New York Tribune*, July 21, 1877.

3. "Gen. G. H. Sharpe Dead," *New York Times*, January 15, 1900.

4. "Workingmen for Protection: William H. Grace Writes to the Senate Finance Committee Reasons Why the Wilson Bill Should Be Defeated," *New-York Tribune*, February 14, 1984.

5. "A Little Case Which Only Begins a Big One," *Brooklyn Daily Eagle*, September 25, 1877; "Four Months in Jail. Ex-Inspector Grace Convicted," *New York Times*, September 25, 1877.

6. "A Custom House Fracas."

7. "A Sweet Fraud: The Alleged Rascalities in Sugar," *Brooklyn Daily Eagle*, January 23, 1879.

8. "The Schmidt Defalcation," *New York Times*, September 2, 1877; John Sherman, *Recollections of Forty Years in the House, Senate and Cabinet: An Autobiography* (Werner, 1895), 680; "Grace's Vindication," *Brooklyn Daily Eagle*, December 18, 1877.

9. "Grace's Vindication."

10. H. V. Boynton, "The Whiskey Ring," *North American Review*, October 1, 1876.

11. Charles [Carlos] Rebello, *The Pith of the Sugar Question* (Macgowan and Slipper, 1879); "Ex-Inspector William H. Grace. An Order Revoking His Appointment Received at the Custom-House," *New York Times*, July 3, 1879; Hayes to Sherman, June 30, 1879, in *Diary and Letters of Rutherford Birchard Hayes: Nineteenth President of the United States*, ed. Charles Richard Williams, vol. 3 (Ohio State Archæological and Historical Society, 1924)., 563. See also the article from the New York *Sun*, November 20, 1878, reprinted in Rebello, *The Pith of the Sugar Question*. In 1879, Grace's friends attempted to get him reappointed to the customhouse based on his supposed good work exposing fraud. President Hayes was unhappy to learn from George Curtis, of the New York Civil Service Reform Association, that "a man named Grace, who was convicted a year or two ago of a flagrant assault upon Surveyor Sharpe," had been put forward again. Hayes was less concerned about Grace's violent temper than he was about the idea that Grace might get the position ahead of candidates recommended through the newly instituted and still-fragile civil service examination system, so Sherman made clear he was not to be reappointed.

12. "A Little Case Which Only Begins a Big One."

13. Stephen Skowronek, *Building a New American State: The Expansion of National Administrative Capacities, 1877–1920* (Cambridge University Press, 1982), 47–84.

14. Alfred S Eichner, *The Emergence of Oligopoly: Sugar Refining as a Case Study* (Johns Hopkins University Press, 1969), 51.

15. *Annual Report of the Secretary of the Treasury on the State of the Finances for the Year 1877*, xxvi; *Review of the Efforts of the Forty-fifth Congress of the United States for the Revision of the Sugar Tariff, Together with Speeches Made by Members of the House of Representatives*, 3–4.

16. Harvey Washington Wiley, *An Autobiography* (Bobbs-Merrill, 1930), 139.

17. *Review of the Efforts of the Forty-fifth Congress of the United States for the Revision of the Sugar Tariff, Together with Speeches Made by Members of the House of Representatives*, 4–5.

18. Charles A. Browne, "A History of the New York Sugar Trade Laboratory," 3. Container 33, Charles Albert Browne Papers, Manuscript Division, Library of Congress, Washington, DC (hereafter CAB papers, LOC).

19. Garfield, *Speech of Hon. James A. Garfield, of Ohio, Delivered in the House of Representatives, Wednesday, February 26, 1879*, 14.

20. Garfield, *Sugar Tariff*, 14–17.

21. Joshua Specht, "A Failure to Prohibit: New York City's Underground Bob Veal Trade," *Journal of the Gilded Age and Progressive Era* 12, no. 4 (2013): 475–501.

22. Cohen, *Pure Adulteration*, 38, 43.

23. Cohen, *Pure Adulteration*, chap. 8; Alan I. Marcus, "Setting the Standard: Fertilizers, State Chemists, and Early National Commercial Regulation, 1880–1887," *Agricultural History* 61, no. 1 (1987): 48.

24. Chandler to Ceballos & Co.; New York, April 18, 1878, Charles F. Chandler Papers, box 260, folder 7, Rare Book and Manuscript Library, Columbia University, New York.

25. Moore, *Statement Relative to the Artificial Coloring of Imported Sugars*, 4.

26. *Testimony in Relation to the Sugar Frauds, Taken by the Subcommittee of the Committee of Ways and Means of the House of Representatives, New York, September, 1878*, 44.

27. Henry A. Brown, *Sugar Frauds and the Tariff: Their Relations to Home Product, Consumption, Industry, Imports, Duties and Revenue, Analyzed and Exhibited. Duty & Drawback, Sampling & Grading. Adulterations and the Polariscope. Official Statistics, Remedies, &c.* (Saxonville, MA, 1879), 16.

28. *Memorial Presented to the Committee of Ways and Means on the Sugar Tariff, Advocating a Uniform Rate of Duty up to No. 13 Dutch Standard, with Arguments in Favor of Same by John E. Searles, Jr., And Edward P. Eastwick*, 16.

29. *Testimony in Relation to the Sugar Frauds, Taken by the Subcommittee of the Committee of Ways and Means of the House of Representatives, New York, September, 1878*, 44.

30. Biot, "Sur l'utilité que pourraient offrir les caractères optiques dans l'exploitation des sucreries et des raffineries," 265.

31. William Booth to Chandler, January 16, 1878, Charles F. Chandler Papers, box 260, folder 7, Rare Book and Manuscript Library, Columbia University, New York.

32. Sherman, *John Sherman's Recollections of Forty Years in the House, Senate and Cabinet*, 676.

33. Sherman, *John Sherman's Recollections* 678.

34. *Testimony in Relation to the Sugar Frauds, Taken by the Subcommittee of the Committee of Ways and Means of the House of Representatives, New York, September, 1878*, 44.

35. "The Sugar Tariff. Large Meeting of Sugar Men." *New York Times*, February 8, 1878.

36. *Testimony in Relation to the Sugar Frauds, Taken by the Subcommittee of the Committee of Ways and Means of the House of Representatives, New York, September, 1878*, 98.

37. Invitation in Chandler Papers, Charles F. Chandler Papers, box 260, Rare Book and Manuscript Library, Columbia University, New York.

38. "The Sugar Men's Quarrel," *New York Times*, January 7, 1879.

39. "The Sugar Tariff Ring," *New York Times*, April 14, 1880.

40. Moore, *Statement Relative to the Artificial Coloring of Imported Sugars*, 7.

41. Deborah Jean Warner, "How Sweet It Is: Sugar, Science, and the State," *Annals of Science* 64, no. 2 (2007): 158–59.

42. "Five Hours Spent in Talk, the Senate Objecting to the Polariscope Test for Sugar," *New York Times*, July 26, 1882.

43. Eichner, *The Emergence of Oligopoly*, 62–64; César J. Ayala, *American Sugar Kingdom: The Plantation Economy of the Spanish Caribbean, 1898-1934*. (University of North Carolina Press, 1999), 4, 256; "A Million-Dollar Fire: Havemeyer's Refinery a Huge Mass of Ruins," *New York Times*, June 12, 1887.

44. César J. Ayala, *American Sugar Kingdom: The Plantation Economy of the Spanish Caribbean, 1898–1934* (University of North Carolina Press, 1999), 4, 33; Eichner, *The Emergence of Oligopoly*, 343.

45. Naomi R. Lamoreaux, *The Great Merger Movement in American Business, 1895–1904* (Cambridge University Press, 1985), 187–88.

46. Eichner, *The Emergence of Oligopoly*, 17.

47. Richard B. Sheridan, *Sugar and Slavery: An Economic History of the British West Indies, 1623–1775* (Johns Hopkins University Press, 1973), 56.

48. Daniel Catlin, *Good Work Well Done: The Sugar Business Career of Horace Havemeyer, 1903–1956* (D. Catlin, 1988), 7–12; George M. Rolph, *Something About Sugar: Its History, Growth, Manufacture and Distribution* (J. J. Newbegin, 1917), 44–90.

49. Eichner, *The Emergence of Oligopoly*, 158.

50. Lamoreaux, *The Great Merger Movement in American Business*, 99–102.

51. Eichner, *The Emergence of Oligopoly*, 343.

52. Eichner, *The Emergence of Oligopoly*, 70–92.

53. Eichner, *The Emergence of Oligopoly*, 160.

54. For recipes, see Mary Randolph, *Virginia Housewife; Or, Methodical Cook* (John Plaskitt, 1836).

55. Kathleen Brady, *Ida Tarbell: Portrait of a Muckraker* (University of Pittsburgh Press, 1989), 121.

56. Eichner, *The Emergence of Oligopoly*, 120.

57. Charles W. McCurdy, "The Knight Sugar Decision of 1895 and the Modernization of American Corporation Law, 1869–1903," *Business History Review* 53, no. 3 (1979): 328.

58. Alfred D. Chandler, *The Visible Hand: The Managerial Revolution in American Business* (Belknap Press of Harvard University Press, 1977), 328–29.

59. Uwe Spiekermann, "Claus Spreckels: Robber Baron and Sugar King," *Immigrant Entrepreneurship, 1720 to the Present*, June 7, 2011, https://www.immigrantentrepreneurship.org/entries/claus-spreckels-robber-baron-and-sugar-king/; Carol A. MacLennan, *Sovereign Sugar: Industry and Environment in Hawai'i* (University of Hawai'i Press, 2014).

60. Eichner, *The Emergence of Oligopoly*, 153; "Claus Spreckels and the Sugar Industry," *Frank Leslie's Illustrated Newspaper*, August 9, 1890.

61. David Genesove and Wallace P. Mullin, "Predation and Its Rate of Return: The Sugar Industry, 1887–1914," *RAND Journal of Economics* 37, no. 1 (2006): 47–69; Jeremiah Whipple Jenks, *The Trust Problem* (Doubleday, Page, 1909), 15. The Progressive Era antitrust economist Jeremiah Jenks even argued that the polariscope drove refiners to combination, because the industry had the power to ensure that "quality is uniform and may be easily tested." Thus sugar became part of a class of goods for which firms could only compete on price.

62. United States v. E. C. Knight Co., 156 U.S. 1, 18 (1895).

63. "Byrne Again in Office," *Boston Daily Globe*, September 13, 1893; US Customs Service, *A History of Enforcement in the United States Customs Service, 1789–1875*, chap. 5. "Special Agent of the Treasury" could be a fluid term.

64. Letter from the Secretary of the Treasury, Transmitting, in Response to Senate Resolution of January 8, 1889, Information Relative to the Sugar Frauds, 50th Congress, 1st session, January 20, 1889, 2–5.

65. Browne, "History," 2, in Browne Papers, LOC.

66. Obituary of Ira Ayer, *San Francisco Call*, February 4, 1903.

67. Letter from the Secretary of the Treasury, Transmitting, in Response to Senate Resolution of January 8, 1889, Information Relative to the Sugar Frauds, 45–50.

68. "Sugar Frauds Alleged: Somebody Has Been Tampering with Dr. Leary's Polariscope," February 26, 1889; Warner, "How Sweet It Is," 161.

69. Letter from the Secretary of the Treasury, Transmitting in Response to a Senate Resolution, July 14, 1890, a Report of Special Agents Who Were Instructed to Investigate the Manner in Which Sugars Are Classified, 51st Congress, 1st session, July 18, 1890, 5.

70. Letter from the Secretary of the Treasury, Transmitting in Response to a Senate Resolution, July 14, 1890, a Report of Special Agents Who Were Instructed to Investigate the Manner in Which Sugars Are Classified," 5–6.

71. "Crafty Plot," *Boston Daily Globe*, February 25, 1889.

72. Letter from the Secretary of the Treasury, Transmitting in Response to a Senate Resolution, July 14, 1890, a Report of Special Agents Who Were Instructed to Investigate the Manner in Which Sugars Are Classified, 7–10.

73. For personal equations, see Simon Schaffer, "Astronomers Mark Time: Discipline and the Personal Equation," *Science in Context* 2 (1988): 115–43.

74. "Will Not Test Sugar Any More," *Brooklyn Daily Eagle*, June 27, 1890.

75. "Appraisers Get to Work," *New York Times*, July 25, 1890.

76. "No Sand in This Sugar: How Uncle Sam Tests the Imported Article. The Functions of the Polariscope and the Nicety to Which It Determines the Grade of Sugar, July 27, 1890," *New York Times*, July 27, 1890.

77. Eichner, *The Emergence of Oligopoly*, 107–8,

78. Sugar Frauds. Witness, Charles R. Heike (June 15, 1911), 25.

79. Letter from the Secretary of the Treasury, Transmitting in Response to a Senate Resolution, July 14, 1890, a Report of Special Agents Who Were Instructed to Investigate the Manner in Which Sugars Are Classified," 10.

80. Douglas A. Irwin, *Clashing over Commerce: A History of US Trade Policy* (University of Chicago Press, 2017).

81. "The True Issue," *Harper's Weekly*, September 8, 1894.

82. T. J. Jackson Lears, *Rebirth of a Nation: The Making of Modern America, 1877–1920* (HarperCollins, 2009), 3.

83. Richard L. McCormick, "The Discovery That Business Corrupts Politics: A Reappraisal of the Origins of Progressivism," *American Historical Review* 86, no. 2 (1981): 247–74.

84. Bruce E. Baker and Barbara Hahn, *The Cotton Kings: Capitalism and Corruption in Turn-of-the-Century New York and New Orleans* (Oxford University Press, 2015).

85. Richard White, *Railroaded: The Transcontinentals and the Making of Modern America* (W. W. Norton, 2012), xxxiii–xxxiv.

86. *Corruption and Reform: Lessons from America's Economic History*, ed. Edward L. Glaeser

and Claudia Goldin, National Bureau of Economic Research Conference Report (University of Chicago Press, 2008), 19–20; William J. Novak, "The Myth of the 'Weak' American State," *American Historical Review* 113, no. 3 (2008): 752–72.

87. Andrew Wender Cohen, *Contraband: Smuggling and the Birth of the American Century* (W. W. Norton, 2015); Peter Andreas, *Smuggler Nation: How Illicit Trade Made America* (Oxford University Press, 2014).

88. David Roth Singerman, "Science, Commodities, and Corruption in the Gilded Age," *Journal of the Gilded Age and Progressive Era* 15, no. 3 (2016): 278–93.

89. Cohen, *Contraband*, 249–50; Warner, "How Sweet It Is"; Jerome Sternstein, "Corruption in the Gilded Age Senate: Nelson W. Aldrich and the Sugar Trust," *Capitol Studies*, Spring 1978, 14–38; *Testimony in Relation to the Sugar Frauds, Taken by the Subcommittee of the Committee of Ways and Means of the House of Representatives, New York, September, 1878, 52.*

90. See, for instance, Theodore M. Porter, *Trust in Numbers: The Pursuit of Objectivity in Science and Public Life* (Princeton University Press, 1996); William Deringer, "For What It's Worth: Historical Financial Bubbles and the Boundaries of Economic Rationality," *Isis* 106, no. 3 (2015): 646–56.

91. "Contemporary Humor," *Brooklyn Daily Eagle*, September 27, 1894.

92. "New York's Sugar Ring," *Boston Daily Globe*, March 30, 1888.

93. Brown, *Sugar Frauds and the Tariff*, 16.

Chapter 7

1. "A Sweet Story," *Brooklyn Daily Eagle*, January 5, 1889.

2. "Warrants Out: English Dupes Want to Arrest the Sugar Swindlers," *New York World* January 5, 1889 (evening edition).

3. David A. Curtis, "The Electric Sugar Swindle," *Harper's Weekly*, March 30, 1889.

4. Eichner, *The Emergence of Oligopoly*, 65 and appendix D.

5. Curtis, "The Electric Sugar Swindle."

6. Cohen, *Pure Adulteration*, 143–46. I'm indebted to Ben Cohen for years of conversations helping each other make sense of the electric sugar swindle.

7. David E. Nye, *Electrifying America: Social Meanings of a New Technology, 1880–1940* (MIT Press, 1990), 4.

8. Nye, *Electrifying America*, 166.

9. "News of the Day," *Birmingham Daily Post*, January 7, 1889.

10. Nye, *Electrifying America*, 2; "Professor Thomson Takes 500,000 Volts," *Washington Evening Star*, June 9, 1905.

11. Cohen, *Pure Adulteration*, 1–8, 144.

12. "Big Sugar Swindle," *New York Times*, January 8, 1889.

13. "A Sweet Story."

14. Pamela H. Smith, *The Body of the Artisan: Art and Experience in the Scientific Revolution* (University of Chicago Press, 2006), 167.

15. Neil Harris, *Humbug: The Art of P. T. Barnum* (Little, Brown, 1973); James W. Cook, *The Arts of Deception: Playing with Fraud in the Age of Barnum* (Harvard University Press, 2001).

16. Amy Reading, *The Mark Inside: A Perfect Swindle, a Cunning Revenge, and a Small History of the Big Con* (Knopf, 2013), 8.

17. "Big Sugar Swindle."

18. "Friend's Chicago Deal," *Chicago Tribune*, January 6, 1889; Electric Sugar Refining Company, June–July 1885, Charles Chandler Papers, box 260, folder 10, Rare Book and Manuscript Library, Columbia University, New York.

19. "Electric Sugar Company: Reported Collapse. Story of the Enterprise," *Birmingham Daily Post*, January 4, 1889.

20. "Sugar by Electricity," *Birmingham Daily Post*, November 12, 1886.

21. Curtis, "The Electric Sugar Swindle." See also the additional item in that day's newspaper.

22. William Lloyd Fox, "Harvey W. Wiley's Search for American Sugar Self-Sufficiency," *Agricultural History* 54, no. 4 (1980): 516–26; Deborah Jean Warner, *Sweet Stuff: An American History of Sweeteners from Sugar to Sucralose* (Smithsonian Institution Scholarly Press in cooperation with Rowman & Littlefield, 2011), chap. 9.

23. Sterling Syrup Works to Thomas Edison, December 23, 1888, Thomas A. Edison Papers, Rutgers University, folder D-88-22, edison.rutgers.edu.

24. "Electric Sugar Company," and see also the additional item in that day's newspaper.

25. Curtis, "The Electric Sugar Swindle."

26. "A Sweet Story."

27. "Warrants Out."

28. "That Alleged Humbug: The Electric Sugar Refining Company's Fiasco," *Semi-Weekly Miner*, January 9, 1889, https://chroniclingamerica.loc.gov/lccn/sn84036034/1889-01-09/ed-1/seq-1/.

29. Jeremy Zallen, "'Dead Work,' Electric Futures, and the Hidden History of the Gilded Age," *Montana: The Magazine of Western History* 66, no. 2 (2016): 39–65.

30. "Now Serving Time," *Semi-Weekly Miner*, December 14, 1889, https://chroniclingamerica.loc.gov/lccn/sn84036034/1889-12-14/ed-1/seq-3/.

31. "Big Sugar Swindle."

32. "Big Sugar Swindle"; [untitled article.], *Brooklyn Daily Eagle*, January 7, 1889.

33. Friend's draft, census, and marriage records, as well as his arrival manifest, are available on Ancestry.com.

34. "Friend's Chicago Deal."

35. Reading, *The Mark Inside*, 104.

36. "Friend's Chicago Deal."

37. "The Electric Sugar Bubble: Interview with Mr. Roberts," *Birmingham Daily Post*, January 30, 1889.

38. "Electric Sugar Company"; "The Electric Sugar Panic," *London Morning Post*, January 5, 1889.

39. "A Victim of the Electric Sugar Refining Company," *Huddersfield Daily Chronicle*, May 2, 1889 edition.

40. "'Electric Sugar': A Startling Disclosure," *Nottingham Evening Post*, January 5, 1889.

41. "The Electric Sugar Frauds Interesting to Local Victims," *Tamworth Herald*, July 19, 1890.

42. "The Electric Sugar Bubble."

43. "The Electric Sugar Refining Fraud," *Preston Guardian*, July 27, 1889.

44. "The Electric Sugar Bubble."

45. "The Electric Sugar Bubble"; "Friend's Chicago Deal."

46. Curtis, "The Electric Sugar Swindle."

47. "Friend's Chicago Deal."

48. "The Ihpetonga," *Brooklyn Daily Eagle*, January 17, 1889.

49. Alfred Ord Tate to Sterling Syrup Works, Thomas Edison Papers, Rutgers University, General letterbook Series: LB-027 (November 1888–January 1889) [LB027783; TAEM 138:728], edison.rutgers.edu.

Chapter 8

1. Pérez, *Cuba*, 112–13.

2. Irwin, *Clashing over Commerce*, 266–67.

3. Transcript of Record, *American Sugar Refining Co, New York v. United States*, 211 U.S. 155, 192 (1908), https://link.gale.com/apps/doc/DW0106394994/SCRB. For instance, the Treasury Department's regulations governing the sugar bounty appended a list of thirty-three types of blank books and forms needed to administer payouts to producers. There were forms for a company to notify the department of its methods of production and forms to apply for a license, but these were different depending on whether the applicant intended to produce sugar from maple or from beet, cane, or sorghum. Many of the regulations themselves concerned which forms and blanks needed to be filled out when, with what information, and kept where and for how long, along with how they would be checked. The prescribed formats organized the information within those books to make them easy for inspectors to inspect, and specified where to add particular details such as the specific hydrometer used to measure the density of the juice.

4. Income Tax Act of 1894 ("Wilson-Gorman Tariff"), 53rd Congress, 2nd Sess., 349 (1894), 521; Irwin, *Clashing over Commerce*, 291.

5. Judson C. Welliver, "The Secret of the Sugar Trust's Power," *Hampton's Magazine*, May 1910, 720.

6. Transcript of Record, *American Sugar Refining Co, New York v. United States*, 211 U.S. 155 (1908), 86.

7. F. W. Taussig, "The Tariff Act of 1897," *Quarterly Journal of Economics* 12, no. 1 (1897): 42; The Dingley Act of 1897, chap 11, 30 Stat. 151 (1897), 168.

8. Harvey W. Wiley, "The True Meaning of the New Sugar Tariff," *The Forum*, February 1898, https://www.unz.com/print/Forum-1898feb-00689/, 693.

9. Clement W. Andrews, "The Influence of Temperature on the Specific Rotation of Cane Sugar," *Technology Quarterly* 2 (1889): 367.

10. Biot, "Examen comparatif du sucre de maïs et du sucre de betterave."

11. Harvey W. Wiley, *Principles and Practice of Agricultural Analysis*, vol. 3, *Agricultural Projects* (Chemical Publishing Company, 1897), 117.

12. Andrews, "The Influence of Temperature on the Specific Rotation of Cane Sugar."

13. Harvey W. Wiley, "The Influence of Temperature on the Specific Rotation of Sucrose and Method of Correcting Readings of Compensating Polariscopes Therefor," *Journal of the American Chemical Society* 21, no. 7 (1899): 595.

14. C. A. Browne, "A Constant Temperature Laboratory for the Polarization of Sugars," in *Eighth International Congress of Applied Chemistry, Washington and New York, September 4 to 13, 1912.* (Rumford Press, 1912), 526.

15. Transcript of Record, *American Sugar Refining Co., New York v. United States*, 211 U.S. 155 (1908), 100.

16. Charles Albert Browne, *A Handbook of Sugar Analysis: A Practical and Descriptive Treatise for Use in Research, Technical and Control Laboratories* (John Wiley and Sons, 1912), 28.

17. Wiley, "The Influence of Temperature on the Specific Rotation of Sucrose," 589.

18. Andrews, "The Influence of Temperature on the Specific Rotation of Cane Sugar," 375.

19. Wiley, "The Influence of Temperature on the Specific Rotation of Sucrose," 573.

20. Sibum, "Reworking the Mechanical Value of Heat."

21. Wiley, "The Influence of Temperature on the Specific Rotation of Sucrose and Method of Correcting Readings of Compensating Polariscopes Therefor," 596.

22. Board of General Appraisers, "Sharpe, George Henry," Federal Judicial Center, accessed January 5, 2024, https://www.fjc.gov/history/courts/board-general-appraisers-sharpe-george-henry; "Howell, William Barberie," Federal Judicial Center," accessed January 5, 2024, https://www.fjc.gov/history/judges/howell-william-barberie.

23. Transcript of Record, *American Sugar Refining Co., New York v. United States*, 211 U.S. 155 (1908), 79–83.

24. Transcript of Record, *American Sugar Refining Co., New York v. United States*, 95–115; US Department of the Treasury, *Treasury Decisions Under Tariff, Internal Revenue, Immigration, Navigation Laws, Etc., January 1 to June 30, 1899*, 2 vols. (Government Printing Office, 1899), 2:539.

25. US Department of the Treasury, *Synopsis of the Decisions of the Treasury Department on the Construction of the Tariff, Navigation, and Other Laws for the Year Ended December 31, 1880* (Government Printing Office, 1881), 293–94.

26. Warner, "How Sweet It Is," 159.

27. Transcript of Record, *American Sugar Refining Co., New York v. United States*, 113.

28. Paolo Brenni, "Some Considerations about the Prices of Physics Instruments in the Nineteenth Century," in *How Scientific Instruments Have Changed Hands*, ed. A. D. Morrison-Low and Sara J. Schechner (Brill, 2017), 77.

29. US Department of the Treasury, *Treasury Decisions Under Tariff, Internal Revenue, Immigration, Navigation Laws, Etc., January 1 to June 30, 1899*, 2:344–47.

30. Transcript of Record, *American Sugar Refining Co., New York v. United States*, 134.

31. Transcript of Record, *American Sugar Refining Co., New York v. United States*, 398.

32. *Report of the Superintendent of the U.S. Coast and Geodetic Survey Showing the Progress of the Work from July 1, 1897, to June 30, 1898* (Government Printing Office, 1899), 166–71.

33. Transcript of Record, American Sugar Refining Co., New York v. United States 211 U.S. 155 (1908) 175.

34. *Bartram Bros. v. United States*, 123 F. 327 (Cir. Court, S.D.N.Y, 1903), rev'd, 131 F. 833 (Cir. Court of Appeals, 1904), cert. denied, 195 U.S. 635 (1904), appeal dismissed, 211 U.S. 155 (1908). Transcript of Record, *American Sugar Refining Co., New York v. United States*, 5.

35. R. Elberton Smith, *Customs Valuation in the United States: A Study in Tariff Administration* (University of Chicago Press, 1948), 131.

36. Michael Lynch, "The Discursive Production of Uncertainty: The OJ Simpson 'Dream Team' and the Sociology of Knowledge Machine," *Social Studies of Science* 28, no. 5/6 (1998): 829–68.

37. For numbers as arguments, see William Deringer, *Calculated Values: Finance, Politics, and the Quantitative Age* (Harvard University Press, 2018).

38. US Department of the Treasury, *Synopsis of the Decisions of the Treasury Department on the Construction of the Tariff, Navigation, and Other Laws for the Year Ended December 31, 1897* (Government Printing Office, 1897), 990 (emphasis in the original).

39. US Department of the Treasury, *Treasury Decisions Under Tariff, Internal Revenue, Immigration, Navigation Laws, Etc., January 1 to June 30, 1899*, 542.

40. Transcript of Record, *American Sugar Refining Co., New York v. United States*, 70.

41. US Department of the Treasury, *Treasury Decisions Under Tariff, Internal Revenue, Immigration, Navigation Laws, Etc., January 1 to June 30, 1899*, 544.

42. George Stade, "Modern Polariscopes," *International Sugar Journal*, February 1899; see also *International Sugar Journal* 2, no. 6 (1900), 282, and "U.S.A. Polariscope Case Decided," *International Sugar Journal* 6, no. 7 (1905), 311.

43. United States v. Bartram Bros., 131 F. 833 (Cir. Court of Appeals, S.D.N.Y, 1904).

44. United States v. Bartram Bros., 131 F. 833.

45. "New Test by Sugar Trust," *New York Times*, November 16, 1908, https://www.nytimes.com/1908/11/16/archives/new-test-by-sugar-trust-brings-a-suit-to-defeat-governments-method.html.

Chapter 9

1. Harold J. Howland, "The Case of the Seventeen Holes," *The Outlook*, May 1, 1909.

2. *Hearings Before the Committee on Expenditures in the Treasury Department: Sugar Frauds: Witness, Charles R. Heike*, 62nd Congress, 1st Sess. (Government Printing Office, June 15, 1911), https://www.google.com/books/edition/Hearings_Before_the_Committee_on_Expendi/q133POwxz3cC?hl=en&gbpv=1.

3. Howland, "The Case of the Seventeen Holes," 38.

4. Eichner, *The Emergence of Oligopoly*, 293; Howland, "The Case of the Seventeen Holes."

5. Catlin, *Good Work Well Done*, 35–36.

6. *Sugar Frauds: Witness, Charles R. Heike*, 18–26 .

7. David Genesove and Wallace P. Mullin, "Rules, Communication, and Collusion: Narrative Evidence from the Sugar Institute Case," *American Economic Review* 91, no. 3 (2001): 379–98.

8. Charles A. Browne, "A History of the New York Sugar Trade Laboratory" (typescript), 24. In Charles Albert Browne Papers, Library of Congress, box 33.

9. Sugar Research Foundation, "New York Sugar Trade Laboratory: A Respected Scientific Institution Rounds Out 40 Years of Service to the Industry," *Sugar Molecule*, January 1949, 10.

10. Ortiz, *Cuban Counterpoint*, 39.

11. Sugar Association of London, *Rules and Regulations for Beet and Cane Sugar Contracts* (Published for the Association by Wm. O'Toole, Secretary, 1928), 73.

12. César J. Ayala, *American Sugar Kingdom: The Plantation Economy of the Spanish Caribbean, 1898–1934*. (University of North Carolina Press, 1999) 4, 41–45.

13. Browne, "A History of the New York Sugar Trade Laboratory," 9.

14. Herbert S. Klickstein and Henry M. Leicester, "Charles Albert Browne as an Historian of Chemistry." *Journal of Chemical Education* 25, no. 6 (1948): 315. In addition to his analytical career, Browne was also a tireless historian of his own field. Each night for several years in the early 1920s he wrote a journal of the day's chemistry-related conversations and happenings, leaving future scholars with a disproportionately, even dizzyingly, detailed record of what went on within the walls of his own laboratory. He even recorded the early history of the Division of the History of Chemistry of the American Chemical Society, a unit that he had founded, and then, at an anniversary meeting of the division, he related that history back to itself.

15. Browne, "A History of the New York Sugar Trade Laboratory," 15 (emphasis added).

16. Browne, "A History of the New York Sugar Trade Laboratory," 18, 22.

17. Robert G. W. Anderson, "Chemistry Laboratories, and How They Might Be Studied," *Studies in History and Philosophy of Science Part A* 44, no. 4 (2013): 669–75.

18. US Department of the Treasury, *Treasury Decisions Under Tariff, Internal Revenue, Immigration, Navigation Laws, Etc., January 1 to June 30, 1898*, 2 vols. (Government Printing Office, 1898), 1:1, 982; "No Sand in This Sugar" July 27, 1890.

19. Clayton Anderson Coppin and Jack C. High, *The Politics of Purity: Harvey Washington Wiley and the Origins of Federal Food Policy* (University of Michigan Press, 1999), 86.

20. Browne, "A History of the New York Sugar Trade Laboratory," 8.

21. Browne, "A History of the New York Sugar Trade Laboratory," 18, 26.

22. F. G. Wiechmann, "Review: The Polariscope in the Chemical Laboratory, an Introduction to Polarimetry and Related Methods by George William Rolfe," *Science*, n.s. 23, no. 590 (1906): 627–28.

23. George W. Rolfe, "Temperature Corrections of Sugar Polarization," *Science*, n.s. 24, no. 610 (1906): 307–8.

24. H. W. Wiley, "The Polariscope in the Chemical Laboratory." *Journal of the American Chemical Society* 27, no. 12 (1905): 1572–73.

25. Browne, "A History of the New York Sugar Trade Laboratory," 41–44.

26. Bartram Bros. v. United States, 123 F. 330.

27. Browne, "A Constant Temperature Laboratory for the Polarization of Sugars."

28. C. A. Browne, "The Use of Temperature Corrections in the Polarization of Raw Sugars and Other Products Upon Quartz Wedge Saccharimeters," *Journal of Industrial & Engineering Chemistry* 1, no. 8 (1909): 567–80.

29. Browne, *A Handbook of Sugar Analysis*, 255.

30. Browne, "The Use of Temperature Corrections in the Polarization of Raw Sugars and Other Products upon Quartz Wedge Saccharimeters," 569.

31. Frederick J. Bates, *Circular of the Bureau of Standards* No. 44: *Polarimetry* (US Department of Commerce, National Institute of Standards and Technology, 1918), 166.

32. Sidney J. Osborn, *Methods of Analysis and Laboratory Control of the Great Western Sugar Company* (Great Western Sugar Company, 1920), 173.

33. "Laboratory building," CGu/0300/P0001-P0015, Colección Central Guánica, Archivo de Arquitectura y Construcción de la Universidad de Puerto Rico.

34. Anderson, "Chemistry Laboratories, and How They Might Be Studied," 669.

35. George W. Rolfe, "Raw Sugar Polarizations," *Louisiana Planter and Sugar Manufacturer*, February 12, 1910, 99.

36. Pavel Hánek and Pavel Hánek Sr., "Tradition of Geodetic Instruments Production in the Czech Republic," *History of Geo- and Space Sciences* 12, no. 2 (2021): 171–78; E. H. Sargent and Company, *Price List No. 20 of Scientific Laboratory Apparatus* (1914), 236; David Singerman, "'A Doubt Is at Best an Unsafe Standard': Measuring Sugar in the Early Bureau of Standards," *Journal of Research of the National Institute of Standards and Technology* 112, no. 1 (2007): 63; Warner, "How Sweet It Is."

37. For the "geography of precision," see Simon Schaffer, "Golden Means: Assay Instruments and the Geography of Precision in the Guinea Trade," in *Instruments, Travel and Science: Itineraries of Precision from the Seventeenth to the Twentieth Century*, ed. Marie-Noëlle Bourguet et al. (Routledge, 2002), 20–50.

38. *International Sugar Journal* 2 (June 1900): 282.

39. Ferdinand G. Wiechmann, "Third International Congress of Applied Chemistry, Vienna, 1898," *Science* 8, no. 194 (1898): 360–62; C. E. Munroe, "Third International Congress of Applied Chemistry," *Journal of the American Chemical Society* 21, no. 1 (1899): 73–102.

40. Ferdinand G. Wiechmann, *Sugar Analysis for Cane-Sugar and Beet-Sugar Houses, Refineries and Experimental Stations and as a Handbook of Instruction in Schools of Chemical Technology*, 3rd. ed. (John Wiley & Sons, 1914), 218.

41. Browne, *A Handbook of Sugar Analysis*, 108.

42. Michael Zakim, "Paperwork," *Raritan: A Quarterly Review* 33, no. 4 (Spring 2014): 34–56.

43. Frederick J. Bates, *Circular of the Bureau of Standards No. 440: Polarimetry, Saccharimetry and the Sugars* (National Bureau of Standards, 1942), 73–80.

44. Ferdinand G. Wiechmann, *Sugar Analysis for Cane-Sugar and Beet-Sugar Houses, Refineries and Experimental Stations and as a Handbook of Instruction in Schools of Chemical Technology*, 3rd ed. (John Wiley & Sons, 1914), 12.

45. Joseph O'Connell, "Metrology: The Creation of Universality by the Circulation of Particulars," *Social Studies of Science* 23, no. 1 (1993): 129–73.

46. Transcript of Record, American Sugar Refining Co. v. United States, 211 U.S. 155 (1908), 346.

47. Transcript of Record, American Sugar Refining Co. v. United States, 211 U.S. 155 (1908), 347.

48. Ferdinand G. Wiechmann, "The Question of Temperature-Influence on the Specific Rotation of Sucrose," *International Sugar Journal* 2 (1900): 491.

49. Wiechmann, *Sugar Analysis for Cane-Sugar and Beet-Sugar Houses*, 221.

50. John Harrison to Charles Browne, [date unknown], Browne Papers, box 7.

51. Transcript of Record, American Sugar Refining Co vs. United States, 211 U.S. 155 (1908), 357–58.

52. C. A. Browne, "The Polariscope Situation and the Need of an International Saccharimetric Scale," *Journal of Industrial & Engineering Chemistry* 10, no. 11 (1918): 916–18.

53. C. A. Browne, "Industrial and Agricultural Chemistry in British Guiana; with a Review of the Work of Professor J. B. Harrison," *Journal of Industrial & Engineering Chemistry* 11, no. 9 (1919): 881.

54. John Harrison, "Letter from the Government Laboratory, Georgetown, British Guiana," *International Sugar Journal*, September 12, 1900, 610.

55. Transcript of Record, American Sugar Refining Co. v. United States, 211 U.S. 155 (1908), 357–58.

56. César J. Ayala, *American Sugar Kingdom: The Plantation Economy of the Spanish Caribbean, 1898–1934* (University of North Carolina Press, 1999), 4, 79–85.

57. Guilford Spencer to Charles Browne, April 10, 1909. In Browne Papers, Library of Congress, box 25.

58. Spencer, *General Instructions and Methods of Analysis and Chemical Control*.

59. Guilford Spencer to Charles Browne, April 10, 1909. In Browne Papers, Library of Congress, box 25.

60. Rolfe, "Raw Sugar Polarizations."

61. Rolfe, "Temperature Corrections of Sugar Polarization."

62. Rolfe, "Raw Sugar Polarizations."

63. Charles A. Browne to George W. Rolfe, March 18, 1910, in Browne Papers, box 16, Library of Congress.

64. Central Mercedita Library, Colección Central Mercedita Archivo General de Puerto Rico, San Juan, Puerto Rico.

65. Spencer, *General Instructions and Methods of Analysis and Chemical Control*, 25.

66. Graeme Gooday, "The Premises of Premises: Spatial Issues in the Historical Construction of Laboratory Credibility," in *Making Space for Science: Territorial Themes in the Shaping of Knowledge*, ed. Crosbie Smith and Jon Agar (Macmillan, 1998), 239; Latour, *Science in Action*.

67. García-Muñiz, "Louisiana's 'Sugar Tramps' in the Caribbean Sugar Industry."

68. Sugar Research Foundation, "New York Sugar Trade Laboratory," 8.

69. Sugar Research Foundation, 11.

Chapter 10

1. See 7 USC § 609(d)(6).

2. Deerr, *The History of Sugar*, 2:449.

3. John Orval Ellsworth, "Sugar Marketing, with Special Reference to the Functions of the New York Coffee and Sugar Exchange" (PhD diss., Cornell University, 1926), 6.

4. Wells, *The Sugar Industry of the United States*, 109.

5. Atkins letter, II.7.79, and Atkins to Terry, II.8.471, December 31, 1889, MHS.

6. US Department of Commerce (Bureau of Foreign and Domestic Commerce), *The Cane Sugar Industry. Agricultural, Manufacturing, and Marketing Costs in Hawaii, Porto Rico, Louisiana, and Cuba*, misc. ser. no. 53 (Government Printing Office, 1917.

7. Moreno Fraginals, "Plantation Economies and Societies in the Spanish Caribbean, 1860–1930," 192–93.

8. Mark J. Smith, "Creating an Industrial Plant: The Biotechnology of Sugar Production in Cuba," in *Industrializing Organisms: Introducing Evolutionary History*, ed. Philip Scranton and Susan R. Schrepfer (Routledge, 2004), 92.

9. Moreno Fraginals, "Plantations in the Caribbean," 3.

10. Browne, "The Use of Temperature Corrections in the Polarization of Raw Sugars and Other Products Upon Quartz Wedge Saccharimeters."

11. Eichner, *The Emergence of Oligopoly*, 107–8.

12. E. M. Brunn, "The New York Coffee and Sugar Exchange," *Annals of the American Academy of Political and Social Science* 155 (1931): 112.

13. "Federal Attack on Sugar Trust," *New York Times*, November 29, 1910, https://www.nytimes.com/1910/11/29/archives/federal-attack-on-sugar-trust-american-sugar-refining-co-30.html.

14. Ellsworth, "Sugar Marketing, with Special Reference to the Functions of the New York Coffee and Sugar Exchange," 88–89; Allen, *A Story of the Growth of E. Atkins & Co. and the Sugar Industry in Cuba*, 19.

15. United States v. New York Coffee and Sugar Exchange, Inc. 263 U.S. 616.

16. Tim Paulson, "Paper Steaks: Live Cattle Futures Markets and the Financial Revolution of 1964," *Enterprise & Society*, May 31, 2024, 14–15.

17. Jonathan Levy, *Freaks of Fortune: The Emerging World of Capitalism and Risk in America* (Harvard University Press, 2012), chap. 7; Morton Rothstein, "The International Market for Agricultural Commodities, 1850–1873," in *Economic Change in the Civil War Era*, ed. David T. Gilchrist and W. David Lewis (Eleutherian Mills–Hagley Foundation, 1975), 62–82.

18. Nortz & Co., *Coffee and Sugar Facts: The New York Coffee and Sugar Exchange* (New York, 1925), 46; Brunn, "The New York Coffee and Sugar Exchange," 113–14.

19. United States v. New York Coffee and Sugar Exchange et al., 263 U.S. at 616.

20. Benjamin Wheeler Dyer, *Modern Methods of Marketing Cuban Raw Sugar: How You Can Lessen Business Risks by Using Sugar Futures: A Booklet for Sugar Centrals and Colonos* (Lamborn & Company, 1923), 12.

21. Annual Reports to the Directors of the New York Sugar Trade Laboratory, 1923 and 1924, in Browne Papers, Library of Congress.

22. United States v. New York Coffee and Sugar Exchange et al., 263 U.S. at 610.

23. Dyer, *Modern Methods of Marketing Cuban Raw Sugar*, 12.

24. Appellee's brief in United States v New York Coffee & Sugar Exchange, 263 U.S. 611 (1924), 5.

25. Cronon, *Nature's Metropolis*; Baker and Hahn, *The Cotton Kings*.

26. In April 2020, the price for May crude oil briefly went below zero, because sellers had no place to keep it, and they needed to pay buyers to take it off their hands. Stanley Reed and Clifford Krauss, "Too Much Oil: How a Barrel Came to Be Worth Less Than Nothing," *New York Times*, April 20, 2020, https://www.nytimes.com/2020/04/20/business/oil-prices.html.

27. Dyer, *Modern Methods of Marketing Cuban Raw Sugar*, 17.

28. Levy, *Freaks of Fortune*, 243; Appellee's brief in United States v New York Coffee & Sugar Exchange, 263 U.S. 611 (1924), 5.

29. Rolfe, "Temperature Corrections of Sugar Polarization," 308.

30. Nortz & Co., *Coffee and Sugar Facts*, 43–47; Francis Maxwell, *Economic Aspects of Cane Sugar Production* (N. Rodger, 1927), 103.

31. Hurford Janes and H. J. Sayers, *The Story of Czarnikow* (Harley, 1963), 50; Ellsworth, "Sugar Marketing."

32. Sedgewick, *Coffeeland*, 88.

33. Appellee's brief in United States v New York Coffee & Sugar Exchange, 263 U.S. 611 (1924), 24.

34. *Charter, By-Laws, and Rules of the New York Coffee and Sugar Exchange* (New York, 1927), 49.

35. Ellsworth, "Sugar Marketing,"52.

36. *Charter, By-Laws, and Rules of the New York Coffee and Sugar Exchange* 54; Sedgewick, *Coffeeland*.

37. Brunn, "The New York Coffee and Sugar Exchange," 115.

38. *Charter, By-Laws, and Rules of the New York Coffee and Sugar Exchange, Inc.*, 142.

39. Julius Bernard Baer and Olin Glenn Saxon, *Commodity Exchanges and Futures Trading* (Harper, 1949), 26; see also Aashish Velkar, "Measurement Standards and Market Governance: London Corn Trade Association and International Grain Markets (1880–1914)," *Histoire & Mesure* 38, no. 1(2023): 65–92.

40. See New York Sugar Trade Laboratory report for 1923, Box 33, Browne Papers, LOC.

41. Transcript of Record, American Sugar Refining Co., New York v. United States, 211 U.S. 155 (1908), 453.

42. Nevers & Callaghan to E. H. Costello, January 16, 1928. In Sugar (Polarization)—Refiners' Committee on Sugar Differentials, Sugar Institute—Correspondence. Series 10: Manuel Rionda y Polledo subject files, 1902–1946. Braga Brothers collection, University of Florida George A. Smathers Libraries. Consulted via University of Florida Digital Collections. https://original-ufdc.uflib.ufl.edu/AA00010154/00024.

43. US Industrial Commission, *Preliminary Report on Trusts and Industrial Combination*, vol. 1 (Government Printing Office, 1900), 44, 65, 70.

44. *Charter, By-Laws, and Rules of the New York Coffee and Sugar Exchange*, 49.

45. Raw Sugar Contract, American Sugar Refining Company, 1920s. In Sugar (Polarization)—Refiners' Committee on Sugar Differentials, Sugar Institute—Correspondence, UFL.

46. See New York Sugar Trade Laboratory report for 1927, Box 33, Browne Papers, LOC.

47. Sam Anderson, "The Weather God of Oklahoma City," *New York Times Magazine*, August 9, 2013, https://www.nytimes.com/2013/08/11/magazine/the-weather-god-of-oklahoma-city.html.

48. F. E. Sullivan to James H. Post, January 30, 1928. In Sugar (Polarization)—Refiners' Committee on Sugar Differentials, Sugar Institute—Correspondence, UFL.

49. Alfred S. Eichner, "Monopoly, the Emergence of Oligopoly and the Case of Sugar Refining: A Reply," *Journal of Law & Economics* 14, no. 2 (1971): 523.

50. David Genesove and Wallace Mullin, "The Sugar Institute Learns to Organize Information Exchange," in *Learning by Doing in Markets, Firms, and Countries*, ed. Naomi R. Lamoreaux et al. (University of Chicago Press, 1999), 103–44; Genesove and Mullin, "Rules, Communication, and Collusion."

51. Report of Committee on Raw Sugar Polarization Allowances. In Sugar (Polarization)—Refiners' Committee on Sugar Differentials, Sugar Institute—Correspondence, UFL.

52. C. J. Welch to E. H. Costello, January 28, 1928. In Sugar (Polarization)—Refiners' Committee on Sugar Differentials, Sugar Institute—Correspondence, UFL.

53. Central Aguirre Superintendent's Report for the Year 1928, Colección Mercedita, Archivo General de Puerto Rico (AGPR).

54. C. J. Welch to E. H. Costello, January 28, 1928. In Sugar (Polarization)—Refiners' Committee on Sugar Differentials, Sugar Institute—Correspondence, UFL.

55. Frank C. Lowry to E. H. Costello, February 21, 1928. In Sugar (Polarization)—Refiners' Committee on Sugar Differentials, Sugar Institute—Correspondence, UFL.

56. Report of Committee on Raw Sugar Polarization Allowances. In Sugar (Polarization)—Refiners' Committee on Sugar Differentials, Sugar Institute—Correspondence, UFL.

57. C. J. Welch to E. H. Costello, January 28, 1928. In Sugar (Polarization)—Refiners' Committee on Sugar Differentials, Sugar Institute—Correspondence, UFL.

58. See New York Sugar Trade Laboratory report for 1925, box 33, Browne Papers, LOC.

59. Appellee's brief, 5, United States v New York Coffee & Sugar Exchange, 263 U.S. 611 (1924).

60. Browne, History of the New York Sugar Trade Laboratory, 63, in Browne Papers, LOC.

61. Browne, Laboratory Journal of interviews, July 1919, New York Public Library.

62. F. E. Sullivan to James H. Post, January 30, 1928. In Sugar (Polarization)—Refiners' Committee on Sugar Differentials, Sugar Institute—Correspondence, UFL.

63. National Bureau of Economic Research, "Retail Price of Sugar for New York, NY, 1911–1943," *Federal Reserve Economic Data* (Federal Reserve Bank of St. Louis), updated August 16, 2012, accessed December 8, 2023, https://fred.stlouisfed.org/series/M04031US35620M267NNBR.

64. Transcript of Record, vol. 1 (1935), 93, Sugar Institute v. United States 297 U.S. 553 (1936), https://link.gale.com/apps/doc/DW0108066207/SCRB, 93.

65. Albert A. Hopkins, "From Sugar Cane to Crystal Cube," *Scientific American* 144, no. 2 (1931): 106.

66. Eichner, *The Emergence of Oligopoly*, 71.

67. Eichner, 102–6. See Eichner, *The Emergence of Oligopoly*, chapter 5, for more on the Trust's mechanisms for controlling inputs and outputs.

68. Genesove and Mullin, "Rules, Communication, and Collusion," 385.

69. United States v. New York Coffee and Sugar Exchange, Inc., et al., 263 U.S. 618–20.

70. Alfred Best memorandum to Mr. [?] Carpenter, undated. In Sugar (Polarization)—Refiners' Committee on Sugar Differentials, Sugar Institute—Correspondence, UFL.

71. C. J. Welch to E. H. Costello, January 28, 1928. In Sugar (Polarization)—Refiners' Committee on Sugar Differentials, Sugar Institute—Correspondence, UFL.

72. Transcript of Record, vol. 1 (1935): 233, Sugar Institute v. United States, 297 U.S. 553 (1936).

73. Transcript of Record, vol. 1 (1935): 233, Sugar Institute v. United States, 297 U.S. 553 (1936).

74. Brief for United States (1936), Sugar Institute v. United States, 297 U.S. 553 (1936). https://link.gale.com/apps/doc/DW0104166844/SCRB.

75. C. J. Welch to E. H. Costello, January 28, 1928; Alfred Best memorandum to Mr. [?] Carpenter (undated), In Sugar (Polarization)—Refiners' Committee on Sugar Differentials, Sugar Institute—Correspondence, UFL.

76. See New York Sugar Trade Laboratory report for 1934, box 33, Browne Papers, LOC.

77. "Sugar No. 16 Futures" (n.d.); author's interview with Charley Richard.

Chapter 11

1. Alfred Best memorandum to Mr. [?] Carpenter (undated) In Sugar (Polarization)—Refiners' Committee on Sugar Differentials, Sugar Institute—Correspondence, UFL.

2. "The Valuation of Sugars," In Sugar (Polarization)—Refiners' Committee on Sugar Differentials, Sugar Institute—Correspondence, UFL.

3. Ortiz, *Cuban Counterpoint*, 23.

4. "The Valuation of Sugars," In Sugar (Polarization)—Refiners' Committee on Sugar Differentials, Sugar Institute—Correspondence, UFL.

5. George A. Akerlof, "'The Market for Lemons': Quality Uncertainty and the Market Mechanism," *Quarterly Journal of Economics* 84, no. 3 (1970): 488–500. It would be a classic case of the market for lemons, but sugar is not supposed to be sour.

6. Ellsworth, "Sugar Marketing, with Special Reference to the Functions of the New York Coffee and Sugar Exchange," 54.

7. US Industrial Commission, *Preliminary Report on Trusts and Industrial Combination*, 93–100.

8. Eichner, *The Emergence of Oligopoly*, 71, 218, 226; Genesove and Mullin, "Predation and Its Rate of Return."

9. William Stout, *The Autobiography of William Stout of Lancaster, 1665–1752* (Manchester, 1967), 145.

10. Rolph, *Something About Sugar*, 66.

11. George Martineau, *Sugar, Cane and Beet: An Object Lesson.*, 4th ed. (Sir I. Pitman & Sons, 1918), 142.

12. American Sugar Refining Co. V. United States, 181 U.S. 610, 611 (1901). Morton to Chandler, April 11, 1878, Charles F. Chandler Papers, box 260, folder 7, Rare Book and Manuscript Library, Columbia University, New York.

13. Stobart, *Sugar and Spice*, 72.

14. Moreno Fraginals, "Plantations in the Caribbean," 8–9. Morton to Chandler, April 11, 1878, Charles F. Chandler Papers, box 260, folder 7, Rare Book and Manuscript Library, Columbia University, New York.

15. Baer and Saxon, *Commodity Exchanges and Futures Trading*, 118.

16. Kula, *Measures and Men*, 103. Kula notes that many premodern European societies featured different "measures of entry" and "measures of exit."

17. Genesove and Mullin, "Predation and Its Rate of Return," 68; Myer Lynsky, *Sugar Economics, Statistics, and Documents* (US Cane Sugar Refiners' Association, 1938), 12–13; National Bureau of Economic Research, "Raw Sugar Stocks at Four Ports for United States," *Federal Reserve Economic Data* (Federal Reserve Bank of St. Louis), updated August 17, 2012, accessed September 27, 2022, https://fred.stlouisfed.org/series/M0502AUSM576NNBR. Calculated using capacity data from Genesove and Mullin and stock data from NBER.

18. US Department of Commerce, Bureau of Foreign and Domestic Commerce, *The Cane Sugar Industry*, 425.

19. Roy A. Ballinger, "A History of Sugar Marketing Through 1974" (US Department of Agriculture, March 1978), 21.

20. Annual Report of the Punta Alegre Sugar Company for the Year Ending May 31, 1919, p. 4. Braga Brothers Collection, University of Florida Digital Collections. https://ufdc.ufl.edu/aa00069061/00001.

21. C. A. Browne, "The Deterioration of Raw Cane Sugar: A Problem in Food Conservation." *Journal of Industrial & Engineering Chemistry* 10, no. 3 (1918): 178–90.

22. Report by George W. Goethals and Company, Inc., on the Cuba Cane Sugar Corporation, July 11, 1919, Braga Brothers online collection, University of Florida Libraries, https://ufdc.ufl.edu/AA00010155/00001/.

23. Report of the New York Sugar Trade Laboratory for 1923, in Box 33, Browne Papers, LOC.

24. John D. Kerr, "Walter Maxwell (1854–1931)," in *Australian Dictionary of Biography* (National Centre of Biography, Australian National University), accessed December 13, 2023, https://adb.anu.edu.au/biography/maxwell-walter-7535/text13143; C. Allan Jones and Robert V. Osgood, *From King Cane to the Last Sugar Mill: Agricultural Technology and the Making of Hawai'i's Premier Crop* (Honolulu: University of Hawai'i Press, 2015), 68–69.

25. Walter Maxwell, *Report of Work of the Experiment Station of the Hawaiian Sugar Planters' Association for the Year 1895* (Reprint), Division of Agriculture and Chemistry Bulletins 1 (Honolulu, 1905), 39.

26. Maxwell, *Report of Work of the Experiment Station*, 43–46.

27. Noël Deerr and R. S. Norris, *The Deterioration of Sugars in Storage*, Division of Agriculture and Chemistry Bulletins 24 (Honolulu, 1908).

28. "The Valuation of Sugars," no date or author given. In Sugar (Polarization)—Refiners' Committee on Sugar Differentials, Sugar Institute—Correspondence, UFL.

29. Rolph, *Something About Sugar*, 88.

30. Harrison to Browne, February 25, 1919, Browne Papers LOC.

31. Browne, *A Handbook of Sugar Analysis*, 14; C. A. Browne and F. W. Zerban, *Physical and Chemical Methods of Sugar Analysis: A Practical and Descriptive Treatise for Use in Research, Technical, and Control Laboratories*, 3rd ed. (John Wiley & Sons, 1941), 18–23.

32. Browne to Directors, June 4, 1919, Browne Papers, LOC. He suggested that Rolfe could be an independent investigator, if the trustees wished to establish a commission.

33. Browne, "The Deterioration of Raw Cane Sugar"; Nicholas Kopeloff and Lillian Kopeloff, "The Deterioration of Manufactured Cane Sugar by Molds." *Journal of Industrial & Engineering Chemistry* 11, no. 9 (1919): 845–50; Nicholas Kopeloff et al., *The Prevention of Sugar Deterioration*, Bulletin of the Agricultural Experiment Station of the Louisiana State University and A. & M. College 175 (Ramires-Jones, 1920); William Ludwell Owen, *Deterioration of Cane Sugars in Storage: Its Causes and Suggested Measures for Its Control*, Bulletin of the Agricultural Experiment Station of the Louisiana State University and A. & M. College 162 (Ramires-Jones, 1918).

34. Browne, "The Deterioration of Raw Cane Sugar," 180.

35. Noël Deerr, *Cane Sugar: A Textbook on the Agriculture of the Sugar Cane, the Manufacture of Cane Sugar, and the Analysis of Sugar-House Products*, 2nd ed. (Norman Rodger, 1921), 436.

36. Reports of the New York Sugar Trade Laboratory for 1923–1927, box 33, Browne Papers, LOC.

37. Rolfe to Browne, August 31, 1939, box 16, Browne Papers, LOC.

38. Eimer & Amend, *Illustrated Wholesale Catalogue with Prices Current of Chemical & Physical Apparatus and Assay Goods* (New York, 1892), 261.

39. Rolfe to Browne, February 25, 1920, box 16, Browne Papers, LOC.

40. C. A. Browne, "The First High-School Chemistry Laboratory?" *Journal of Chemical Education* 18, no. 8 (1941): 395.

41. Rolfe to Browne, February 25, 1920, box 16, Browne Papers, LOC.

42. Abstracts of Custom Court Decisions—Classification, 3 U.S. Cust. Ct. Rep. 321 (n.d.), 364.

43. Rolfe to Browne, November 4, 1919, box 16, Browne Papers, LOC.

44. Singerman, "'A Doubt Is at Best an Unsafe Standard,'" 65.

45. R. W. Plews, *The History of ICUMSA: The First 100 Years, 1897–1997* (International Commission for Uniform Methods of Sugar Analysis, 1997).

46. Browne to Directors, June 4, 1919, box 33, Browne Papers, LOC.

47. Browne, Journal of Laboratory Interviews, January 13, 1920, New York Public Library.

48. Rolfe, *The Polariscope in the Chemical Laboratory*, 142.

49. Rolfe to Browne, August 31, 1939, box 16, Browne Papers, LOC.

50. Rolfe to Browne, February 25, 1920, box 16, Browne Papers, LOC.

51. Rolfe to Browne, August 31, 1939, box 16, Browne Papers, LOC.

52. Shapin, "The Invisible Technician."

53. Schaffer, "Easily Cracked."

54. Browne, "Industrial and Agricultural Chemistry in British Guiana; with a Review of the Work of Professor J. B. Harrison," 881.

55. US Department of the Treasury, *Treasury Decisions Under Tariff, Internal Revenue, Immigration, Navigation Laws, Etc., January 1 to June 30, 1898*, 992.

56. Browne to Rolfe, February 19, 1920, box 16, Browne Papers, LOC.

57. Rolfe to Browne, January 19, 1930, box 16, Browne Papers, LOC.

58. Bates, "Circular of the Bureau of Standards No. 440," 75.

59. American Sugar Refining Company, *A Century of Sugar Refining in the United States, 1816–1916* (De Vinne Press, 1916), 16–18.

60. "The Vanished Sugar Barrel," *Sugar*, January 1946.

61. "Short Interviews on the Tariff Question, no. 4: Sugar," *Puck*, October 3, 1888, 90.

62. National Bureau of Economic Research, "Raw Sugar Stocks at All Ports for United States," *Federal Reserve Economic Data* (Federal Reserve Bank of St. Louis), updated August 17, 2012, accessed December 12, 2023, https://fred.stlouisfed.org/series/M0502BUSM576NNBR.

63. Nortz & Co., *Coffee and Sugar Facts*, 11.

64. "New No. 5 Contract Advantageous," *Sugar*, August 1947.

65. Browne and Zerban, *Physical and Chemical Methods of Sugar Analysis*, 21.

66. Hall et al., "Production Control in the Revere Sugar Refinery," 89.

67. *Testimony in Relation to the Sugar Frauds, Taken by the Subcommittee of the Committee of Ways and Means of the House of Representatives, New York, September 1878*, 44.

Selected Bibliography

Archives and Manuscript Sources Consulted

BERKELEY, CALIFORNIA

Bancroft Library, University of California, Berkeley
- Papers of Dan Gutleben
- Records of the Western Sugar Refinery

BOSTON, MASSACHUSETTS

Massachusetts Historical Society
- Atkins Family Papers

GLASGOW, SCOTLAND

Glasgow School of Art
University of Glasgow Archives and Special Collections
- Records of Duncan Stewart & Co.
- Records of Mirrlees Watson & Co.
- Records of Smith Mirrlees and associated companies

University of Strathclyde Archives and Special Collections

HONOLULU, HAWAII

University of Hawai'i at Manoa Hawaiian and Pacific Collections
University of Hawai'i at Manoa Manuscripts Collections
- Hawaiian Sugar Planters Association
- Honolulu Iron Works

NEW YORK, NEW YORK

Columbia University Rare Book and Manuscript Library
- Charles Chandler Papers

New York Public Library
Charles A. Browne Papers
Moses Taylor Papers

PHILADELPHIA, PENNSYLVANIA

American Philosophical Society
Pennsylvania Historical Society
Science History Institute

PUERTO RICO

Archivo General de Puerto Rico, San Juan
Colección Central Mercedita
Departemento de Estado
Fondo Oficina del Gobernador
Records of the Spanish Governors of Puerto Rico
Universidad de Puerto Rico, Rio Piedras
Archivo de Arquitectura y Construcción
Colección Puertorriqueña

WASHINGTON, DC

Library of Congress Manuscript Division
Papers of Charles A. Browne
Papers of James A. Garfield
Papers of John Sherman
Papers of David Ames Wells
Papers of Harvey W. Wiley
National Archives and Records Administration
RG 36, Records of the U.S. Customs Service
National Institute for Standards and Technology
United States Department of the Treasury, Treasury Library

Selected Newspapers and Periodicals

American Sugar Industry and Beet Sugar Gazette
Boston Daily Globe
Brooklyn Daily Eagle
Frank Leslie's Illustrated Newspaper
Harper's Weekly
International Sugar Journal
Journal of the American Chemical Society
Journal of Chemical Education
Journal of Industrial & Engineering Chemistry
Louisiana Planter and Sugar Manufacturer

New York Herald
The New York Times
New-York Tribune
New York World
The Sugar Cane

US Government Documents

EXECUTIVE BRANCH

Report of the Superintendent of the U.S. Coast and Geodetic Survey Showing the Progress of the Work from July 1, 1897, to June 30, 1898. Government Printing Office, 1899.

Tenth Census of the United States, Report on the Manufactures of the United States. 22 vols. Government Printing Office, 1883. https://catalog.hathitrust.org/Record/000441415.

United States Customs Service, ed. *A History of Enforcement in the United States Customs Service, 1789–1875.* Special Customs bicentennial reissue. Customs Publication, no. 522. U.S. Customs Service, 1988.

United States Department of Commerce (Bureau of Foreign and Domestic Commerce). *The Cane Sugar Industry. Agricultural, Manufacturing, and Marketing Costs in Hawaii, Porto Rico, Louisiana, and Cuba.* Miscellaneous Series, No. 53. Government Printing Office, 1917. https://catalog.hathitrust.org/Record/100559910.

United States Department of the Treasury. *Annual Report of the Secretary of the Treasury on the State of the Finances for the Year 1877.* Government Printing Office, 1877.

United States Department of the Treasury. *Annual Report of the Secretary of the Treasury on the State of the Finances for the Year 1878.* Government Printing Office, 1878.

United States Department of the Treasury. *Annual Report of the Secretary of the Treasury on the State of the Finances for the Year 1879.* Government Printing Office, 1879.

United States Department of the Treasury. *Letter from the Secretary of the Treasury, Transmitting the Report of Certain Commissioners Appointed to Examine Custom-houses, and Recommending Appropriations for Their Pay.* 45th Congress, 1st Session, October 25, 1877. https://doi.org/10.1017/CBO9780511665004.

United States Department of the Treasury. *Letter from the Secretary of the Treasury, Transmitting in Response to a Senate Resolution, July 14, 1890, a Report of Special Agents Who Were Instructed to Investigate the Manner in Which Sugars Are Classified.* 51st Congress, 1st session, July 18, 1890.

United States Department of the Treasury. *Letter from the Secretary of the Treasury, Transmitting, in Response to Senate Resolution of January 8, 1889, Information Relative to the Sugar Frauds.* 50th Congress, 1st session, January 20, 1889.

United States Department of the Treasury. *Reports from the Secretary of the Treasury, of Scientific Investigations in Relation to Sugar and Hydrometers, Made Under the Superintendence of Professor R. S. McCulloh.* Wendell and Van Benthuysen, 1848.

United States Department of the Treasury. *Synopsis of the Decisions of the Treasury Department on the Construction of the Tariff, Navigation, and Other Laws for the Year Ended December 31, 1880.* Government Printing Office, 1881. https://catalog.hathitrust.org/Record/010424084.

United States Department of the Treasury. *Synopsis of the Decisions of the Treasury Department on the Construction of the Tariff, Navigation, and Other Laws for the Year Ended December 31, 1897.* Government Printing Office, 1897. https://catalog.hathitrust.org/Record/010424084.

United States Department of the Treasury. *Treasury Decisions Under Tariff, Internal Revenue, Immigration, Navigation Laws, Etc., January 1 to June 30, 1898*. Vol. 1.1. Government Printing Office, 1898. https://catalog.hathitrust.org/Record/007516750.

United States Department of the Treasury. *Treasury Decisions Under Tariff, Internal Revenue, Immigration, Navigation Laws, Etc., January 1 to June 30, 1899*. Vol. 2.1. Government Printing Office, 1899. https://catalog.hathitrust.org/Record/007516750

United States Industrial Commission. *Preliminary Report on Trusts and Industrial Combination*. Government Printing Office, 1900.

JUDICIAL BRANCH

Abstracts of Custom Court Decisions—Classification, 3 U.S. Cust. Ct. Rep. 321 (n.d.)

American Sugar Refining Co. v. United States, 181 U.S. 610 (1901)

American Sugar Refining Co. v. United States, 211 U.S. 155 (1908)

Bartram Bros. v. United States, 123 F. 327 (Cir. Court, S.D.N.Y, 1903), rev'd, 131 F. 833 (Cir. Court of Appeals, S.D.N.Y, 1904), cert. denied, 195 U.S. 635 (1904), appeal dismissed, 211 U.S. 155 (1908)

Sugar Institute v. United States, 297 U.S. 553 (1936)

United States v. Bartram Bros., 131 F. 833 (Cir. Court of Appeals, S.D.N.Y., 1904), cert. denied, 195 U.S. 635 (1904), appeal dismissed, 211 U.S. 155 (1908)

United States v. E. C. Knight Co., 156 U.S. 1 (1895)

United States v. New York Coffee & Sugar Exchange, 263 U.S. 611 (1924)

LEGISLATIVE BRANCH

Tariff of 1861 (Morrill Tariff), chap. 48, 12 Stat. 178 (March 2, 1861).

Revenue Act of 1861, chap. 45, 12 Stat. 292 (August 5, 1861)

Tariff of 1894 (Wilson-Gorman Tariff), chap. 349, 28 Stat. 570 (August 27, 1894)

Tariff of 1897 (Dingley Tariff), chap. 11, 30 Stat. 151 (July 24, 1897)

Books and Articles

Achard, Franz Carl. *Traité complet sur le sucre européen de betteraves: Culture de cette plante considérée sous le rapport agronomique et manufacturier*. Edited by Ch. Derosne, translated by D. Angar. Paris, 1812. http://hdl.handle.net/2027/nyp.33433006676716

Achim, Miruna. "The Art of the Deal, 1828: How Isidro Icaza Traded Pre-Columbian Antiquities to Henri Baradère for Mounted Birds and Built a National Museum in Mexico City in the Process." *West 86th: A Journal of Decorative Arts, Design History, and Material Culture*, July 19, 2015. https://doi.org/10.1086/662519.

Adams, Henry. *The Education of Henry Adams*. Houghton Mifflin, 1918.

Akerlof, George A. "The Market for 'Lemons': Quality Uncertainty and the Market Mechanism." *Quarterly Journal of Economics* 84, no. 3 (1970): 488–500. https://doi.org/10.2307/1879431.

Alasheev, Sergei. "On a Particular Kind of Love and the Specificity of Soviet Production." In *Management and Industry in Russia: Formal and Informal Relations in the Period of Transition*, ed. Simon Clarke, 69–98. Edward Elgar, 1995.

Alder, Ken. *The Measure of All Things: The Seven-Year Odyssey and Hidden Error That Transformed the World*. Free Press, 2002.

Alder, Ken. "Scientific Conventions: International Assemblies and Technical Standards from the Republic of Letters to Global Science." In *Nature Engaged: Science in Practice from the Renaissance to the Present*, ed. Mario Biagioli and Jessica Riskin, 19–39. Palgrave Studies in Cultural and Intellectual History. Palgrave Macmillan, 2012.

Allen, Benjamin. *A Story of the Growth of E. Atkins & Co. and the Sugar Industry in Cuba*, 1925.

American Sugar Refining Company. *A Century of Sugar Refining in the United States, 1816–1916*. De Vinne Press, 1916. https://babel.hathitrust.org/cgi/pt?id=mdp.39015067165111&seq=1.

Anderson, Robert G. W. "Chemistry Laboratories, and How They Might Be Studied." *Studies in History and Philosophy of Science, Part A*, 44, no. 4 (2013): 669–75. https://doi.org/10.1016/j.shpsa.2013.07.003.

Andreas, Peter. *Smuggler Nation: How Illicit Trade Made America*. Oxford University Press, 2014.

Andrews, Clement W. "The Influence of Temperature on the Specific Rotation of Cane Sugar." *Technology Quarterly* 2 (1889): 367–71. https://catalog.hathitrust.org/Record/008616579.

Atkins, Edwin Farnsworth. *Sixty Years in Cuba: Reminiscences of Edwin F. Atkins*. Privately printed at the Riverside Press, 1926.

Ayala, César J. *American Sugar Kingdom: The Plantation Economy of the Spanish Caribbean, 1898–1934*. University of North Carolina Press, 1999.

Baer, Julius Bernard, and Olin Glenn Saxon. *Commodity Exchanges and Futures Trading*. Harper, 1949.

Baeza, Miguel de. *Los quatro libros del arte de la confitería*. 1592. Repr. Antonio Pareja Editor, 2014.

Baker, Bruce E., and Barbara Hahn. *The Cotton Kings: Capitalism and Corruption in Turn-of-the-Century New York and New Orleans*. Oxford University Press, 2015.

Baker, John Proculus. *An Essay on the Art of Making Moscovado Sugar Wherein a New Process Is Proposed*. Joseph Weatherby, 1775. http://link.gale.com/apps/doc/CW0106823254/ECCO.

Ballinger, Roy A. *A History of Sugar Marketing Through 1974*. US Department of Agriculture, March 1978. http://www.ers.usda.gov/publications/pub-details/?pubid=40540.

Bates, Frederick J. *Circular of the Bureau of Standards No. 44: Polarimetry*. US Department of Commerce, National Institute of Standards and Technology, 1918. https://books.google.com?id=cKQ_swEACAAJ.

Beachey, R. W. *The British West Indies Sugar Industry in the Late 19th Century*. Blackwell, 1957.

Beckert, Sven, et al. "Commodity Frontiers and the Transformation of the Global Countryside: A Research Agenda." *Journal of Global History* 16, no. 3 (2021): 435–50. https://doi.org/10.1017/S1740022820000455.

Beckert, Sven. *Empire of Cotton: A Global History*. Alfred A. Knopf, 2014.

Belisle, Donica. "Eating Clean: Anti-Chinese Sugar Advertising and the Making of White Racial Purity in the Canadian Pacific." *Global Food History* 6, no. 1 (2020): 41–59. https://doi.org/10.1080/20549547.2020.1712577.

Bensel, Richard Franklin. *The Political Economy of American Industrialization, 1877–1900*. Cambridge University Press, 2000. https://doi.org/10.1017/CBO9780511665004.

Berg, Maxine. *The Machinery Question and the Making of Political Economy, 1815–1848*. Cambridge University Press, 1980.

Bessemer, Henry. *On a New System of Manufacturing Sugar from the Cane*. Privately printed, 1852. http://www.globalcommodities.amdigital.co.uk/Documents/Details/UL_GL_1852a_New System.

Bigelow, Allison Margaret. *Mining Language: Racial Thinking, Indigenous Knowledge, and Colonial Metallurgy in the Early Modern Iberian World*. University of North Carolina Press, 2020. http://muse.jhu.edu/book/74318.

Bigelow, Allison, and Pablo Cruz. "*Ingenios* and Ingenuity: Rethinking Indigenous Histories of Silver in the Colonial Andean Mining Industry." *Colonial Latin American Review* 30, no. 4 (2021): 520–44. https://doi.org/10.1080/10609164.2021.1996989.

Biot, Jean-Baptiste. "Comparaison du sucre et de la gomme ara bigue dans leur action sur la lumière polarisée." *Bulletin des sciences par la société philomatique de Paris*, 1816, 125–27.

Biot, Jean-Baptiste. "Examen comparatif du sucre de maïs et du sucre de betterave, soumis aux épreuves de la polarisation circulaire." *Comptes rendus hebdomadaires des séances de l'Académie des Sciences* 2 (1836): 464–67. https://doi.org/10.1259/jrs.1912.0009.

Biot, Jean-Baptiste. "Sur l'application des propriétés optiques à l'analyse quantitatives des mélanges liquides ou solides, dans lesquels le sucre de canne cristallisable est associé à des sucres incristallisables." *Comptes rendus hebdomadaires des séances de l'Académie des Sciences* 16 (1843): 619–39. https://doi.org/10.1259/jrs.1912.0009.

Biot, Jean-Baptiste. "Sur l'utilité que pourraient offrir les caractères optiques dans l'exploitation des sucreries et des raffineries." *Comptes rendus hebdomadaires des séances de l'Académie des Sciences* 10 (1840): 264–66. https://doi.org/10.1259/jrs.1912.0009.

Blum, Deborah. *The Poison Squad: One Chemist's Single-Minded Crusade for Food Safety at the Turn of the Twentieth Century*. Penguin Press, 2018.

Booker, Peter Jeffrey. *A History of Engineering Drawing*. Chatto & Windus, 1963.

Bookers Sugar Estates, Ltd. *Bookers Sugar: Supplement to the Accounts of Booker Brothers, McConnell & Co., Limited*. Georgetown, British Guiana, 1954.

Bosma, Ulbe. *The World of Sugar: How the Sweet Stuff Transformed Our Politics, Health, and Environment over 2,000 Years*. Belknap Press of Harvard University Press, 2023.

Bouillet, Marie Nicolas. *Atlas universel d'histoire et de géographie*. Hachette et Cie, 1865. https://books.google.com?id=fOgOAAAAQAAJ.

Bourguet, Marie-Noëlle, et al., Introduction. In *Instruments, Travel and Science: Itineraries of Precision from the Seventeenth to the Twentieth Century*, edited by Marie- Noëlle Bourguet et al., 1–19. Routledge, 2002.

Boyle, Robert. *Some Considerations Touching the Usefulnesse of Experimental Naturall Philosophy: The Second Part*. Hen. Hall, Printer to the University, for Ric. Davis, 1663. http://name.umdl.umich.edu/A29031.0001.001.

Boynton, H. V. "The Whiskey Ring" *North American Review*, October 1, 1876, http://www.jstor.org/stable/25109986.

Brady, Kathleen. *Ida Tarbell: Portrait of a Muckraker*. University of Pittsburgh Press, 1989.

Brenni, Paolo. "Some Considerations about the Prices of Physics Instruments in the Nineteenth Century." In *How Scientific Instruments Have Changed Hands*, ed. A. D. Morrison-Low and Sara J. Schechner, 57–87. Brill, 2017.

Brock, William. *Justus von Liebig: The Chemical Gatekeeper*. Cambridge University Press, 1997.

Brookings Institution. *The Sugar Problem of Puerto Rico*. Association of Sugar Producers of Puerto Rico, 1936.

Brown, Henry A. *Statements Made Before the Committee of Ways and Means on the Sugar Question in the Interests of American Consumers, Home Industries, and Revenue*. Judd & Detweiler, 1880. https://books.google.com?id=eg5BnQEACAAJ.

Brown, Henry A. *Sugar Frauds and the Tariff: Their Relations to Home Product, Consumption, Industry, Imports, Duties and Revenue, Analyzed and Exhibited. Duty & Drawback, Sampling & Grading. Adulterations and the Polariscope. Official Statistics, Remedies, &c.* Saxonville, MA, 1879.

Browne, C. A. "A Constant Temperature Laboratory for the Polarization of Sugars." In *Eighth International Congress of Applied Chemistry, Washington and New York, September 4 to 13, 1912*, 519–29. Rumford Press, 1912. https://catalog.hathitrust.org/Record/001042582.

Browne, C. A. "The Deterioration of Raw Cane Sugar: A Problem in Food Conservation." *Journal of Industrial & Engineering Chemistry* 10, no. 3 (1918): 178–90. https://doi.org/10.1021/ie50099a010.

Browne, C. A. "The First High-School Chemistry Laboratory?" *Journal of Chemical Education* 18, no. 8 (1941): 395. https://doi.org/10.1021/ed018p395.

Browne, C. A. *A Handbook of Sugar Analysis: A Practical and Descriptive Treatise for Use in Research, Technical and Control Laboratories.* John Wiley and Sons, 1912.

Browne, C. A. "Industrial and Agricultural Chemistry in British Guiana; with a Review of the Work of Professor J. B. Harrison." *Journal of Industrial & Engineering Chemistry* 11, no. 9 (1919): 874–81. https://doi.org/10.1021/ie50117a012.

Browne, C. A. "The Polariscope Situation and the Need of an International Saccharimetric Scale." *Journal of Industrial & Engineering Chemistry* 10, no. 11 (1918): 916–18. https://doi.org/10.1021/ie50107a015.

Browne, C. A. "Sugar Industry in Foreign Lands." *Industrial & Engineering Chemistry* 25, no. 1 (1933): 61–68. https://doi.org/10.1021/ie50277a015.

Browne, C. A. "The Use of Temperature Corrections in the Polarization of Raw Sugars and Other Products upon Quartz Wedge Saccharimeters." *Journal of Industrial & Engineering Chemistry* 1, no. 8 (1909): 567–80. https://doi.org/10.1021/ie50008a014.

Browne, C. A., and F. W. Zerban. *Physical and Chemical Methods of Sugar Analysis: A Practical and Descriptive Treatise for Use in Research, Technical, and Control Laboratories.* 3rd ed. John Wiley & Sons, 1941.

Browne, E. J. *Charles Darwin: A Biography.* 2 vols. Princeton University Press, 1996–2002.

Browne, Edith A. *Peeps at Industries: Sugar.* A. & C. Black, 1911. http://www.biodiversitylibrary.org/bibliography/24261.

Brunn, E. M. "The New York Coffee and Sugar Exchange." *Annals of the American Academy of Political and Social Science* 155 (1931): 110–18. https://www.jstor.org/stable/1018010.

Buchanan, R. A. "The Diaspora of British Engineering." *Technology and Culture* 27, no. 3 (1986): 501–24. https://doi.org/10.2307/3105383.

Cabrera Salcedo, Lizette. *De los bueyes al vapor: Caminos de la tecnología del azúcar en Puerto Rico y El Caribe.* Editorial Universidad de Puerto Rico, 2010.

Callon, Michel. *Markets in the Making: Rethinking Competition, Goods, and Innovation.* Edited by Martha Poon. Translated by Olivia Custer. Zone Books, 2021.

Carroll, Charles. *Journal of Charles Carroll of Carrollton During His Visit to Canada in 1776, as One of the Commissioners from Congress: With a Memoir and Notes.* J. Murphy for the Maryland Historical Society, 1876.

Case, Holly. *The Age of Questions, or, a First Attempt at an Aggregate History of the Eastern, Social, Woman, American, Jewish, Polish, Bullion, Tuberculosis, and Many Other Questions over the Nineteenth Century, and Beyond.* Princeton University Press, 2018.

Casid, Jill H. *Sowing Empire: Landscape and Colonization.* University of Minnesota Press, 2004.

Catlin, Daniel. *Good Work Well Done: The Sugar Business Career of Horace Havemeyer, 1903–1956*. D. Catlin, 1988.

Chamberlin, S. E., H. J. Abbott, and F. M. Endlich. *Report on the Methods of Manufacturing Sugars in West India Islands and British Guiana*. Government Printing Office, 1880.

Chandler, Alfred D. *The Visible Hand: The Managerial Revolution in American Business*. Belknap Press of Harvard University Press, 1977.

Chandler, Charles. "Sugar." In *Johnson's New Universal Cyclopædia: A Scientific and Popular Treasury of Useful Knowledge*, 4:622–43. New York, 1877. https://babel.hathitrust.org/cgi/pt?id=uc1.c2755795&seq=7.

Chang, Hasok. *Is Water H_2O? Evidence, Realism, and Pluralism*. Springer Netherlands, 2012.

Chaptal, Jean-Antoine. *Mémoire sur le sucre de betteraves*. Huzard, 1818. https://gallica.bnf.fr/ark:/12148/bpt6k9757832s.

Charlevoix, Pierre-François-Xavier de. *Histoire et description générale de la nouvelle France, avec le journal historique d'un voyage fait par ordre du Roi dans l'Amérique septentrionnale*. Vol. 5. Pierre-François Giffart, 1744. https://gallica.bnf.fr/ark:/12148/bpt6k53240290.

Charter, By-Laws, and Rules of the New York Coffee and Sugar Exchange, Inc. New York, 1927.

Checkland, S. G., and Anthony Slaven, eds. *Dictionary of Scottish Business Biography, 1860–1960*. Aberdeen University Press, 1986.

Child, David Lee. *The Culture of the Beet, and Manufacture of Beet Sugar*. Weeks, Jordan & Co., 1840. https://catalog.hathitrust.org/Record/006519368.

Cochrane, Raymond. *Measures for Progress: A History of the National Bureau of Standards*. MP275. National Bureau of Standards, US Department of Commerce, 1966.

Cohen, Andrew Wender. *Contraband: Smuggling and the Birth of the American Century*. W. W. Norton, 2015.

Cohen, Andrew Wender. "Smuggling, Globalization, and America's Outward State, 1870–1909." *Journal of American History* 97, no. 2 (2010): 371–98. http://www.jstor.org/stable/40959765.

Cohen, Benjamin R. *Pure Adulteration: Cheating on Nature in the Age of Manufactured Food*. University of Chicago Press, 2022.

Commerell, François de. *An Account of the Culture and Use of the Mangel Wurzel, or Root of Scarcity*. C. Dilly and J. Phillips, 1787. https://books.google.com?id=vGMmAQAAMAAJ.

Cook, Harold John. *Matters of Exchange: Commerce, Medicine, and Science in the Dutch Golden Age*. Yale University Press, 2007.

Cook, James W. *The Arts of Deception: Playing with Fraud in the Age of Barnum*. Harvard University Press, 2001.

Coons, George H. "The Sugar Beet: Product of Science." *Scientific Monthly*. Business Press (Lancaster, PA), 1949. https://catalog.hathitrust.org/Record/000519252.

Cooper, William. "Letter on the Manufacture of Maple Sugar." *Transactions of the Society for the Promotion of Agriculture, Arts and Manufactures, Instituted in the State of New York*, 1:83. New York, 1792. https://catalog.hathitrust.org/Record/008606675.

Coppin, Clayton Anderson, and Jack C. High. *The Politics of Purity: Harvey Washington Wiley and the Origins of Federal Food Policy*. University of Michigan Press, 1999.

Cronon, William. *Nature's Metropolis: Chicago and the Great West*. W. W. Norton, 1991.

Crosland, Maurice P. *The Society of Arcueil: A View of French Science at the Time of Napoleon I*. Harvard University Press, 1967.

Crowley, John E. "Sugar Machines: Picturing Industrialized Slavery." *American Historical Review* 121, no. 2 (2016): 403–36. https://doi.org/10.1093/ahr/121.2.403.

Curry-Machado, Jonathan. *Cuban Sugar Industry: Transnational Networks and Engineering Migrants in Mid-Nineteenth-Century Cuba*. Palgrave Macmillan, 2011.

Curtin, Philip D. *The Rise and Fall of the Plantation Complex: Essays in Atlantic History*. 2nd ed. Studies in Comparative World History. Cambridge University Press, 1998. https://doi.org/10.1017/CBO9780511819414.

Curtis, David A. "The Electric Sugar Swindle." *Harper's Weekly*, March 30, 1889.

Daniels, Christian, and Nicholas K. Menzies. *Science and Civilisation in China*, vol. 6, *Biology and Biological Technology*, part 3, *Agro-Industries and Forestry, Agro-Industries: Sugarcane Technology, Forestry*. Cambridge University Press, 2010.

Daniels, John, and Christian Daniels. "The Origin of the Sugarcane Roller Mill." *Technology and Culture* 29, no. 3 (1988): 493–535. https://doi.org/10.2307/3105272.

Daniels, John, and Christian Daniels. "Sugarcane in Prehistory." *Archaeology in Oceania* 28, no. 1 (1993): 1–7. https://doi.org/10.1002/j.1834-4453.1993.tb00309.x.

Daston, Lorraine, and Peter Galison. *Objectivity*. Zone Books, 2007.

Daston, Lorraine, and Katharine Park. *Wonders and the Order of Nature, 1150–1750*. Rev. ed. Zone Books, 2001.

Deerr, Noël. *Cane Sugar: A Textbook on the Agriculture of the Sugar Cane, the Manufacture of Cane Sugar, and the Analysis of Sugar-House Products*. 2nd ed. Norman Rodger, 1921.

Deerr, Noël. *The History of Sugar*. 2 vols. Chapman and Hall, 1950.

Deerr, Noël. *Sugar and the Sugar Cane: An Elementary Treatise on the Agriculture of the Sugar Cane and on the Manufacture of Cane Sugar*. N. Rodger, 1905.

Deerr, Noël. *Sugar House Notes and Tables: A Reference Book for Planters, Factory Managers, Chemists, Engineers, and Others Employed in the Manufacture of Cane Sugar*. E. & F. N. Spon, 1900.

Deerr, Noël, and R. S. Norris. *The Deterioration of Sugars in Storage*. Division of Agriculture and Chemistry Bulletins 24. Honolulu, 1908.

Delbourgo, James. *Collecting the World: The Life and Curiosity of Hans Sloane*. Allen Lane, 2017.

Deringer, William. *Calculated Values: Finance, Politics, and the Quantitative Age*. Harvard University Press, 2018.

Deringer, William. "For What It's Worth: Historical Financial Bubbles and the Boundaries of Economic Rationality." *Isis* 106, no. 3 (2015): 646–56. https://doi.org/10.1086/683529.

Devine, T. M., and Gordon Jackson, eds. *Glasgow*. Vol. 1, *Beginnings to 1830*. Manchester University Press, 1995.

Dew, Nicholas. "Vers la ligne: Circulating Measurements Around the French Atlantic." In *Science and Empire in the Atlantic World*, edited by James Delbourgo and Nicholas Dew, 53–72. Routledge, 2008.

Dillon, Elizabeth Maddock. "The Cost of Sugar: Narratives of Loss of Life and Limb." Paper presented at the Conference *Beyond Sweetness: New Histories of Sugar in the Early Atlantic World*, John Carter Brown Library, Brown University, October 24–25, 2013.

Dunn, Richard S. *Sugar and Slaves: The Rise of the Planter Class in the English West Indies, 1624–1713*. Published for the Institute of Early American History and Culture at Williamsburg, VA, by the University of North Carolina Press, 1972.

Dye, Alan. "Avoiding Holdup: Asset Specificity and Technical Change in the Cuban Sugar Industry, 1899–1929." *Journal of Economic History* 54, no. 3 (1994): 628–53. http://www.jstor.org/stable/2123871.

Dye, Alan. *Cuban Sugar in the Age of Mass Production: Technology and the Economics of the Sugar Central, 1899–1929*. Stanford University Press, 1998.

Dyer, Benjamin Wheeler. *Modern Methods of Marketing Cuban Raw Sugar: How You Can Lessen Business Risks by Using Sugar Futures: A Booklet for Sugar Centrals and Colonos*. Lamborn & Company, 1923. https://books.google.com?id=YtFMAAAAYAAJ.

E. H. Sargent and Company. *Price List No. 20 of Scientific Laboratory Apparatus*. The Company, 1914. https://books.google.com?id=8ccJAAAAIAAJ.

Edgerton, David. *The Shock of the Old: Technology and Global History Since 1900*. Oxford University Press, 2006.

Edson, Hubert. *Sugar, from Scarcity to Surplus*. Chemical Publishing Co., 1958.

Eichner, Alfred S. *The Emergence of Oligopoly: Sugar Refining as a Case Study*. Johns Hopkins University Press, 1969. https://www.jstor.org/stable/724964.

Eichner, Alfred S. "Monopoly, the Emergence of Oligopoly and the Case of Sugar Refining: A Reply." *Journal of Law & Economics* 14, no. 2 (1971): 521–27.

Eimer & Amend. *Illustrated Wholesale Catalogue with Prices Current of Chemical & Physical Apparatus and Assay Goods*. New York, 1892. http://catalog.hathitrust.org/Record/008631161.

Ellsworth, John Orval. "Sugar Marketing, with Special Reference to the Functions of the New York Coffee and Sugar Exchange." PhD diss., Cornell University, 1926.

Fakhri, Michael. *Sugar and the Making of International Trade Law*. Cambridge Studies in International and Comparative Law. Cambridge University Press, 2014. https://doi.org/10.1017/CBO9781139629089.

Fernández-de-Pinedo, Nadia, and David Pretel. "Circuits of Knowledge: Foreign Technology and Transnational Expertise in Nineteenth-Century Cuba." In *The Caribbean and the Atlantic World Economy: Circuits of Trade, Money and Knowledge, 1650–1914*, edited by Adrian Leonard and David Pretel, 263–89. Cambridge Imperial and Post-Colonial Studies.: Palgrave Macmillan, 2015.

Fox, Celina. *The Arts of Industry in the Age of Enlightenment*. Yale University Press for The Paul Mellon Centre for Studies in British Art, 2009.

Fox, William Lloyd. "Harvey W. Wiley's Search for American Sugar Self-Sufficiency." *Agricultural History* 54, no. 4 (1980): 516–26. http://www.jstor.org/stable/3742486.

Galison, Peter Louis. *Einstein's Clocks, Poincaré's Maps: Empires of Time*. W. W. Norton, 2003.

Galloway, J. H. *The Sugar Cane Industry: An Historical Geography from Its Origins to 1914*. Cambridge University Press, 1989.

García-Muñiz, Humberto. "Louisiana's 'Sugar Tramps' in the Caribbean Sugar Industry." *Revista/Review Interamericana* 29, no. 1–4 (1999).

García-Muñiz, Humberto. *Sugar and Power in the Caribbean: The South Porto Rico Sugar Company in Puerto Rico and the Dominican Republic, 1900–1921*. Editorial Universidad de Puerto Rico, 2010.

García Rodríguez, Gervasio L., and Emma Aurora Dávila Cox, eds. *Puerto Rico en la mirada extranjera: La correspondencia de los cónsules norteamericanos, franceses e ingleses, 1869–1900*. Centro de Investigaciones Históricas, Decanato de Estudios Graduados e Investigación, Universidad de Puerto Rico, Recinto de Río Piedras, 2005.

Garfield, James A. *Sugar Tariff: Speech of Hon. James A. Garfield, of Ohio, Delivered in the House of Representatives, Wednesday, February 26, 1879*. Washington, DC, 1879. http://catalog.hathitrust.org/Record/006103504,

Gayer, Arthur D., et al. *The Sugar Economy of Puerto Rico*. Columbia University Press, 1938.

Geerligs, H. C. Prinsen. *The World's Cane Sugar Industry, Past and Present*. Norman Rodger, 1912. https://catalog.hathitrust.org/Record/100559910.

Genesove, David, and Wallace P. Mullin. "Predation and Its Rate of Return: The Sugar Industry, 1887–1914." *RAND Journal of Economics* 37, no. 1 (2006): 47–69. https://www.jstor.org/stable/25046226.

Genesove, David, and Wallace P. Mullin. "Rules, Communication, and Collusion: Narrative Evidence from the Sugar Institute Case." *American Economic Review* 91, no. 3 (2001): 379–98. https://www.jstor.org/stable/2677870.

Genesove, David, and Wallace P. Mullin. "The Sugar Institute Learns to Organize Information Exchange." In *Learning by Doing in Markets, Firms, and Countries*, edited by Naomi R. Lamoreaux et al., 103–44. University of Chicago Press, 1999. https://doi.org/10.3386/w5981.

Glaeser, Edward L., and Claudia Goldin, eds. *Corruption and Reform: Lessons from America's Economic History*. National Bureau of Economic Research Conference Report. University of Chicago Press, 2008.

Gooday, Graeme. "The Premises of Premises: Spatial Issues in the Historical Construction of Laboratory Credibility." In *Making Space for Science: Territorial Themes in the Shaping of Knowledge*, edited by Crosbie Smith and Jon Agar, 216–45. Macmillan, 1998.

Gough, J. B. "Achard, Franz Karl." In *Dictionary of Scientific Biography*, edited by Charles Coulston Gillispie. Vol. 1. New York: Scribner, 1980.

Gould, Lewis L. "Tariffs and Markets in the Gilded Age." *Reviews in American History* 2, no. 2 (1974): 266–71. https://doi.org/10.2307/2701666.

Grandville, J. J., and Taxile Delord. *Un autre monde; Transformations, visions, incarnations, ascensions, locomotions, explorations, pérégrinations, excursions, stations, cosmogenies, fantasmagories, réveries, lubies, facéties, folatreries, métamorphoses, boomorphoses, lithomorphoses, métempsycoses, apothéoses, et autres choses*. Paris, H. Fournier, 1844. http://archive.org/details/unautremondetran00gran_0.

Guerra y Sánchez, Ramiro. *Sugar and Society in the Caribbean: An Economic History of Cuban Agriculture*. Translated by Marjory M. Urquidi. Yale University Press, 1964.

Hacking, Ian. *Historical Ontology*. Harvard University Press, 2002.

Hacking, Ian. "Making Up People." *London Review of Books*, August 17, 2006.

Hall, Reginald Sexton, et al. "Production Control in the Revere Sugar Refinery." Massachusetts Institute of Technology, Department of Engineering Administration, 1922.

Hall, William Garvie. "Future of Sugar Is in the Orient." *The Trans-Pacific: A Magazine of International Service Covering the Far East and Australasia*, September 1921. https://books.google.com?id=aiE10AEACAAJ,

Hánek, Pavel, and Pavel Hánek Sr. "Tradition of Geodetic Instruments Production in the Czech Republic." *History of Geo- and Space Sciences* 12, no. 2 (2021): 171–78. https://doi.org/10.5194/hgss-12-171-2021.

Harris, Neil. *Humbug: The Art of P. T. Barnum*. Little, Brown, 1973.

Harvey, Robert. *Early Days of Engineering in Glasgow*. Aird & Coghill, 1919.

Harvey, Robert. "The History of the Sugar Machinery Industry in Glasgow." *International Sugar Journal*, 1917.

Hassall, Arthur Hill. *Food and Its Adulterations; Comprising the Reports of the Analytical Sanitary Commission of "the Lancet" for the Years 1851 to 1854 Inclusive, Revised and Extended*. London, 1855. https://babel.hathitrust.org/cgi/pt?id=uc1.$b31726&seq=9.

Havemeyer, Harry W. *Merchants of Williamsburgh: Frederick C. Havemeyer, Jr., William Dick, John Mollenhauer, Henry O. Havemeyer*. H.W. Havemeyer, 1989.

Hayes, Rutherford B. *Diary and Letters of Rutherford Birchard Hayes: Nineteenth President of the United States*. Edited by Charles Richard Williams. Vol. 3, *1865–1881*. Ohio State Archaeological and Historical Society, 1922.

Hecht, Gabrielle. *Being Nuclear: Africans and the Global Uranium Trade*. MIT Press, 2012.

Heitmann, John Alfred. *The Modernization of the Louisiana Sugar Industry*, 1830–1910. Louisiana State University Press, 1987.

Helot, Jules. *Le sucre de betterave en France de 1800 à 1900*. Fernand et Paul Deligne, 1900. https://babel.hathitrust.org/cgi/pt?id=coo.31924013882513&seq=258.

Heriot, Thomas H. P. *Science in Sugar Production: An Introduction to Methods of Chemical Control*. N. Rodger, 1907.

Heriot, Thomas H. P. "The Sugar Industry After the War." *Proceedings of the Royal Society of Glasgow* 49 (1918): 31.

Heriot, Thomas H. P. "Technical Training for the Sugar Industry." *International Sugar Journal* 28 (1916): 173–79.

Higman, B. W. "The Sugar Revolution." *Economic History Review* 53, no. 2 (2000): 213–36. https://doi.org/10.1111/1468-0289.00158.

Hoiem, Elizabeth Massa. "The Progress of Sugar: Consumption as Complicity in Children's Books about Slavery and Manufacturing, 1790–2015." *Children's Literature in Education* 52, no. 2 (2021): 162–82. https://doi.org/10.1007/s10583-020-09411-y.

Hopkins, Albert A. "From Sugar Cane to Crystal Cube." *Scientific American* 144, no. 2 (1931): 106–9. https://www.jstor.org/stable/24975622.

Howe, Robert. *The Labor Side of the Great Sugar Question*. New York, 1878. https://babel.hathitrust.org/cgi/pt?id=uc1.b3110426&seq=3.

Howland, Harold J. "The Case of the Seventeen Holes." *The Outlook*, May 1, 1909. http://www.UNZ.org/Pub/Outlook-1909may01-00025,

Hull, Matthew S. *Government of Paper: The Materiality of Bureaucracy in Urban Pakistan*. University of California Press, 2012.

Humboldt, Alexander von, and Aimé Bonpland. *Relation historique du voyage aux régions équinoxiales du nouveau continent*. Vol. 3. F. A. Brockhaus, 1825. https://www.biodiversitylibrary.org/item/95419.

Irwin, Douglas A. *Clashing over Commerce: A History of US Trade Policy*. University of Chicago Press, 2017.

James, Bill. *Popular Crime: Reflections on the Celebration of Violence*. Scribner, 2011.

Janes, Hurford, and H. J. Sayers. *The Story of Czarnikow*. Harley, 1963.

Jenks, Jeremiah Whipple. *The Trust Problem*. New Doubleday, Page, 1909. http://link.gale.com/apps/doc/F0151067605/MOML.

Johnson, Walter. *River of Dark Dreams: Slavery and Empire in the Cotton Kingdom*. Belknap Press of Harvard University Press, 2013.

Jones, C. Allan, and Robert V. Osgood. *From King Cane to the Last Sugar Mill: Agricultural Technology and the Making of Hawaii's Premier Crop*. University of Hawaiʻi Press, 2015. https://muse.jhu.edu/pub/5/monograph/book/39348.

Jung, Moon-Ho. *Coolies and Cane: Race, Labor, and Sugar in the Age of Emancipation*. Johns Hopkins University Press, 2006.

Kerr, John D. "Walter Maxwell (1854–1931)." In *Australian Dictionary of Biography*. Canberra:

National Centre of Biography, Australian National University. Accessed December 13, 2023. https://adb.anu.edu.au/biography/maxwell-walter-7535/text13143.

Kershaw, Michael. “The International Electrical Units: A Failure in Standardisation?” *Studies in History and Philosophy of Science* 38, no. 1 (2007): 108–31. https://doi.org/10.1016/j.shpsa.2006.12.012.

Kessler, Lawrence Helfgot. “Planter’s Paradise: Nature, Culture, and Hawai‘i’s Sugarcane Plantations.” PhD diss., Temple University, 2016. https://search.proquest.com/pqdtglobal/docview/1794167570/abstract/A0D7A82497234473PQ/1.

Kingsbury, Noël. *Hybrid: The History and Science of Plant Breeding*. University of Chicago Press, 2009.

Klein, Ursula. “Apothecary Shops, Laboratories and Chemical Manufacture in Eighteenth-Century Germany.” In *The Mindful Hand: Inquiry and Invention from the Late Renaissance to Early Industrialisation*, edited by Lissa Roberts et al., 247–79. Koninkliijke Nederlandse Akademie van Wetenschappen, 2007.

Klein, Ursula. “Contexts and Limits of Lavoisier’s Analytical Plant Chemistry: Plant Materials and Their Classification.” *AMBIX* 52, no. 2 (2005): 107–57. https://doi.org/10.1179/000269805X70102.

Klein, Ursula, and Wolfgang Lefèvre. *Materials in Eighteenth-Century Science: A Historical Ontology*. MIT Press, 2007.

Klickstein, Herbert S., and Henry M. Leicester. “Charles Albert Browne as an Historian of Chemistry.” *Journal of Chemical Education* 25, no. 6 (1948): 315. https://doi.org/10.1021/ed025p315.

Kopeloff, Nicholas, and Lillian Kopeloff. “The Deterioration of Manufactured Cane Sugar by Molds.” *Journal of Industrial & Engineering Chemistry* 11, no. 9 (1919): 845–50. https://doi.org/10.1021/ie50117a007.

Kopeloff, Nicholas, et al. *The Prevention of Sugar Deterioration*. Bulletin of the Agricultural Experiment Station of the Louisiana State University and A. & M. College 175. Ramires-Jones, 1920.

Kostelnick, Charles. “Visualizing Technology and Practical Knowledge in the *Encyclopédie*’s Plates: Rhetoric, Drawing Conventions, and Enlightenment Values.” *History & Technology* 28, no. 4 (2012): 443–54. https://doi.org/10.1080/07341512.2012.771465.

Kuhn, Mary. *The Garden Politic: Global Plants and Botanical Nationalism in Nineteenth-Century America*. America and the Long Nineteenth Century. New York University Press, 2023.

Kuhn, Thomas S. *The Structure of Scientific Revolutions*. University of Chicago Press, 1962.

Kula, Witold. *Measures and Men*. Translated by R. Sretzer. Princeton University Press, 1986.

Kumar, Prakash. *Indigo Plantations and Science in Colonial India*. Cambridge University Press, 2012.

Kusch, Martin. “Objectivity and Historiography.” *Isis* 100, no. 1 (2009): 127–31, https://doi.org/10.1086/597564.

Lamoreaux, Naomi R. *The Great Merger Movement in American Business, 1895–1904*. Cambridge University Press, 1985. https://doi.org/10.1017/CBO9780511665042.

Larsen, Esther Louise, and Peter Kalm. “Peter Kalm’s Description of How Sugar Is Made from Various Types of Trees in North America.” *Agricultural History* 13, no. 3 (1939): 149–56. https://www.jstor.org/stable/3739306.

Latour, Bruno. “Give Me a Laboratory and I Will Raise the World.” In *Science Observed: Perspectives on the Social Study of Science*, edited by Karen D. Knorr-Cetina and Michael Mulkay, 141–70. Sage, 1983.

Latour, Bruno. *Science in Action: How to Follow Scientists and Engineers Through Society*. Harvard University Press, 1987.

Laws of Puerto Rico Annotated. San Juan, Puerto Rico: Butterworth Legal Publishers, Equity Pub. Division, 1954.

Le Romain, Jean-Baptiste-Pierre. "Sucre." In *Encyclopédie ou dictionnaire raisonné des sciences, des arts et des métiers*, 608b–614a. Paris, 1765.

Lears, T. J. Jackson. *Rebirth of a Nation: The Making of Modern America, 1877–1920*. HarperCollins, 2009.

Leland, E. R. "Mites, Ticks, and Other Acari." *Popular Science Monthly*, February 1879.

Levitt, Theresa. *Elixir: A Parisian Perfume House and the Quest for the Secret of Life*. Harvard University Press, 2023.

Levitt, Theresa. *The Shadow of Enlightenment: Optical and Political Transparency in France, 1789–1848*. Oxford: Oxford University Press, 2009.

Levy, Jonathan. *Freaks of Fortune: The Emerging World of Capitalism and Risk in America*. Harvard University Press, 2012.

Liebig, Justus von. *Chemistry, and Its Application to Physiology, Agriculture, and Commerce*. New York, 1847. https://babel.hathitrust.org/cgi/pt?id=osu.32435016165052&seq=3.

Liebig, Justus von. *Familiar Letters on Chemistry, in Its Relations to Physiology, Dietetics, Agriculture, Commerce, and Political Economy*. Edited by John Blyth. 4th ed. Walton and Maberly, 1859. https://babel.hathitrust.org/cgi/pt?id=pst.000057583136&seq=5.

Ligon, Richard. *A True and Exact History of the Island of Barbados*. Humphrey Moseley, 1657. http://hdl.handle.net/2047/D20235165.

Lincoln, Mervyn David. "The Culture of the South African Sugarmill: The Impress of the Sugarocracy." PhD diss., University of Cape Town, 1985.

Littleton, Edward. *The Groans of the Plantations, or, a True Account of Their Grievous and Extreme Sufferings by the Heavy Impositions Upon Sugar and Other Hardships Relating More Particularly to the Island of Barbados*. London, 1689. https://www.proquest.com/eebo/docview/2240910972/citation/83DD64FE9F0A4CD4PQ/1.

Locke, John. *An Essay Concerning Humane Understanding*. London, 1694.

Lucier, Paul. *Scientists & Swindlers: Consulting on Coal and Oil in America, 1820–1890*. Johns Hopkins Studies in the History of Technology. Johns Hopkins University Press, 2008.

Lynch, Michael. "The Discursive Production of Uncertainty: The OJ Simpson 'Dream Team' and the Sociology of Knowledge Machine." *Social Studies of Science* 28, no. 5/6 (1998): 829–68. https://www.jstor.org/stable/285519.

Lynsky, Myer. *Sugar Economics, Statistics, and Documents*. US Cane Sugar Refiners' Association, 1938.

Macherey, Pierre. *The Object of Literature*. Cambridge University Press, 1995. https://books.google.com?id=oKfG4rFyGbsC.

MacKenzie, Donald A. *Inventing Accuracy: A Historical Sociology of Nuclear Missile Guidance*. MIT Press, 1990.

Mainardi, Patricia. "Grandville, Visions, and Dreams." *Public Domain Review*, September 26, 2018. https://publicdomainreview.org/essay/grandville-visions-and-dreams/.

Marcet, Jane Haldimand. *Conversations on Chemistry, in Which the Elements of That Science Are Familiarly Explained and Illustrated by Experiments and Plates*. Sidney's Press for Increase Cooke & Co., 1813. http://hdl.handle.net/2027/osu.32435018311860.

Marcus, Alan I. "Setting the Standard: Fertilizers, State Chemists, and Early National Commercial Regulation, 1880–1887." *Agricultural History* 61, no. 1 (1987): 47–73. https://www.jstor.org/stable/3743917.

Marggraf, Andreas Sigismund. "Expériences chymiques faites dans le dessein de tirer un véritable sucre de diverses plantes, qui croissent dans nos contrées." *Mémoires de l'Académie royale des Sciences et Belles Lettres*, 1747, 79–90.

Martin, Samuel. *An Essay on Plantership: Inscribed to Sir George Thomas, Bart. As a Monument to Ancient Friendship. The Seventh Edition, with All the Additions from the Author's Experiments to the Time of His Death. By Colonel Martin of Antigua.* Robert Mearns, 1785

Martineau, George. *Sugar, Cane and Beet: An Object Lesson.* 4th ed. Sir I. Pitman & Sons, 1918. https://www.biodiversitylibrary.org/item/60594.

Martínez Vergne, Teresita. *Capitalism in Colonial Puerto Rico: Central San Vicente in the Late Nineteenth Century.* University Press of Florida, 1992.

Mata, Juan de la. *Arte de reposteria: En que se contiene todo gènero de hacer dulces secos y en lìquido, vizcochos, turrones, natas, bebidas heladas de todos generos, rosolis, mistelas [et]c. con una breve instruccion para conocer las frutas y servirlas crudas. y diez mesas con su explicacion.* Imprenta de Josef Herrera, 1786. https://books.google.com?id=XlIuKxgFnIkC.

Maxwell, Francis. *Economic Aspects of Cane Sugar Production.* N. Rodger, 1927.

Maxwell, Walter. *Report of Work of the Experiment Station of the Hawaiian Sugar Planters' Association for the Year 1895* (Reprint). Division of Agriculture and Chemistry Bulletins 1. Honolulu, 1905.

Mayer, John. *Notices of Some of the Principal Manufacturers of the West of Scotland.* Glasgow: Blackie & Son, 1876.

Mazumdar, Sucheta. "China and the Global Atlantic: Sugar from the Age of Columbus to Pepsi-Coke and Ethanol." *Food and Foodways* 16, no. 2 (2008): 135–47. https://doi.org/10.1080/07409710802086070.

Mazumdar, Sucheta. *Sugar and Society in China.* Peasants, Technology, and the World Market, Harvard-Yenching Institute Monograph Series 45. Harvard University Asia Center, 1998.

McCormick, Richard L. "The Discovery That Business Corrupts Politics: A Reappraisal of the Origins of Progressivism." *American Historical Review* 86, no. 2 (1981): 247–74. https://doi.org/10.2307/1857438.

McCulloh, R. S., and A. D. Bache. *Reports from the Secretary of the Treasury, of Scientific Investigations in Relation to Sugar and Hydrometers, Made Under the Superintendence of Professor R. S. McCulloh.* Wendell and Van Benthuysen, 1848. https://catalog.hathitrust.org/Record/011609712.

McCurdy, Charles W. "The Knight Sugar Decision of 1895 and the Modernization of American Corporation Law, 1869–1903." *Business History Review* 53, no. 3 (1979): 304–42. https://doi.org/10.2307/3114089.

McGee, David. "From Craftsmanship to Draftsmanship: Naval Architecture and the Three Traditions of Early Modern Design." *Technology and Culture* 40, no. 2 (1999): 209–36. http://www.jstor.org/stable/25147307.

McGillivray, Gillian. *Blazing Cane: Sugar Communities, Class, and State Formation in Cuba, 1868–1959.* American Encounters / Global Interactions. Duke University Press, 2009.

McLean, Angus. *Local Industries of Glasgow and the West of Scotland.* British Association for the Advancement of Science, 1901.

McMahon, Benjamin. *Jamaica Plantership*. E. Wilson, 1839.

Memorial Presented to the Committee of Ways and Means on the Sugar Tariff, Advocating a Uniform Rate of Duty up to No. 13 Dutch Standard, with Arguments in Favor of Same by John E. Searles, Jr., and Edward P. Eastwick. T. McGill & Co., 1880.

Merleaux, April. *Sugar and Civilization: American Empire and the Cultural Politics of Sweetness*. University of North Carolina Press, 2015.

Miller, William Allen. *Elements of Chemistry: Theoretical and Practical*. Part 3, *Organic Chemistry*. J. W. Parker and Son, 1857.

Mintz, Sidney W. "The Plantation as Socio-Cultural Type." In *Plantation Systems of the New World; Papers and Discussion Summaries*. Social Science Monographs. Pan American Union, 1959.

Mintz, Sidney W. *Sweetness and Power: The Place of Sugar in Modern History*. Viking, 1985.

Mintz, Sidney W. *Tasting Food, Tasting Freedom: Excursions Into Eating, Culture, and the Past*. Beacon Press, 1996.

Mitsein, Rebekah. "Humanism and the Ingenious Machine: Richard Ligon's True and Exact History of the Island of Barbados." *Journal for Early Modern Cultural Studies* 16, no. 1 (2016): 95–122.

Munroe, C. E. "Third International Congress of Applied Chemistry." *Journal of the American Chemical Society* 21, no. 1 (1899): 73–102. https://doi.org/10.1021/ja02051a013.

Montgomery, David. *The Fall of the House of Labor: The Workplace, the State, and American Labor Activism, 1865–1925*. Cambridge University Press, 1987.

Moore, Gideon E. *Statement Relative to the Artificial Coloring of Imported Sugars*. Government Printing Office, 1881.

Moore, Jason W. "Madeira, Sugar, and the Conquest of Nature in the 'First' Sixteenth Century, Part 2: From Regional Crisis to Commodity Frontier, 1506—1530." *Review* (Fernand Braudel Center) 33, no. 1 (2010): 1–24. https://www.jstor.org/stable/41427556.

Moore, Jason W. "Sugar and the Expansion of the Early Modern World-Economy: Commodity Frontiers, Ecological Transformation, and Industrialization." *Review* (Fernand Braudel Center) 23, no. 3 (2000): 409–33. https://www.jstor.org/stable/40241510.

Moore, Paul H., et al. "Sugarcane: The Crop, the Plant, and Domestication." In *Sugarcane: Physiology, Biochemistry, and Functional Biology*, 1–17. John Wiley & Sons, 2013.

Moreno Fraginals, Manuel. "Plantation Economies and Societies in the Spanish Caribbean, 1860–1930." In *The Cambridge History of Latin America*. Volume 4, *c. 1870 to 1930*, edited by Leslie Bethell, 187–232. Cambridge University Press, 1986. https://doi.org/10.1017/CHOL9780521232258.007

Moreno Fraginals, Manuel. "Plantations in the Caribbean: Cuba, Puerto Rico, and the Dominican Republic in the Late Nineteenth Century." In *Between Slavery and Free Labor: The Spanish-Speaking Caribbean in the Nineteenth Century*, edited by Manuel Moreno Fraginals, et al. Johns Hopkins University Press, 1985.

Moreno Fraginals, Manuel. *The Sugarmill: The Socioeconomic Complex of Sugar in Cuba, 1760–1860*. Translated by Cedric Belfrage. New York: Monthly Review Press, 1976.

Morgan, Mary S., and Marcel Boumans. "Secrets Hidden by Two-Dimensionality: The Economy as a Hydraulic Machine." In *Models: The Third Dimension of Science*, edited by Soraya de Chadarevian and Nick Hopwood, 369–401. Stanford University Press, 2004. https://doi.org/10.1515/9781503618992-016.

Moss, Michael S., and John R. Hume. *Workshop of the British Empire: Engineering and Shipbuilding in the West of Scotland*. Heinemann, 1977.

Mullen, Stephen. *The Glasgow Sugar Aristocracy: Scotland and Caribbean Slavery, 1775–1838*. New Historical Perspectives. University of London Press, 2023.

Müller-Wille, Staffan. "Linnaean Paper Tools." In *Worlds of Natural History*, edited by Emma C. Spary et al., 205–20. Cambridge University Press, 2018. 205–20. https://doi.org/10.1017/9781108225229.013.

Munroe, C. E. "Third International Congress of Applied Chemistry." *Journal of the American Chemical Society* 21, no. 1 (1899): 73–102.

Napoleon Bonaparte, Louis. "Analysis of the Sugar Question." In *The Political and Historical Works of Louis Napoleon Bonaparte*, 2:1–91. London: Illustrated London Library, 1852. https://catalog.hathitrust.org/Record/000645354.

National Bureau of Economic Research. "Raw Sugar Stocks at All Ports for United States." Federal Reserve Bank of St. Louis: FRED. Accessed December 12, 2023. https://fred.stlouisfed.org/series/M0502BUSM576NNBR.

National Bureau of Economic Research. "Raw Sugar Stocks at Four Ports for United States." Federal Reserve Bank of St. Louis: FRED. Accessed September 27, 2022. https://fred.stlouisfed.org/series/M0502AUSM576NNBR.

National Bureau of Economic Research. "Retail Price of Sugar for New York, NY." Federal Reserve Bank of St. Louis: FRED. Accessed December 8, 2023. https://fred.stlouisfed.org/series/M04031US35620M267NNBR.

Nearing, Helen, and Scott Nearing. *The Maple Sugar Book: Together With Remarks on Pioneering as a Way of Living in the Twentieth Century*. Schocken Books, 1970.

Nellist, George F. *The Story of Hawaii and Its Builders, with Which Is Incorporated Volume III Men of Hawaii*. Honolulu Star-Bulletin, 1925. http://evols.library.manoa.hawaii.edu/handle/10524/56596.

Nguyen, Sophia. "Kara Walker 'Sweet Talks' Harvard." *Harvard Magazine*, December 11, 2014. https://harvardmagazine.com/2014/12/kara-walker-sweet-talk-harvard.

Niccol, Robert. *The Sugar Insect: "Acarus Sacchari," Found in Raw Sugar*. De Armond & Goodrich, 1868. https://catalog.hathitrust.org/Record/011545138.

Nortz & Co. *Coffee and Sugar Facts: The New York Coffee and Sugar Exchange*. New York, 1925.

Novak, William J. "The Myth of the 'Weak' American State." *American Historical Review* 113, no. 3 (2008): 752–72. https://www.jstor.org/stable/30223051.

Nye, David E. *Electrifying America: Social Meanings of a New Technology, 1880–1940*. MIT Press, 1990.

O'Connell, Joseph. "Metrology: The Creation of Universality by the Circulation of Particulars." *Social Studies of Science* 23, no. 1 (1993): 129–73. http://www.jstor.org/stable/285692.

Ogborn, Miles. "Vegetable Empire." In *Worlds of Natural History*, edited by Emma C. Spary et al., 271–86. Cambridge University Press, 2018. https://doi.org/10.1017/9781108225229.017.

Ortiz, Fernando. *Cuban Counterpoint: Tobacco and Sugar*. Translated by Harriet De Onís. Duke University Press, 1995.

Osborn, Sidney J. *Methods of Analysis and Laboratory Control of the Great Western Sugar Company*. Great Western Sugar Company, 1920. https://books.google.com?id=wBBLAAAAMAAJ.

Otremba, Eric. "Inventing Ingenios: Experimental Philosophy and the Secret Sugar-Makers of the Seventeenth-Century Atlantic." *History and Technology* 28, no. 2 (2012): 119–47. https://doi.org/10.1080/07341512.2012.694204.

Owen, William Ludwell. *Deterioration of Cane Sugars in Storage: Its Causes and Suggested Measures for Its Control*. Bulletin of the Agricultural Experiment Station of the Louisiana State University and A. & M. College 162. Ramires-Jones, 1918.

Pacione, Michael. *Glasgow: The Socio-Spatial Development of the City*. J. Wiley, 1995.

Papers Relative to Drawback Rates on Exported Sugar. New York: Evening Post, 1877.

Pares, Richard. *The Historian's Business, and Other Essays*. Clarendon Press, 1961.

Pares, Richard. "The London Sugar Market, 1740–1769." *Economic History Review* 9, no. 2 (1956): 254–70. https://doi.org/10.2307/2591745.

Pares, Richard. *A West-India Fortune*. Longmans, Green and Co., 1950.

Parrillo, Nicholas R. *Against the Profit Motive: The Salary Revolution in American Government, 1780–1940*. Yale University Press, 2013.

Paulson, Tim. "Paper Steaks: Live Cattle Futures Markets and the Financial Revolution of 1964." *Enterprise & Society*, May 31, 2024, 1–32. https://doi.org/10.1017/eso.2024.16.

Peñaranda, Carlos. *Cartas Puertorriqueñas: Dirigidas al célebre poeta Don Ventura Ruiz Aguilera, 1878–1880*. Tip. Sucesores de Rivadeneyra, 1885.

Pennington, Neil L., and Charles W. Baker. *Sugar: A User's Guide to Sucrose*. Van Nostrand Reinhold, 1990.

Pennsylvania Sugar Company. "The Story of Cane Sugar." Pennsylvania Sugar Company, 1925.

Pérez, Bernadette. "Before the Sun Rises: Contesting Power and Cultivating Nations in the Colorado Beet Fields." PhD diss., University of Minnesota, 2017.

Pérez, Louis A. *Cuba: Between Reform and Revolution*. 3rd ed. Latin American Histories.] Oxford University Press, 2006.

Perkins, John. "The Political Economy of the Sugar Beet in Imperial Germany." In *Crisis and Change in the International Sugar Economy 1860–1914*, edited by Bill Albert and Adrian Graves. ISC Press, 1984.

Piqueras, José Antonio. "The Discovery of Progress in Cuba: Machines, Slaves, Businesses." Translated by Paul Edgar. In *New Frontiers of Slavery*, edited by Dale W. Tomich, 49–75. State University of New York Press, 2016. https://doi.org/10.1353/book.83858.

Playfair, Lyon, and Wemyss Reid. *Memoirs and Correspondence of Lyon Playfair*. Cassell & Co., 1899. https://catalog.hathitrust.org/Record/101708740.

Plews, R. W. *The History of ICUMSA: The First 100 Years, 1897–1997*. International Commission for Uniform Methods of Sugar Analysis, 1997.

Porter, Theodore M. *Trust in Numbers: The Pursuit of Objectivity in Science and Public Life*. Princeton University Press, 1996.

Portuondo, María M. "Plantation Factories: Science and Technology in Late-Eighteenth-Century Cuba." *Technology and Culture* 44, no. 2 (2003): 231–57. http://www.jstor.org/stable/25148106.

Poskett, James. *Horizons: The Global Origins of Modern Science*. Mariner Books, 2022.

Proust, Joseph. "Lettre du Professeur Proust à Jean-Christophe Delamétherie sur le sucre du raisin." *Journal de physique, de chimie et d'histoire naturelle*, 1802, 113

Pugliano, Valentina. "Natural History in the Apothecary's Shop." In *Worlds of Natural History*, edited by Emma C. Spary et al., 44–60. Cambridge University Press, 2018. https://doi.org/10.1017/9781108225229.004.

Quélus, D. *Histoire naturelle du cacao et du sucre*. Paris: d'Houry, 1719. https://gallica.bnf.fr/ark:/12148/bpt6k15193845.

Raj, Kapil. *Relocating Modern Science: Circulation and the Construction of Knowledge in South Asia and Europe, 1650–1900*. Palgrave Macmillan, 2008.

Ramos Mattei, Andrés. *La sociedad del azúcar en Puerto Rico, 1870–1910*. Universidad de Puerto Rico, Recinto de Río Piedras, 1988.

Ramos Mattei, Andrés. "Technical Innovations and Social Change in the Sugar Industry of Puerto Rico." In *Between Slavery and Free Labor: The Spanish-Speaking Caribbean in the Nineteenth Century*, edited by Manuel Moreno Fraginals et al., 158–78. Johns Hopkins University Press, 1985.

Randolph, Mary. *Virginia Housewife: or, Methodical Cook*. John Plaskitt, 1836. https://books.google.com?id=vDjmpfFUsFcC.

Ratekin, Mervyn. "The Early Sugar Industry in Española." *Hispanic American Historical Review* 34, no. 1 (1954): 1–19. https://doi.org/10.2307/2509696.

Reading, Amy. *The Mark Inside: A Perfect Swindle, a Cunning Revenge, and a Small History of the Big Con*. Knopf, 2013.

Rebello, Charles [Carlos]. *The Pith of the Sugar Question*. Macgowan and Slipper, 1879.

A Review of the Case of the United States vs. 712 Bags of Dark Demerara Centrifugal Sugars, A.W. Perot & Co., Claimants: Tried at the September Term of the District Court of the United States for the District of Maryland. Lucas Brothers, 1878.

Review of the Efforts of the Forty-fifth Congress of the United States for the Revision of the Sugar Tariff, Together with Speeches Made by Members of the House of Representatives. New York, 1879.

Rickman, John. *Journal of Captain Cook's Last Voyage, to the Pacific Ocean: On Discovery: Performed in the Years 1776, 1777, 1778, 1779, and 1780*. E. Newbery, 1785. http://name.umdl.umich.edu/004895036.0001.000.

Ripley, George, and Charles A. Dana, eds. "Sugar." In *The New American Cyclopædia: A Popular Dictionary of General Knowledge*, 15:156–66. D. Appleton & Co., 1869.

Roberts, Lissa. "The Death of the Sensuous Chemist: The 'New' Chemistry and the Transformation of Sensuous Technology." *Studies in History and Philosophy of Science, Part A*, 26, no. 4 (1995): 503–29. https://doi.org/10.1016/0039-3681(95) 00013-5.

Robertson, Frances. "Delineating a Rational Profession: Engineers and Draughtsmen as 'Visual Technicians' in Early Nineteenth Century Britain." Paper presented at the 3 Societies Meeting, Philadelphia, July 11–14, 2012.

Robertson, Frances. "Manufacturing the Visual Economy in Nineteenth-Century Britain." Paper presented at the International Society for Cultural History, Lunéville, July 2–5, 2012.

Roche, Daniel. *A History of Everyday Things: The Birth of Consumption in France, 1600–1800*. Translated by Brian Pearce. Cambridge University Press, 2000.

Rodney, Walter. *Guyanese Sugar Plantations in the Late Nineteenth Century: A Contemporary Description from the "Argosy."* Release Publishers, 1979.

Rodney, Walter. *A History of the Guyanese Working People, 1881–1905*. Johns Hopkins Studies in Atlantic History and Culture. Johns Hopkins University Press, 1981.

Rolfe, George "Temperature Corrections of Sugar Polarization." *Science*, n.s. 24, no. 610 (September 7, 1906): 307–8. http://www.jstor.org/stable/1632113.

Rolfe, George W. "Raw Sugar Polarizations." *The Louisiana Planter and Sugar Manufacturer*, February 12, 1910, 99.

Rolfe, George William. *The Polariscope in the Chemical Laboratory: An Introduction to Polarimetry and Related Methods*. Macmillan, 1905.

Rolph, George M. *Something About Sugar: Its History, Growth, Manufacture and Distribution*. J. J. Newbegin, 1917.

Rood, Daniel B. *The Reinvention of Atlantic Slavery: Technology, Labor, Race, and Capitalism in the Greater Caribbean*. Oxford University Press, 2017.

Rood, Daniel B. Review of *De los bueyes al vapor: Caminos de la tecnología en Puerto Rico y El Caribe*. *Caribbean Studies* 39, no. 1/2 (2011): 243–48. https://www.jstor.org/stable/41495098.

Rosenthal, Caitlin. *Accounting for Slavery: Masters and Management*. Harvard University Press, 2018.

Rothstein, Morton. "The International Market for Agricultural Commodities, 1850–1873." In *Economic Change in the Civil War Era*, edited by David T. Gilchrist and W. David Lewis, 62–82. Eleutherian Mills–Hagley Foundation, 1975.

Roy, Rai Bahadur Joges Chandra. "Sugar Industry in Ancient India." *Journal of the Bihar and Orissa Research Society* 4, no. 4 (December 1918): 435–54.

Rush, Benjamin. *An Account of the Sugar Maple-Tree, of the United States, and of the Methods of Obtaining Sugar from It* [. . .]. R. Aitken & Son, 1791. http://name.umdl.umich.edu/N19030.0001.001.

Russell, Andrew, and Lee Vinsel. "Hail the Maintainers." *Aeon*, April 7, 2016. https://aeon.co/essays/innovation-is-overvalued-maintenance-often-matters-more.

Scarlett, Sarah Fayen. "The Craft of Industrial Patternmaking." *Journal of Modern Craft* 4, no. 1 (2011): 27–48. https://doi.org/10.2752/174967811X12949160069018.

Schaffer, Simon. "Accurate Measurement Is an English Science." In *The Values of Precision*, edited by M. Norton Wise, 135–72. Princeton University Press, 1995.

Schaffer, Simon. "Astronomers Mark Time: Discipline and the Personal Equation." *Science in Context* 2 (1988): 115–43.

Schaffer, Simon. "Easily Cracked: Scientific Instruments in States of Disrepair." *Isis* 102, no. 4 (2011): 706–17. https://doi.org/10.1086/663608.

Schaffer, Simon. "Golden Means: Assay Instruments and the Geography of Precision in the Guinea Trade." In *Instruments, Travel and Science: Itineraries of Precision from the Seventeenth to the Twentieth Century*, edited by Marie-Noëlle Bourguet et al., 20–50. Routledge, 2002.

Schaffer, Simon. "Late Victorian Metrology and Its Instrumentation: A Manufactory of Ohms." In *Invisible Connections: Instruments, Institutions, and Science : 1–2 April 1991, London, United Kingdom*, edited by Robert Bud and Susan E. Cozzens. Proceedings of SPIE –the International Society for Optical Engineering. SPIE, 1991. https://doi.org/10.1117/12.2283709.

Schaffer, Simon. "Late Victorian Metrology and Its Instrumentation: A Manufactory of Ohms." In *The Science Studies Reader*, edited by Mario Biagioli, 457–78. Routledge, 1999.

Schaffer, Simon. "Newton on the Beach: The Information Order of Principia Mathematica." *History of Science* 47, no. 3 (2009): 243–76. https://doi.org/10.1177/007327530904700301.

Scherer, M. A. N., and Jean-Baptiste Van Mons. "Note sur l'extraction du sucre de la betterave: Extraite d 'une lettre de M. A. N. Scherer, au Citoyen Van Mons." *Annales de Chimie, ou, Recueil de mémoires concernant la chimie et les arts qui en dépendent* 30 (1799): 299–302. https://catalog.hathitrust.org/Record/006954248.

Schnakenbourg, Christian. "From the Sugar Estate to Central Factory: The Industrial Revolution in the Caribbean (1840–1905)." In *Crisis and Change in the International Sugar Economy, 1860–1914*, edited by Bill Albert and Adrian Graves. ISC Press,1984.

Scott, James C. *Seeing Like a State: How Certain Schemes to Improve the Human Condition Have Failed*. Yale University Press, 1999.

Scott, Rebecca J. *Degrees of Freedom: Louisiana and Cuba After Slavery*. Belknap Press of Harvard University Press, 2005.

Scott, Rebecca J. *Slave Emancipation in Cuba: The Transition to Free Labor, 1860–1899*. University of Pittsburgh Press, 2000.

Scurr, Ruth. *Napoleon: A Life Told in Gardens and Shadows*. Liveright, 2021.

Secord, James A. "Knowledge in Transit." *Isis* 95, no. 4 (2004): 654–72. https://doi.org/10.1086/isis.2004.95.issue-4.

Sedgewick, Augustine. *Coffeeland: One Man's Dark Empire and the Making of Our Favorite Drug*. Penguin Press, 2020.

Sellers, Charles Grier. *The Market Revolution: Jacksonian America, 1815–1846*. Oxford University Press, 1991.

Serres, Olivier de. *Le théâtre d'agriculture et mesnage des champs*. Nouvelle édition conforme au texte, augmentée de notes et d'un vocabulaire. Vol. 2, 1805. https://gallica.bnf.fr/ark:/12148/bpt6k9616450p.

Shammas, Carole. *The Pre-Industrial Consumer in England and America*. Clarendon Press, 1990.

Shapin, Steven, and Barry Barnes. "Science, Nature and Control: Interpreting Mechanics' Institutes." *Social Studies of Science* 7, no. 1 (1977): 31–74. http://www.jstor.org/stable/284633.

Shapin, Steven. "The Invisible Technician." *American Scientist* 77, no. 6 (1989): 554–63. http://www.jstor.org/stable/27856006.

Shapin, Steven, and Simon Schaffer. *Leviathan and the Air-Pump: Hobbes, Boyle, and the Experimental Life*. Princeton University Press, 1985.

Shaw, Nate, and Theodore Rosengarten. *All God's Dangers: The Life of Nate Shaw*. Knopf, 1974.

Sheridan, Richard B. *Sugar and Slavery: An Economic History of the British West Indies, 1623–1775*. Johns Hopkins University Press, 1973.

Sherman, John. *Recollections of Forty Years in the House, Senate and Cabinet: An Autobiography*. Chicago: Werner, 1895.

Sibum, H. O. "Reworking the Mechanical Value of Heat: Instruments of Precision and Gestures of Accuracy in Early Victorian England." *Studies in History and Philosophy of Science* 26, no. 1 (1995): 73–106.

Singerman, David. "'A Doubt Is at Best an Unsafe Standard': Measuring Sugar in the Early Bureau of Standards." *Journal of Research of the National Institute of Standards and Technology* 112, no. 1 (2007): 53. https://doi.org/10.6028/jres.112.004.

Singerman, David Roth. "Science, Commodities, and Corruption in the Gilded Age." *The Journal of the Gilded Age and Progressive Era* 15, no. 3 (2016): 278–93.

Skaife, Wilfrid. "The Field for Chemical Improvement in the Manufacture of Sugar." *Journal of the Franklin Institute* 127, no. 3 (1889): 215–26.

Skaife, Wilfrid. "Future Industrial Opportunities in Cuba." *Engineering Magazine* 15, no. 3 (June 1898): 363–70.

Skaife, Wilfrid. "Sugar Producing Plants." *Canadian Record of Science* 3, no. 8 (1889): 455–75.

Skowronek, Stephen. *Building a New American State: The Expansion of National Administrative Capacities, 1877–1920*. Cambridge University Press, 1982.

Sloane, Hans. *A Voyage to the Islands Madera, Barbados, Nieves, S. Christophers and Jamaica, with the Natural History of the Herbs and Trees, Four-Footed Beasts, Fishes, Birds, Insects, Reptiles, &c. Of the Last of Those Islands* [...] Vol. 1. London: Printed by B. M. for the author, 1707. https://catalog.hathitrust.org/Record/001490666.

Smith, Mark J. "Creating an Industrial Plant: The Biotechnology of Sugar Production in Cuba." In *Industrializing Organisms: Introducing Evolutionary History*, edited by Philip Scranton and Susan R. Schrepfer, 85–106. Routledge, 2004.

Smith, P. S. "Electricity Applied to the Manufacture of Sugar." *General Electric Review*, August 1913.

Smith, Pamela H. *The Body of the Artisan: Art and Experience in the Scientific Revolution*. University of Chicago Press, 2006.

Smith, R. Elberton. *Customs Valuation in the United States: A Study in Tariff Administration*. University of Chicago Press, 1948.

Solá, José. "'The Funnel System in Which His Is the Little End.' The Technological Transformation of the Sugar Industry and American Protectionism in the Emergence of the Colonos in Caguas, Puerto Rico, 1898–1928." PhD diss., University of Connecticut, 2004.

Solnit, Rebecca. *River of Shadows: Eadweard Muybridge and the Technological Wild West*. Viking, 2003.

Specht, Joshua. "A Failure to Prohibit: New York City's Underground Bob Veal Trade." *The Journal of the Gilded Age and Progressive Era* 12, no. 4 (2013): 475–501. https://www.jstor.org/stable/43902976.

Spencer, Guilford Lawson. *General Instructions and Methods of Analysis and Chemical Control: For Use in the Factories of the Cuban-American Sugar Company*. The Cuban-American Sugar Company, 1916.

Spencer, Guilford Lawson. *A Hand-Book for Sugar Manufacturers and Their Chemists*. 2nd ed. New York, 1893. https://babel.hathitrust.org/cgi/pt?id=wu.89047327879&seq=7.

Spiekermann, Uwe. "Claus Spreckels: Robber Baron and Sugar King." Immigrant Entrepreneurship, June 7, 2011. https://www.immigrantentrepreneurship.org/entries/claus-spreckels-robber-baron-and-sugar-king/.

Stade, George. "Modern Polariscopes." *International Sugar Journal*, February 1899.

Staum, Martin S. "Marggraf, Andreas Sigismund." In *Dictionary of Scientific Biography*, edited by Charles Gillispie. Vol. 9. Scribner, 1980.

Sternstein, Jerome. "Corruption in the Gilded Age Senate: Nelson W. Aldrich and the Sugar Trust." *Capitol Studies*, Spring 1978, 14–38.

Stevens, Robert White. *On the Stowage of Ships and Their Cargoes, Freights, Charter-Parties, &c.* Longmans, 1859. http://catalog.hathitrust.org/Record/008606002.

Stewart, Larry. "Assistants to Enlightenment: William Lewis, Alexander Chisholm and Invisible Technicians in the Industrial Revolution." *Notes and Records of the Royal Society of London* 62, no. 1 (2008): 17–29. http://www.jstor.org/stable/20462648.

Stobart, Jon. *Sugar and Spice: Grocers and Groceries in Provincial England, 1650–1830*. Oxford University Press, 2016.

Stols, Eddy. "The Expansion of the Sugar Market in Western Europe." In *Tropical Babylons: Sugar and the Making of the Atlantic World, 1450–1680*, edited by Stuart B. Schwartz, 237–88. University of North Carolina Press, 2004.

Stout, William. *The Autobiography of William Stout of Lancaster, 1665–1752*. Manchester, 1967. https://babel.hathitrust.org/cgi/pt?id=mdp.39015003478131.

Sugar Association of London. *Rules and Regulations for Beet and Cane Sugar Contracts*. Published for the Association by Wm. O'Toole, Secretary, 1928.

Sugar Research Foundation. "New York Sugar Trade Laboratory: A Respected Scientific Institution Rounds Out 40 Years of Service to the Industry." *The Sugar Molecule*, January 1949. https://industrydocuments.ucsf.edu/docs/qknk0226.

Sumner, James. "John Richardson, Saccharometry and the Pounds-Per-Barrel Extract: The Construction of a Quantity." *British Journal for the History of Science* 34, no. 3 (2001): 255–73. http://www.jstor.org/stable/4028098.

Taussig, F. W. "The Tariff Act of 1897." *Quarterly Journal of Economics* 12, no. 1 (1897): 42. https://doi.org/10.2307/1882308.

Taussig, F. W. *The Tariff History of the United States*. 5th ed., G. P. Putnam's Sons, 1910, repr. Von Mises Institute, 2010.

Taylor, Alan. *William Cooper's Town: Power and Persuasion on the Frontier of the Early American Republic*. Alfred A. Knopf, 1995.

Testimony in Relation to the Sugar Frauds, Taken by the Subcommittee of the Committee of Ways and Means of the House of Representatives, New York, September, 1878. Brooklyn Daily Times Print, 1878.

Tomich, Dale W. "The Second Slavery and World Capitalism: A Perspective for Historical Inquiry." *International Review of Social History* 63, no. 3 (December 2018): 477–501. https://doi.org/10.1017/S0020859018000536.

Tomich, Dale W., et al. *Reconstructing the Landscapes of Slavery: A Visual History of the Plantation in the Nineteenth-Century Atlantic World*. University of North Carolina Press, 2021.

Treillon, Roland, and Jean Guérin. "La guerre des sucres." *Culture Technique* 16 (1986): 224–35. http://documents.irevues.inist.fr/handle/2042/29799.

Tresch, John. *The Romantic Machine: Utopian Science and Technology After Napoleon*. University of Chicago Press, 2012.

Turner, Gerard L'Estrange. *Nineteenth-Century Scientific Instruments*. University of California Press, 1983.

University of Hawaii (Honolulu). *The Story of Cane Sugar, as Told to an Imaginary Class in Agriculture*. University of Hawaii, 1928.

Van Leeuwenhoek, Antoni. "Other Microscopical Observations, Made by the Same, about the Texture of the Blood, the Sap of Some Plants, the Figure of Sugar and Salt, and the Probable Cause of the Difference of Their Tasts." *Philosophical Transactions of the Royal Society of London* 10, no. 117 (1675): 380–85. https://doi.org/10.1098/rstl.1675.0033.

Vascik, George S. "Sugar Barons and Bureaucrats: Unravelling the Relationship Between Economic Interest and Government in Modern Germany, 1799–1945." *Business and Economic History* 21 (1992): 336–42. https://www.jstor.org/stable/23703236.

Velkar, Aashish. "Measurement Standards and Market Governance: London Corn Trade Association and International Grain Markets (1880–1914)." *Histoire & Mesure* 38, no. 1 (2023): 65–92. https://doi.org/10.4000/histoiremesure.19043.

Vigreux, Pierre. "Aux origines du savoir agro-alimentaire: La création de l'École Nationale des Industries Agricoles (Douai, 1893)." *Revue du Nord* 72, no. 285 (1990): 255–89. https://doi.org/10.3406/rnord.1990.4528.

Vogt, Karl Christoph. *Aus meinem leben: Erinnerungen und rückblicke*. E. Nägele, 1896. https://books.google.com?id=xyESAAAAYAAJ.

Walker, Francis Amasa, ed. *United States Centennial Commission. International Exhibition, 1876: Reports and Awards, Groups III–VII*. Vol. 4. Government Printing Office, 1880. https://books.google.com?id=bf0mAQAAIAAJ.

Walker, Kara. *A Subtlety: Or the Marvelous Sugar Baby, an Homage to the Unpaid and Overworked Artisans Who Have Refined Our Sweet Tastes from the Cane Fields to the Kitchens of the New World on the Occasion of the Demolition of the Domino Sugar Refining Plant*. 2014. http://creativetime.org/projects/karawalker/.

Wallis-Tayler, A. J. *Sugar Machinery: A Descriptive Treatise Devoted to the Machinery and Processes Used in the Manufacture of Cane and Beet Sugars*. London: W. Rider and Son, Ltd., 1895.

Warner, Deborah Jean. "How Sweet It Is: Sugar, Science, and the State." *Annals of Science* 64, no. 2 (2007): 147–70.

Warner, Deborah Jean. *Sweet Stuff: An American History of Sweeteners from Sugar to Sucralose.* Smithsonian Institution Scholarly Press in cooperation with Rowman & Littlefield, 2011.

Welliver, Judson C. "The Secret of the Sugar Trust's Power." *Hampton's Magazine*, May 1910.

Wells, David Ames. *The Sugar Industry of the United States, and the Tariff. Report on the Assessment and Collection of Duties on Imported Sugar* [. . .]. New York, 1878.

White, Richard. *Railroaded: The Transcontinentals and the Making of Modern America.* 1st ed. W. W. Norton, 2012.

White, Richard. *The Republic for Which It Stands: The United States During Reconstruction and the Gilded Age, 1865–1896.* Oxford History of the United States. Oxford University Press, 2017.

Wiechmann, F. G. "Review: The Polariscope in the Chemical Laboratory, an Introduction to Polarimetry and Related Methods by George William Rolfe." *Science*, n.s. 23, no. 590 (1906): 627–28. http://www.jstor.org/stable/1633734.

Wiechmann, Ferdinand G. "The Question of Temperature-Influence on the Specific Rotation of Sucrose." *International Sugar Journal* 2 (1900): 491.

Wiechmann, Ferdinand G. *Sugar Analysis for Cane-Sugar and Beet-Sugar Houses, Refineries and Experimental Stations and as a Handbook of Instruction in Schools of Chemical Technology.* 3rd ed. John Wiley & Sons, 1914.

Wiechmann, Ferdinand G. "Third International Congress of Applied Chemistry, Vienna, 1898." *Science* 8, no. 194 (1898): 360–62. https://doi.org/10.1126/science.8.194.360.

Wiley, Harvey W. *An Autobiography.* Bobbs-Merrill, 1930. https://babel.hathitrust.org/cgi/pt?id=uc1.$b51093&seq=9

Wiley, Harvey W. "The Influence of Temperature on the Specific Rotation of Sucrose and Method of Correcting Readings of Compensating Polariscopes Therefor." *Journal of the American Chemical Society* 21, no. 7 (1899): 568–96. https://doi.org/10.1021/ja02057a002.

Wiley, Harvey W. "The Polariscope in the Chemical Laboratory." *Journal of the American Chemical Society* 27, no. 12 (1905): 1572–73. https://doi.org/10.1021/ja01990a017

Wiley, Harvey W. *Principles and Practice of Agricultural Analysis.* Vol. 3: Agricultural Projects. Chemical Publishing Company, 1897. https://books.google.com?id=ohBJAAAAMAAJ.

Wiley, Harvey W. "The True Meaning of the New Sugar Tariff." *The Forum*, February 1898.

Williams, Eric Eustace. *Capitalism & Slavery.* University of North Carolina Press, 1944.

Williams, Jeffrey C. "The Origin of Futures Markets." *Agricultural History* 56, no. 1 (1982): 306–16. https://www.jstor.org/stable/3742318.

Woloson, Wendy A. *Refined Tastes: Sugar, Confectionery, and Consumers in Nineteenth-Century America.* Johns Hopkins University Press, 2002.

The Women's Petition Against Coffee: Representing to Publick Consideration the Grand Inconveniencies Accruing to Their Sex from the Excessive Use of That Drying, Enfeebling Liquor [. . .]. London, 1674.

Woods, Rebecca J. H. *The Herds Shot Round the World: Native Breeds and the British Empire, 1800–1900.* Flows, Migrations, and Exchanges. University of North Carolina Press, 2017.

Worden, William L. *Cargoes: Matson's First Century in the Pacific.* University Press of Hawai'i, 1981.

Wright, Carroll Davidson. *Comparative Wages, Prices, and Cost of Living: From the Sixteenth Annual Report of the Massachusetts Bureau of Statistics of Labor, for 1885.* Wright & Potter, 1889. https://catalog.hathitrust.org/Record/005856297.

Yarrington, Jonna. "Droits and Frontières: Sugar and the Edge of France, 1800–1860." Master's thesis, University of Arizona, 2014.

Zakim, Michael. "Paperwork." *Raritan: A Quarterly Review* 33, no. 4 (2014): 34–56.

Zallen, Jeremy. *American Lucifers: The Dark History of Artificial Light, 1750–1865*. University of North Carolina Press, 2019.

Zallen, Jeremy. "'Dead Work,' Electric Futures, and the Hidden History of the Gilded Age." *Montana: The Magazine of Western History* 66, no. 2 (2016): 39–65. http://www.jstor.org/stable/26322796.

Zanetti Lecuona, Oscar, et al. "Nöel Deerr en la Guayana Británica, Cuba y Puerto Rico (1897–1921). Memorándum para la historia del azúcar en El Caribe." *Revista Mexicana Del Caribe* 6, no. 11 (2001): 57–154.

Zeitlin, Jonathan. "Between Flexibility and Mass Production: Strategic Ambiguity and Selective Adaptation in the British Engineering Industry, 1830–1914." In *World of Possibilities: Flexibility and Mass Production in Western Industrialization*, edited by Charles F. Sabel and Jonathan Zeitlin, 241–72. Cambridge University Press, 1997.

Index

Note: Page numbers in italics indicate figures.

A. & W. Smith, 61, 66
Abbott, Henry, 111, 114–15
Académie royale des sciences et belles-lettres, Berlin, 25–28
accounting, 46, 72
Achard, Franz Karl, 27–31, 35–39
Achim, Miruna, 243n82
Adams, Henry, 8
Adams, John Quincy, 28
Adeline Sugar Factory, Louisiana, 66
Aguirre (factory), 79–82, 92, 191–92, 221–23
American Chemical Society, 271n14
American Sugar Refining Company, 65, 134–35, 160, 168–70, 172–74, 182, 208, 221, 225; contract for sugar delivery, *205*; receiving wharf, *213*
Anderson's Institution, Glasgow, Scotland, 59
Andreas, Peter, 141
Anti-Trust Act (US), 134
appraisal. *See* inspection and appraisal of sugar
Arago, François, 41
Arbuckle, John, 135, 174–76
Arthur, Chester, 131, 132
artificial coloring, 110–13
ash, 213
Association of Sugar Producers (Puerto Rico), 90
Atkins, Edwin, xi–xiii, 1, 15–17, 51–52, 56, 67–68, 74, 76, 83, 84, 86, 87, 195
Audubon Sugar School, Louisiana, 177
Ayer, Ira, 136

Bacon, Francis, 70
bagasse (cane trash), 44, 46, 80–81, 91
Baltimore, sugar trade in, 99–100, 111–12
Balzac, Honoré de, 40
Barbados, 6, 14, 114
Barnum, P. T., 146
barrels. *See* hogsheads
Bates, Frederick, 184, 222, 224
Baxa, Jakob, and Guntwin Bruhns, *Zucker im Leben der Völker*, 37–38, *38*
Beckert, Sven, 10
Beetroot Sugar Association, 175
beets, Marggraf's experiments on, 26. *See also* beet sugar
beet sugar, 21–43; disappointments of, 35–40; early production of, 27–28; in France, 21–23, 27–30, 35–40, 217; international trade law and, 43; invention of, 24–27; meanings and associations of, 19, 38; patriotism linked to, 23; price of, 29, 65; production of, 37–38, *38*, 48; purity of, 35–36; significance of, 23; state support of, 22–23, 28, 29–30, 35–36; taste of, 35–37; taxes on, 41. *See also* cane sugar vs. beet sugar; sugar
Bennett, Robert, 60
Bessemer, Henry, 62
Bigelow, Allison, 242n66
Biot, Jean-Baptiste, 41, 72–73, 82, 130, 161
Blenheim (plantation), 100
Board of General Appraisers, 163, 168–69
boiling of sugar, xi–xiii, 5–6, 31–32, 46, 67, 70–72, *72*, 84–88, *86*, *88*, 115, 255n10; charge, 85. *See also* striking of syrup
Booth, William, 130
Booth & Edgar, 116
Bosma, Ulbe, 5, 13
Boston, sugar trade in, 135–39
Boston Daily Globe (newspaper), 137, 142
Boston Herald (newspaper), 135

Braid, Andrew, 163, 167–68
Brazil, 6, 202, 203
British Guiana, 50, 54, 112, 119, 185, 188–89, 202. *See also* Demerara sugar
Brookings Institution, 90
Brooklyn Daily Eagle (newspaper), 123–24, 126, 142, 155
Brooklyn Sugar Refining Company, 215
Browne, Charles Albert, 176–78, 180–85, 189–93, 212, 216–17, 219–20, 222–23, 271n14; *Handbook of Sugar Analysis*, *105*, *181*, *183*, 192–93, 219; "The Use of Temperature in the Polarization of Raw Sugars and other Products upon Quartz Wedge Saccharimeters," 181–82
Bureau of Chemistry (US), 190
Bureau of Standards (US), 182, 184, 222–23, 225
Butler, Benjamin, 117
Byrne, T. Aubrey, 135–36, 139

Cabrera Salcedo, Lizette, 47
California, 1, 134, 191, 218
Cambalache (factory), 92
Canary Islands, 53–54
cane sugar: history of, 4–7; maple sugar as alternative to, 30–34; marketing of, 4; meanings and associations of, 19; price of, 7, 27–29, 30, 65; taxes/duties on, xi, 22. *See also* cane sugar vs. beet sugar; production of cane sugar; sugar; sugarcane
cane sugar vs. beet sugar: competition of, 21; market competition involving, 6, *7*, 27–30, 42–43; similarities and differences, 19, 23–24, 26–28, *30*, 35–40, 42–43
cane trash. See *bagasse*
capitalism: industrial and technological developments in the nineteenth century, 8–10; social relations transformed into apparently natural ones by, 12, 16; standardization's role in, 8–12. *See also* commodities
Carmichael, James Kennedy Wann, 64
cartels. *See* price cartels
Centennial Exhibition (Philadelphia, 1876), 64
central factories (centrals): colonos in relation to, 89–94; control of, 74, 89–90; emergence and growth of, 6, 67, 68; Glasgow firms and, 61; image of, *75*; slavery in, 84. See also *ingenio*
centrifuges (centrifugals), 61, 64, 65; explained, 48; maintenance and repair of, 60; sugar from, 48, 100, 103, 216–17, *217*
Chamberlin, Simon, 111–12, 114–15
Chance (appraiser), 163, 167
Chandler, Charles, 107–9, 112, 129, 131, 146–48; analysis of sugar sample, *147*
Chang, Hasok, 243n81
Chaparra (factory), 191
Chaptal, Jean-Antoine, 30, 37, 39
Charlevoix, Pierre de, 31
chemistry: and analysis of food and other organic substances, 129; and appraisal of sugar quality, 112, 117, 157; and contracts for sugarcane, 91–94; human knowledge and skill involved in, 18, 20; identification of sucrose by, 14; identification of sugar by, 14, 16, 19, 23–24; nature and, 23; of plants, 24–25, 27; purity in, 16, 23, 24; reports, *79*, *81*; standards in, 157; and sugar production, 72, 74–84
Chicago Board of Trade, 9–10
China: in competition with Western sugar production, 117; early sugar production in, 4; labor from, 67, 83–84; mill design in, 44; technicians from, 224
Christadelphians, 153–54
cistern bottoms, 107
civil service reform, 130
Civil War (US), 102
Clark, William, illustrations of slave plantations, *45*, *72*
clayed sugar, 42, 102, 107
Clerget method, 82
Coast and Geodetic Survey, 163, 167
Cobb, Ned, 12
Coffee Exchange, 10, 198, 201–3, 225
Cohen, Andrew Wender, 141
Collins, Harry, 11
Colonial Sugar Company of Australia, 220
colonos, 89–94
color of sugar: appraisal of, for economic purposes, 101–3, 106–13, 127; artificial manipulation of, 110–13; national preferences concerning, 114; quality in relation to, 110; refiners' consideration of, 213; whiteness, 12, 13, 15, 16, 30, 35–37, 40, 42, 47, 99. *See also* dark sugar
Columbia University, 92, 107
commercial method, 169, 176
commodities: histories of, 16; human factors in the production and sale of, 16–18; standardization of, in the nineteenth century, 8–10, 12, 49; sugar as a paradigm of, 12, 17, 42; sugar as paradigm of, 229; trading in, 199–203, 207–8. *See also* capitalism; standards and grades
company checkers, 172–74
Congress (US). *See* United States Congress
Congressional Record, 99
Constancia (factory), 190
contracts: allowances, 157–58, 203–14, 229; for sugarcane, 89–94; for sugar delivery, 205, *205*, 225, 227–29
Cook, James, 58, 59, 64
Cooper, William, 32–33
corruption. *See* fraud and corruption
cotton, 10
Cotton Brokers' Association, 10

Crampton, Charles, 163–64
creole knowledge, 19, 48
Creoles, 83
Cronon, William, 9–10
crystallization, in sugar production, xi, 2, 4, 46, 70–71, 84–87, 115. *See also* boiling of sugar; striking of syrup
crystals: drying of, 3, 5; salt, 15; shapes of, 15; sugar, *13*, 15, 26
Cuba: anxiety about beet sugar in, 29, 39; as basis for sugar prices, 201–2, 206; central factories in, 89–90; in competition with US sugar production, 116–17; machinery and factories in, 41, 56–57, 64–66; plantations in, 6; production management in, 71, 74; races in, 257n53; slavery in, 67, 83–84, 116; and sugar deterioration, 217–20
Cuba Cane Corporation, 217, 219
Cuban-American Sugar Company, 78, 190–92, 212, 219
customs service: corruption allegations against, 124, 126; inspection and appraisal of sugar by, 100–113, 127; in New York, 95, 121–22, 126, 131, 135–39; oversight of, 107, 111; politics and, 121–23, 130; weighing of sugar by, 172–74
Czarnikow, Macdougall & Co., 176, 177

Dance of the Millions, 202
dark sugar: foreign production of, 115–17; marine transportation of, 106; process of making, 113; sales of, 108, 110–13. *See also* color of sugar
De Castro & Donner, 123–24
Deerr, Noël, 72, 75–78, 87, 195, 238n13, 239n21
Demerara sugar, 100, 109–15, 202, 219
Diderot and d'Alembert, *Encyclopédie*, 12, 37
differentials. *See* allowances
Dingley Act (US, 1897), 160, 163, 168
Division of the History of Chemistry, American Chemical Society, 271n14
Dodd, Thomas, 58, 60
Dominica, 114
Dominican Republic, 202
Dos Hermanos (factory), 57
Doscher, Claus, 215
drawbacks (tax refunds), 108–9
Duncan Stewart (company), 51, 53–54, 61. *See also* Stewart, Duncan
Dutch standard, 102–3, 106, 110, 113–14, 117, 127–29, 131–32, 159–60; jar of no. 16 sugar, *104*
duties. *See* tariffs, taxes, and duties
Dye, Alan, 89–90

Eastern Sugar Associates, 92
Edison, Thomas, 145, 149, 155
Edson, Hubert, 77–78
Eichner, Alfred, 206
electricity, sugar refining fraud based on, 145–55
Electric Sugar Refining Company, 96, 146–54; Chandler's analysis of sugar sample, *147*
enslaved people, overlooked knowledge and skills of, 5–6, 16, 69, 70–73, 255n10. *See also* labor; slavery
equipment. *See* machines
Europe: claims of superiority by, 8; consumption of sugar in, 5

factories. *See* central factories
Fajardo, 92
false crystals, 86–87
feeders, 45–46
fermentation, of sugar, 76, 90, 217–20
flour, inspection of, *105*
France: and beet sugar, 21–23, 27–30, 35–40, 217; and maple sugar, 33–34; sugar empire of, 21–22
fraud and corruption: allegations of, 123–24, 126, 132, 136–39; anxieties over, 95, 96, 99, 103, 107–8, 117; customs service's role in, 111; Friend's electric refinery as example of, 144–55; in the Gilded Age, 140–42; polariscopes and, 128–31; referee laboratory and, 174–85; restraint-of-trade, 158; in weighing of sugar, 172–80, *173*. *See also* price cartels
Frederick William III of Prussia, 28–30
Frić Brothers, 184
Friend, Henry, 144–55
Friend, Olive, 149–52, 154
fructose, 77, 82
Fuller, Lawson, 130, 132, 146–48, 152–54, 229
futures. *See* sugar futures

Galison, Peter, 11
Garfield, James, 99, 115–17, 128–29, 132
Germany, 30, 39–40, 222. *See also* Prussia
Gilbert, Robert, 53–54
Gilded Age, 100, 140, 145, 153
Glasgow, Scotland, 19, 48–66, 250n31
Glasgow and West of Scotland Technical College, 61, 63, *63*
Glasgow School of Art, Scotland, 59
glucose, 77, 82
Gorman, Arthur P., 140
Government School of Design, Glasgow, Scotland, 59
Grace, William H., 121–24, 126, 163, 263n11
grades. *See* standards and grades
grain, 9–10
Grand Army of the Republic, 100
Grandville, J. J., 43; "Combat de deux raffinés," 21–22, *22*; *Un autre monde*, 21
Grant, Ulysses, 121, 124
granulators, 206

grape sugar, 35
Guanica (factory), 76, 182
Guiana. *See* British Guiana
gyrodynat, 161

Hagelberg, G. B., 14
Haiti, 202
Hamilton, Alexander, 32
Harlan, John Marshall, 135
Harper's Weekly (magazine), 140, 148, 149, 154; Sugar Trust cartoon, *141*
Harrison, John, 185–86, 188–90, 219
Harvey, Robert, 58–59
Harvey Engineering Company, 61
Hassall, Arthur, 97
Havemeyer, Henry, 116, 124, 130, 131–33, 136, 139, 142, 144, 174, 175–76, 215
Havemeyer, Theodore, 116, 124, 132, 142
Havemeyer & Elder, 123–24, 131, 133, 144, 146, 151, 172; sugar refinery, *125*
Hawaii, 1, 6, 64–66, 134, 191, 218–19
Hawaiian Sugar Planters' Association, 218
Hayes, Rutherford B., 131, 263n11
hedging, 198–202
Henry, Joseph, 128
Heriot, Thomas, 61–62, 64, 78
hogsheads (barrels), 32, 33, 103, 106, *126*, *141*, 174, 225, *226*
Home of the Friendless, 119
Honolulu Iron Works, 64–66
Hormiguero Central Company, 219
Howe, Robert (nom de plume), "Labor Side of the Great Sugar Question," 116
Howell, William, 163, 170
Humboldt, Alexander von, 39–40

indentured workers, 6, 67, 83–84
India, origins of sugar production in, 4
Industrial Commission (US), 215
ingenio (mechanical equipment/plantation), 6, 7, 20, 44, 68, 82, 249n2. *See also* central factories
ingenios centrales. *See* central factories
inspection and appraisal of sugar: arbitrariness in, 227–29; chemical, 112, 117, 157; controversies over, 100–113, 117, 168–71; Dutch standard used for, 102–3, *104*, 106, 110, 113–14, 117, 127–29, 131–32, 159–60; polariscope used for, 87, 114–15, 127–32, 135–39, 142, 160–71, *164*, *179*, 180–90, *181*; recommendation for refinery criteria in, 212–14; record books for, 269n3; referee laboratory's role in, 174–85, 208; safeguards for, 139; sampling as component of, 103, *105*, 106; uncertainties in, 135–40. *See also* fraud and corruption; standards and grades
Institut de France, 28, 37

intermediate products in cane sugar manufacture: massecuites, 77, 85, 87; melada, 107; meladura, 85
International Commission, 188
International Commission on Uniform Methods of Sugar Analysis, 222
International Exchange, 211
International Sugar Journal, 185
international trade law, 43
Ironside, Martin, 66

Jamaica, 6
Jamaica train, 46, 57, 71
Jefferson, Thomas, 26, 29, 30, 34
Jenks, Jeremiah, 265n61
Jesuits, 44
Johnson's New Universal Cyclopaedia, 107
Joule, James, 240n49
Journal of Industrial and Engineering Chemistry, 181

Kalm, Peter, 31–32
kettles, 46
Kingsbury, Noël, 242n598
knowledge and skill: in chemistry, 18, 20; of enslaved people, 5–6, 16, 70–73, 255n10; in "primitive" areas, 2; standards created by, 11–12, 16–17; in striking of syrup, xi–xii, 71, 82, 84–87, 255n10; in sugar cultivation and production, 5–6, 16, 84–88; technological/capitalist obscuring of the operations of, 12, 16, 18, 20, 49–50, 68–70, 82–84, 133, 142–43, 175, 183–84, 211, 215
Kopeloff, Lillian and Nicholas, 220

La Jalousie (plantation), 115
La Ninfa (factory), 74
labor: control of, 74–79, 83; metrology vs., 228. *See also* enslaved people; human knowledge and skill
"Labor Side of the Great Sugar Question" (pamphlet), 116
Lamborn & Company, 200
Lamoreaux, Naomi, 133
latifundias, 89–90
Latour, Bruno, 60, 193, 240n48
Lavoisier, Antoine-Laurent, 25
Lears, Jackson, 140
Leary, J. T., 135, 137, 139
Leeuwenhoek, Antonie van, 14–15, 26
Leonard, Pat, xi–xiii
Levitt, Theresa, 41
Liebig, Justus von, 23, 40–41
Ligon, Richard, 44, 46
lime, 32, 46, 112, 115
Linnaeus, Carl, 25
Littleton, Edward, 45–46

loaves of sugar, 15, 28, 30, 33, 37–38, 46, 71
Locke, John, 12, 15
Louisiana Planter (newspaper), 191, 222
Louis Napoleon, 22–23; *Analysis of the Sugar Question*, 23, 35
Lynch, Michael, 168

machines, 19, 44–66; advertisements for, *52*; development of, in nineteenth century, 6; disrepair and degradation expected of, 224; effects of, on production process, 73–74; Glasgow, Scotland, as center of production for, 19, 48–66, 250n31; human knowledge and skill obscured by, 16, 18, 20, 49–50, 68–70, 82–84, 133, 142–43, 175, 183–84, 211, 215; longevity of, 54; maintenance of, 54–57, 64; mills, 44–47; in model factories, 61–64, *63*; patterns of, 59–61; plans/drawings for, *57*, 58–61; plantations likened to, 6, 44; and standardization, 49; steam power for, 6, 47; on sugar plantations, 44–50, 67
Madison, James, 33–34
mangel-wurzel (mangold), 26, 37, 39. *See also* beet sugar
maple sugar, 30–34, 37, 246n53
Marggraf, Andreas, 24–28, 35, 242n598
Martin, Samuel, 44, 71, 74
Massachusetts Institute of Technology (MIT), 80, 85, 137, 221–24
massecuites, 77, 85, 87
Matthiessen (family of sugar refiners), 152
Maxwell, Walter, 218
McCarthy (student of Rolfe), 222–23
McKinley, William, 159, 163
McOnie, Peter, 57–58
Mechanics' Institute, Glasgow, Scotland, 59, 61
melada, 107
meladura, 85
Melville, Herman, 151
Mendoza, Victor, 57
mergers, 133–34. *See also* monopoly
Merleaux, April, 99
metrology: conceptions of, 11–12; conflicts involving, 95, 178–80, 187–94; human element in, 11–12, 18, 70, 240n48
microscopes, 26–27
mills, 44–47
Mintz, Sidney, 5, 13–14, 23
Mirrlees, James, 59
Mirrlees Tait, 58–59
Mirrlees Watson, 49–56, 60, 64; order book from, *55*
Mississippi (ship), 99–100, 110–11
MIT. *See* Massachusetts Institute of Technology
mites. *See* sugar mites
molasses: analysis of sucrose in, 82; in beet sugar, 36; crystallizable matter in, 71, 73; as a darkening agent, 114–15; in marine transportation, 103, 106; microscopic image of, *13*; in production process, 15, 71, 85, 103; purging of, 46, 48, 86; sorghum as source for, 149
monoculture, 5, 6
monopoly, in sugar trade, 124, 133–34, 140, 154, 159, 176, 198–99, 210–11. *See also* mergers; price cartels
Montgomery, David, 85
Moreno Fraginals, Manuel, 7, 73–74, 82, 196
Moret Law (Spain, 1870), 83–84
Morton, Henry, 112
Mullen, Stephen, 48
muscovado sugar, 31, 33, 47, 100, 102, 107, 113, 114, 127

Napoleon, 28–30, 35–36, *36*, 40
National Academy of Sciences, 137
National Sugar Refining Company, 177
nature: chemistry and, 23; distinguishing artificial products from those of, 42, 221; impurity/diversity of, 8, 10, 23; and second nature, 9, 211, 221, 229; social relations recast/disguised as, 12, 16; standardization/purification of, 10–11, 18, 23; sugar produced by, 14; theoretical conceptions of, 14
nearby sales, 200, 204, 209
Nearing, Helen and Scott, 246n53
Neilson (company), 58
New American Cyclopedia, The, 97
New Guinea, 4
New York Custom House, 95, 121–22, 126, 131, 135–39
New York Sugar Refinery, 215
New York Sugar Trade Laboratory, 174–85, 191–94, *193*, 196, 200, 203, 207–8, 211, 219, 220, 223, 227; polarization data, *197*, *198*
New York Times (newspaper), 132, 138–39
New York World (newspaper), 151
Newton, Isaac, 8
normal weight, 187–89, 222
N. P. Nathan's & Sons, 53

O. B. Stillman catalog, *xii*, *75*
O'Connell, Joseph, 11
Ortiz, Fernando, 16, 17, 49–50, 214
Otremba, Eric, 70, 242n598
Ottoman Empire, 136, 177

pan men, 68, 84–87, 215
Paraf, Alfred, 145
Pares, Richard, 71
Pasteur, Louis, 42
patronato law, 84
Peeps at Industries (book series), 87
Peñaranda, Carlos, 48

Pennsylvania Sugar Company, 16; barrels on the wharf, *122*; raw sugar in, *17*; *The Story of Cane Sugar*, 1–4, *2*, *3*
Perot, Adolphus William, 100, 117, 119–20
Perot, William Henry, 100, 106, 110–15, 117, 119
Philippines, 115, 202, 207
Pinney, John, 71–72
plantations. *See* sugar plantations
plants: chemical analysis of, 24–25, 27; classification of, 25, 27; commercial availability of products from, 24
polariscopes: accuracy of, 129–30, 136–40, 165, 170, 221–25; arguments for and against the use of, 41–42, 95–96, 127–33, 265n61; development of, 41; how it works, 41, 127, 142–43; inspection of sugar using, 87, 114–15, 130–32, 135–39, 142, 160–71, *164*, *179*, 180–90, *181*; instructions for use of, 163–65, *166*, 167; laboratories for, 128, 136; physical form of, 127–28, *164*; proposed use of, in sugar production, 42, 72–73, 82; quartz plates in, 137, 165, 167, 168, 184, 186; saccharimeters as special type of, 186–88; sources of, 165, 184–85; temperature's effect on, 162–70, 180–86, *181*, *183*, 188–94, 219. *See also* polarization
polarization: detection and measurement of, 42, 91; and deterioration of sugar, 219; of sugar, 80–82, 85; tallies of, *197*, *198*; temperature in relation to, 157, 161–70. *See also* polariscopes; standards and grades: 96° polarization, for sugar
Popular Science (magazine), 99
Portugal, 5
Portuondo, María, 71
price cartels, 132–35, 157–58, 206, 208–9. *See also* monopoly
prices: of beet sugar, 29, 65; of cane sugar, 7, 27–29, 30, 65; of sugar, xi, 5, 27–29, 65, 91, 110, 132–35, 144, 157–58, *199*, 208–9. *See also* price cartels
production of cane sugar: administration and control of, 74–84; basic process of, 4; boiling component of, xi–xiii, 5–6, 46; chemistry's role in, 72, 74–84; compared to production of beet or maple sugars, 37; controlling labor in, 74–79, 83; history of, 4–7, *7*; human knowledge and skills responsible for, 16, 84–88; industrialization of, 73–74; manual components of, 73–74; pamphlet on, 1–4, *2*, *3*; reports, *79*, *81*, *88*; sources of loss in, 76–77; weighing the sugarcane, 76. *See also* intermediate products in cane sugar manufacture; machines; refined sugar; refineries; striking of syrup
proof sticks, *86*
protectionism. *See* tariffs, taxes, and duties
Proust (French chemist), 35
Prussia, 28. *See also* Germany
Puck (magazine), 225
Puerto Rico, 6, 83, 90–94, 202, 206, 220
Pulitzer, Joseph, 150
purgeries, 46
purity: of beet sugar, 35–36; chemical, 16, 23, 24; of maple sugar, 32; meanings of, 4, 14, 16; race linked to, 32; of refined sugar, 4; whiteness equated with, 3, 35–37

race: denigration of non-US sugar production based on, 16, 115–17; purity linked to, 32; supposed inferiority of non-white peoples, 16, 116–17. *See also* enslaved people; slavery
Raj, Kapil, 250n33
Reading, Amy, 151
referee laboratories, 174–85, 207–8
refined sugar: drawbacks and, 108–9; harmful ingredients in, 108–9; meanings of, 14, 21; processes of, 108–9, 133; purity of, 4; raw sugar vs., 195; secrecy surrounding, 108–9. *See also* refineries
refineries: consolidation/mergers of, 133–34; factors in purchasing decisions of, 212–16; fraud scheme using electricity for, 145–55; Havemeyer and, 133, 144; plan section, *196*; and politics, 140, 142; purity of sugar as factor in output of, 208. *See also* fraud and corruption; monopoly; price cartels; refined sugar; Sugar Institute; Sugar Trust; tariffs, taxes, and duties
Republican Party, 100
Revere refinery, 1–2, *226*, *228*
Richardson, John, 241n50
Rilieux, Norbert, 48
Roberts, Lissa, 25
Roberts, Robert, 153–54
Robertson, Frances, 59
Robertson, James, 153
Roche, Daniel, 37–38, 241n55
Rodney, Walter, 101
Rolfe, George, 80, 85, 87, 191–92, 221–25, 257n54; *The Polariscope in the Chemical Laboratory*, 84, 180, 221–22
Rolph, George, 218
Rood, Daniel, 41, 48, 68–69
Rosenthal, Caitlin, 45–46, 72
Ross, Charley, 117–20, *118*
Ross, W. H., 51–52
Royal Society of London, 14
Royal Technical College, 62
rum, 71, 106
Rush, Benjamin, 26, 32–34, 37

saccharimeters, 181–82, *184*, 186–89, 192, 221–25
Saint-Domingue, 6, 21
samplers, 106–7
San Vincente (factory), 48
Schaffer, Simon, 224

Schmidt & Haensch, 137, 221–25; half-shadow polariscope, *164*, 165, 192
Science (magazine), 180
Scientific American (magazine), 208
Scott, James, 243n82
Scott, Rebecca J., 84
second nature, 9, 211, 221, 229
Secord, Jim, 250n33
Sedgewick, Augustine, 10
Seeger, Pete, 246n53
Semi-Weekly Miner (Butte, Montana, newspaper), 150
Serres, Olivier de, 26–27
Sharpe, George, 121–24, 126, 163
Sherer, Edward, 135–39, 146, 163–64, 177
Sherer, John, 136, 146, 168, 177
Sherman, John, 110–12, 114, 127–28, 130–32, 135
Sherman Antitrust Act (US), 210–11
Sibum, Otto, 240n49
Skaife, Wilfrid, xi–xiii, 67, 76, 78–79, 87–88, 257n53
skill. *See* human knowledge and skill
slavery: in Cuba, 67, 83–84, 116; maple sugaring not dependent on, 30, 32, 34; in Puerto Rico, 83; sugar economy based on, 5–6. *See also* enslaved people; sugar plantations
Sloane, Hans, 255n10
Société philomatique, 42
Society for Promoting the Manufacture of Sugar from the Sugar Maple Tree, 34
Solá, José, 91
Soledad (plantation/factory), xi–xiii, 1, 52, 56, 67, 74, 78, 83
sorghum, 148–49
South Porto Rico Sugar Company, 79, 82, 222
Spain, 5
Spanish Antilles, 83
Spaulding, O. Z., 136–39
Spencer, Guilford, 78, 190–92
Spreckels, Claus, 134–35, 154, 213
standards and grades: allowances permitted for, 157–58, 203–14, 229; arbitrariness in, 227–29; chemical, 157; development, implications, and use of, 8–10; embodiments of meter and kilogram standards, 11–12, 241n52; in Europe, 5; human component of, 11–12, 16–17; machinery and, 49; meanings of, 11–12, 196; 96° polarization, for sugar, 4, 81–82, 93, 114, 157, 195–98, 203–4, 206, 209, 211, 212, 214, 218, 229; particular/peculiar features and, 11, 243n82; polarization as basis for, 212; of sugar, 12, 16; universality imputed to, 8, 11, 18; variety of, 107–8. *See also* Dutch standard
steam power, 6, 47
Sterling Syrup Works, 149
Stewart, Duncan, 50–51. *See also* Duncan Stewart
Story of Cane Sugar, The (pamphlet), 1–4, *2*, *3*
strikes, labor, 84
striking of syrup, xi–xii, 46, 71, 82, 84–87, 115, 255n10
sucrose: chemical designation of, 14; measurement of, 80–82, 127, 169; as a molecule, 4, 12–14, *13*; sugar identified with/distinguished from, 4, 13–14, 19, 127, 169, 182, 191
sugar: chemical designation of, 14, 16, 19, 23–24; consumption of, 100, 133; defining, 99, 107; deterioration of, 76, 90, 216–20, *217*; grades of, 12, 16; history of, 4–7; how taste is produced by, 15; human factors in the production and sale of, 17; inspection and appraisal of, 100–113, 117, 127–30; marine transportation of, 103, 106; meanings and associations of, 38; 96 percent standard of, 4, 81–82, 93, 114, 157, 195–98, 203–4, 206, 209, 211, 212, 214, 218, 229; philosophical conceptions of, 12; price of, xi, 5, 27–29, 65, 91, 110, 132–35, 144, 157–59, *199*, 208–9 (*see also* price cartels); profit margins for, 126–27, 133–34, 135, 144; properties of, 7, 42; raw vs. refined, 195; sales of, 108, 110–13, 200, 204, 209; sensory qualities of, 15–17, 28, 213–15; sources of, 6, 27, 35, 39; stability of, 18, 157, 189, 216–20 (*see also* uniformity/universality of); sucrose identified with/distinguished from, 4, 13–14, 169, 182, 191; uniformity/universality of, 12, 16, 23, 26–27, 42, 69, 82, 196–98, 212, 229, 242n598 (*see also* stability of); weight of, purchasing by, 172–85; whiteness as property of, 12, 13, 15, 16, 30, 35–37, 40, 42, 47, 99. *See also* beet sugar; cane sugar; cane sugar vs. beet sugar; clayed sugar; color of sugar; dark sugar; inspection and appraisal of sugar; loaves of sugar; maple sugar; monopoly; muscovado sugar; potted sugar; refined sugar; refineries; sugar, types of
sugar, types of: cucurucho, 107–8; loaves, 15, 28, 30, 33, 37–38, 46, 71; molasses, 107; potted, 107; quebrado, 103
Sugar (journal), 225, 227
Sugar District 13, Brooklyn, New York, 123
Sugar Equalization Board (US), 204
Sugar Exchange, 158, 199–203, 207–9, 211, 225, 227–29
sugar futures, 199–203, 207–8
Sugar Institute, 206, 208–10, 220
Sugar Journal, 189
sugar mites, 97, *98*, 99, 117, 120
Sugar Molecule, The (promotional publication), 193–94
sugar plantations: administration cane, 90; administration land, 89–90; development of the model of, 5; operation of, 44–50, *45*; role of, in sugar production, 5–6
sugar question, 6, 21–24, 35, 39
Sugar Ring, 124, 146

sugar tramps, 193
Sugar Trust, 96, 133–34, 139–40, 142, 154, 157, 159, 169–70, 185, 198–99, 225; cartoon about, *141*
sugarcane: criticisms of, 30–31; cultivation of, 4–5; geographical locations of, 1; origins of, 4. *See also* cane sugar
sulfuric acid, 114, 115
Sullivan, F. E., 205, 208
Supreme Court (US). *See* United States Supreme Court
sweepings of sugar, 124, *126*

Tait, William, 58–59
Taiwan, 6
Tarbell, Ida, 134
tariffs, taxes, and duties: on beet sugar, 41; on cane sugar, xi, 22; drawbacks, 108–9; importance of, 95, 100; politics and, 132, 140; uniform, 131–32; in the United States, 100–103, 108–13, 131–32, 135–42, 159; variety of, 101
taste, physical process of, 14–15
taxes. *See* tariffs, taxes, and duties
technology. *See* machines
temperature, polarization/polariscopes affected by, 157, 161–70, 180–86, *181*, *183*, 188–94, 219
Ten Years' War (1868–1878), 83
testing laboratories. *See* referee laboratories
tobacco, 16, 175
Tomich, Dale W., 242n598
train (machinery). *See* Jamaica train
Trans Pacific, The (journal), 66
transportation, of sugar by ship, 103, 106
triers, 103, *105*, 106

United Fruit Company, 2
United States: and beet sugar, 29; drawbacks (tax refunds) on sugar exports from, 108–9; imports of sugar and molasses to, *101*; and maple sugar, 30–34; sugar consumption in, 100, 133; and sugar production, 65, 115–17, 133–34, 208. *See also* inspection and appraisal of sugar; tariffs, taxes, and duties
United States Congress: and oversight of the sugar industry, 210; and polariscope's use for sugar tariffs, 128, 132, 160, 169; and sugar tariffs, 101–3, 110, 113, 114, 127–29, 136, 160, 197
United States Department of Agriculture, 77, 78, 129, 137, 163, 176
United States Department of Justice, 157–58, 210
United States Department of the Treasury, 95, 107–9, 111, 114, 123–24, 127–29, 135–37, 142, 159–60, 163–65, 167–70, 172, 182, 195, 197, 210
United States Supreme Court, 132, 134, 157, 160–61, 170, 176, 181, 202, 207, 209–10
United States vs. 712 Bags of Dark Demerara Centrifugal Sugars, 113
United States vs. E. C. Knight (1895), 134–35
universality. *See* standards and grades

vacuum pans, xi, *xii*, 19, 47–48, 82, 84–87, *86*
Vergenoegen (plantation), 100–101

Washington, George, 26, 33, 34
Watson, Laidlaw & Co., 60, 61
Weekly Statistical Sugar Trade Journal, 202
weighing the sugarcane, 76
weight of sugar, purchasing by, 172–85
Welch, C. J., 206, 207, 210
Wells, David, 132
West Indies, 28, 32, 48, 50, 57, 63–64
West of Scotland Technical College, 64
Weston, David, 64
Whalley, Richard, 172
Whiskey Ring, 124
White, Richard, 140
whiteness: civilizational hierarchy linked to, 99; of clayed sugar, 102, 107; purity equated with, 3, 35–37; and supposed inferiority of non-white peoples, 16, 116–17. *See also* color of sugar
Wiechmann, Ferdinand, 168, 180, 182, 185–86, 188–90, 222
Wiley, Harvey, 77, 127, 137–38, 160, 161, 163, 167, 169, 174, 176, 177, 180, 182, 185, 190, 192, 222
Willett, Wallace, 146
Willet and Gray, *Weekly Statistical Sugar Trade Journal*, 202
Wilson, William L., 140
Wilson-Gorman Tariff (US), 140, 142, 159; cartoon about, *141*
Woolworth Building, New York, 65
World War I, 65, 199, 217
World War II, 225

zafra (cane harvest season), xi–xiii
Zerban, F. W., 192, 205, 220